Superplastic Forming/Diffusion Bonding Technology of Titanium Alloys

Zhiqiang Li

Superplastic Forming/ Diffusion Bonding Technology of Titanium Alloys

Theories and Applications

国防工業出版社
National Defense Industry Press

Springer

Zhiqiang Li
AVIC Manufacturing Technology Institute
Aviation Industry Corporation of China, Ltd.
Beijing, China

ISBN 978-981-99-3908-4 ISBN 978-981-99-3909-1 (eBook)
https://doi.org/10.1007/978-981-99-3909-1

Jointly published with National Defense Industry Press
The print edition is not for sale in The Mainland of China. Customers from The Mainland of China please order the print book from: National Defense Industry Press.
ISBN of the Co-Publisher's edition: 978-7-118-12541-2

This Springer imprint is published by the registered company Springer Nature Singapore Pte Ltd.
The registered company address is: 152 Beach Road, #21-01/04 Gateway East, Singapore 189721, Singapore

Foreword by Qiao Guan

In the early 1980s, Beijing Aviation Manufacturing Technology Research Institute (the predecessor of AVIC Manufacturing Technology Institute) began to study superplastic forming/diffusion bonding (SPF/DB) technology. In 1986, the Aviation Sheet Metal Forming Technology Laboratory was founded with SPF/DB technology as its major research focus. Zhiqiang Li was one of the first members and has been working for 36 years since then. For more than 40 years, I witnessed the development of SPF/DB technology as well as the growth of pioneers represented by Zhiqiang Li. Recalling the years when we had "nothing but struggling" and looking at the "blooming and thriving" of today, it is because a generation of SPF/DB experts and engineers represented by Zhiqiang Li, the current president of AVIC Manufacturing Technology Institute, sticks to their faith of "choose one thing and do it for a lifetime" and makes dreams come true. The persistence of these people cultivated the SPF/DB technology from a sapling into a towering tree. Their contributions to the development of China's aviation manufacturing technology make me feel very gratified.

After years of continuous research, the AVIC Manufacturing Technology Institute has achieved a large number of innovative results in SPF/DB technology. The material system has expanded from titanium alloys to aluminum alloys, aluminum-lithium alloys, and intermetallic compounds. The structural forms have evolved from single-layer plates with complex shapes and double-layer plate structures to multi-layer hollow integral structures, greatly reducing the number of parts and structural weight. This has provided applicable technology to improve the overall performance of aircraft, while improving material utilization and reducing manufacturing costs. Therefore, a series of thin-walled lightweight structures developed based on SPF/DB technology has been used in various aircrafts, engines, and missiles.

This book is the author's wisdom of over 30 years' research and application of materials' superplasticity, SPF/DB technology. In terms of basic research, it elaborates on the diffusion bonding and superplastic deformation mechanisms. In terms of technical research and development, it systematically introduces SPF/DB structural optimization design, process quality control, and defect control methods. In terms of performance evaluation, it focuses on the fatigue performance of the deformed material and SPF/DB structure. The large amount of data and examples presented in the book are published for the first time, and I believe this book can serve as a source of inspiration for designers and engineers developing new aircraft and solving problems in production. It also has high reference value for scholars engaged in plastic deformation research. As a veteran researcher in the aviation manufacturing field, I hope that more people can read high-quality books like *Superplastic Forming/Diffusion Bonding of Titanium Alloys and Its Applications*, connect the past and future, and make continuous innovations in development, pouring into new motive power for vigorous development of China's advanced manufacturing technology and aviation industry.

May 2022

Qiao Guan
Academician of Chinese Academy of Engineering
Beijing, China

Foreword by Prof. Jianguo Lin

I am very pleased receiving an invitation to write a preface for this extraordinary book authored by Prof. Zhiqiang Li, who have spent over 30 years to the development of new theories, experimental and simulation methods, and innovative processes for superplastic forming and diffusion bonding (SPF/DB) technologies. Particularly, he has rich experience on the manufacturing of hollow, lightweight, and multiple layer engineering components using SPF/DB methods and developed many new processes for different applications. His knowledges and deep scientific understanding will be shared with readers in this book through the introduction of new theories, innovative engineering practice, microstructure of formed parts, and post-forming property assessments.

This book is, for the first time, a systematic summary and conclusion of Prof. Li's research work and engineering practice over the years working with industrialists. With inclusive, systematic, novel, simple but profound, and well-structured contents, this book has high academic values for researchers and also provides ideal references for engineers, who work on superplastic forming and metal forming in general. For example: a mapping relationship between the material microstructure and the product macroscopic performance has been established and introduced; the law of microstructure evolution in material forming processes has been formulated; the design principle of structural parameters and optimization criteria of process parameters have been created; quality control and material characterization methods of lightweight monolithic structure have been established; etc. All the above are the reflection of the author's years of profound academic research, deep scientific understanding, and rich engineering experiences. Publication of this book will provide an important reference and guidance for SPF/DB technology research and development, application promotion, and talent cultivation. In addition, it can be widely used by researchers and engineers engaged in other aspects of metal forming.

I believe this book will be seen as a milestone in superplastic forming and diffusion bonding and plasticity manufacturing technologies in general, such as forging, extrusion, and hot stamping. I would like to sincerely thank Prof. Li's efforts for summarizing his outstanding scientific work and engineering practice experience, with numeral useful examples, as a book, and sharing with us.

May 2022

Prof. Jianguo Lin, FREng, FIoMMM, FIMechE
Imperial College London
London, UK

About This Book

With titanium alloy as the starting point, this book introduces the characteristics, classification, deformation mechanism, and application of titanium alloys in the aviation field. It then summarizes the common forming methods of titanium structures and presents the superplastic forming/diffusion bonding (SPF/DB) technology that can realize the manufacturing of titanium hollow structures. For this technology, the principles, mechanisms, and influencing factors of superplasticity and diffusion bonding are explained, respectively. The structure characteristics and application scenarios of typical SPF/DB structures, such as single-layer plate, double-layer plate, triple-layer plate, and quadruple-layer plate, are proposed. In considering the common features of these four typical structures, the principle and processing route of SPF/DB are introduced in detail. The design criteria of structural parameters based on process feasibility, selection criteria of processing parameters, and mold design methods are proposed. By combining numerical simulation and experimental verification, the control methods of defects such as non-uniform wall thickness, surface wrinkling, local fracture, and surface step difference are investigated in detail. In addition to the dimension control of SPF/DB structure, the microstructure change of material during SPF/DB, the evolution of voids, and DB interface defects are revealed. The static and fatigue properties of material after SPF/DB are characterized to guide the performance optimization of SPF/DB structures during the manufacturing process. Considering the requirements for both dimension and performance of SPF/DB structures, the inspection methods of external and internal structure accuracy, DB interface quality, and residual stress are summarized, and the design and performance evaluation methods of SPF/DB structure are introduced in detail based on application requirements.

This book is recommended to researchers and engineers engaged in SPF/DB technology, and it can also provide guidance for professionals working in material processing.

Introduction

Titanium alloys have been widely used in aerospace industry for the high specific strength, excellent fatigue performance, and good corrosion resistance. When deformed under specific conditions of temperature, strain rate, and microstructure, titanium alloys show exceptional ductility known as superplasticity. Meanwhile, titanium atoms become very active at this temperature, which makes their interface easy for diffusion bonding. In this way, the combination of superplastic forming and diffusion bonding (SPF/DB) is useful for manufacturing products of complex forms in one operation, thereby eliminating unnecessary joints and rivets. Other benefits of this technology include the reduction of residual stresses and the ability to form near net-shape parts that lowers machining costs and material waste. From the perspective of aircraft designing, SPF/DB helps to reduce the weight and the cost of the parts that they may replace.

My understanding of SPF/DB technology has been shaped over 30 years through collaborations, discussions, and correspondence with the research team at AVIC Manufacturing Technology Institute as well as people in universities and industries. I was motivated to write this book by the absence of any recent comprehensive literature on SPF/DB technology. This book aims to provide a modern compendium that addresses both theories and applications of SPF/DB. I would like to extend my sincere gratitude to those colleagues who have continuously worked in the research of SPF/DB and promoted the sustainable development and application of this technology.

I received great support and help from my colleagues and students when writing this book, including Han Xiuquan, Shao Jie, Chen Wei, Qu Haitao, Han Xiaoning, Du Lihua, Deng Ying, Mu Yanhong, Zhang Ning, Liu Yunxi, Li Xiaohua, Zhao Bing, Zhang Yanling, Xu Huiyuan, Fu Mingjie, Zhang Xingzhen, Zhou Lina, Liu Shengjing, Deng Wujing, Zhang Jichun, Wang Jingzhao, Fu Xin, Wu Qiong, Ma Lixia, etc. I hereby would like to express my heartfelt thanks for their contributions in completion of this book.

I hope this book can be a useful means of consultation for researchers, engineers, and students engaged in the design and manufacture of aerospace structures. A representative set of references which provide additional detail for readers interested in specific aspects of SPF/DB is also included. There may be mistakes or inadequacies in the book, and readers' comments are welcomed.

March 2022 Zhiqiang Li

Contents

Chapter 1
Titanium and Titanium Alloy

Next to aluminum, iron and magnesium, titanium (Ti) ranks the forth most abundant structural metal in the Earth's crust, the content of which is about 0.6%. In 1791, the British mineralogist–William Gregor, was the first to discover titanium. Countries around the world have continued to study and improve titanium extraction methods for more than a century since then. In 1932, titanium started commercial production after the Luxembourg chemist Wilhelm Kroll invented the magnesium thermal reduction method. Due to the high specific strength and excellent corrosion resistance, titanium alloys are widely used in aerospace and chemical industries, and then expanded to the fields of metallurgy, electric power, architecture, energy, biomedicine, sports & leisure, transportation and others.

1.1 Titanium Alloy and Its Classification

Today more than 100 titanium alloys are known, of which, however, only about 30 are commonly used. Depending on the material property, titanium alloys are classified as low-strength high-plastic titanium alloy, medium-strength titanium alloy, high-strength titanium alloy and damage-tolerance titanium alloy. According to the application fields, titanium alloys are classified as structural titanium alloy, corrosion resistant titanium alloy, and high-temperature titanium alloy. According to the microstructure characteristics, titanium alloys are classified as α titanium alloy, α + β titanium alloy and β titanium alloy. In summary, different classification methods are all derived from the basic characteristics of α and β phases in titanium alloys.

Z. Li, *Superplastic Forming/Diffusion Bonding Technology of Titanium Alloys*,
https://doi.org/10.1007/978-981-99-3909-1_1

1.1.1 Crystal Structure

For pure titanium, the density is 4.5g/cm^3, the melting point is 1668 °C, and the allotropic transformation temperature (also known as β transformation temperature) is 882.5 °C. Titanium crystallizes at low temperatures in a modified ideally hexagonal close packed (hcp) structure, called α phase. At high temperature (above the β transformation temperature), the body-centered cubic (bcc) structure is stable and is referred to as β phase. The atomic unit cells of the α and β phase are schematically shown in Fig. 1.1. The β → α phase transformation in titanium alloy follows the Burgers orientation relationship, that is $(110)_\beta//(0002)_\alpha$, $[111]_\beta//[1120]_\alpha$. During the phase transformation, $(110)_\beta$ changes to $(0002)_\alpha$ and $[1\ \bar{1}\ 1]_\beta$ changes to $[1\ \bar{1}\ 20]_\alpha$.

As shown in Fig. 1.2, each $(110)_\beta$ plane has two $[1\ \bar{1}\ 1]_\beta$ crystal directions; one $[1\ \bar{1}\ 1]_\beta$ crystal direction is directly parallel to $[\bar{2}\ 110]_\alpha$ direction, while the other $[1\ \bar{1}\ 1]_\beta$ crystal direction has 10.5° angle from $[1\ \bar{1}\ 20]_\alpha$.

a Variant 1; **b** Variant 2

The difference between the crystal structure and lattice parameters of α and β phases leads to lattice distortion at the interface. When the interface range is small, the atoms of the two phase are exactly located at the common positions of the two phase lattices, forming a coherent interface. As the interface range expands, the mismatch of two-phase atoms increases accordingly, and the elastic strain energy caused by lattice distortion increases rapidly. In this case, it is difficult to maintain the coherent interface, and the edge dislocation or structural step will be generated at the interface to reduce the elastic strain energy. The interface of α phase and β phase in titanium alloy is dominated by a step-like semi-coherent interface. During the phase transformation, α lamellae thicken and widen according to the step-like

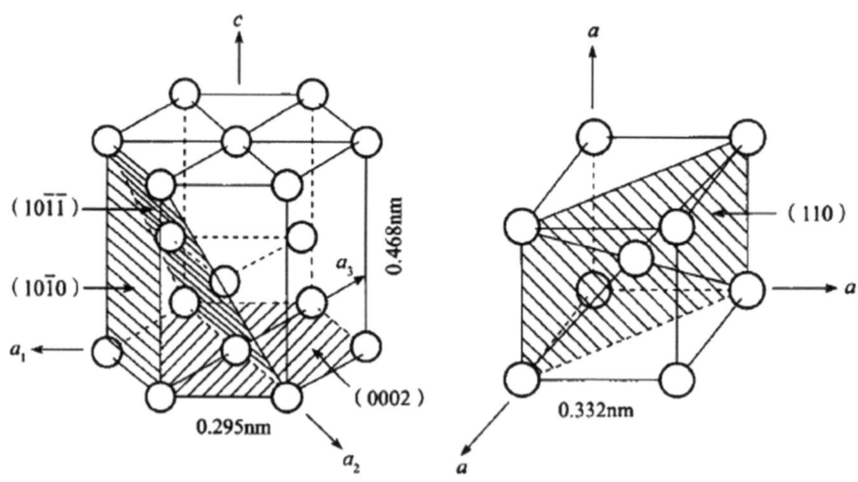

Fig. 1.1 Schematic of unit cells of α phase (hcp) and β phase (bcc)

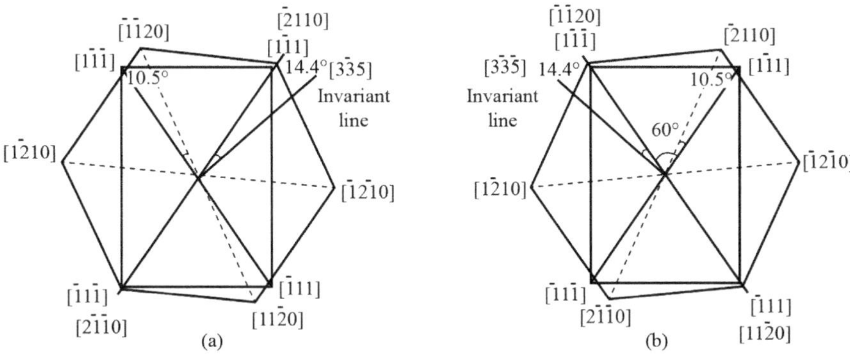

Fig. 1.2 Burgess orientation relationship between α phase and β phase of titanium alloy

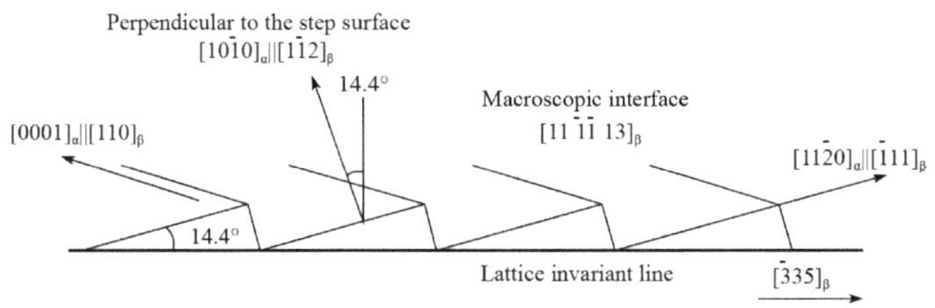

Fig. 1.3 α phase and β phase interface structure of titanium alloy

mechanism, and the steps are gradually lowered along the lattice invariant line $[\bar{3}35]_\beta$ (Fig. 1.3).

1.1.2 Metallurgy

Allotropic transformation and different crystal structures of titanium alloys are considered as the basis of microstructure and property regulation. Depending on their influence on the transformation temperature, the alloying elements of titanium are classified as α-stabilizers, β-stabilizers, or neutral. The α-stabilizers extend the α phase field to higher temperatures, while β-stabilizing elements shift the β phase field to lower temperatures. Neutral elements have only minor influence on the β-transformation temperature. The impact of common alloying elements on the phase diagram of titanium alloy is shown in Fig. 1.4.

α-stable elements can be solidly dissolved into α phase to expand the α phase region and improve the phase transformation point. These elements include Al, Ga, Sb, Ge, Bi, In, B, and impurity elements like C, O, N, etc. Al is the most crucial α-stable element in titanium alloys, and is the only common metal that increases the β transformation temperature. When Al and Ti form a substitutional solid solution,

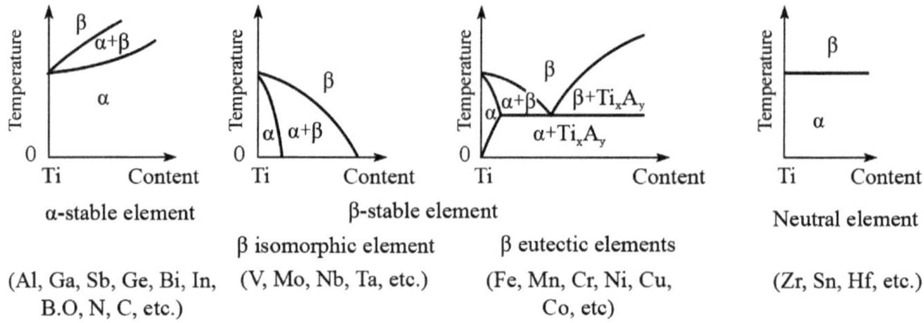

Fig. 1.4 Influence of alloying elements on phase diagrams of titanium alloys (schematically)

they strengthen and reduce the density of alloy. According to the Ti–Al binary phase diagram, the ultimate solubility of Al in Ti is 7.5%. Therefore, the Al content in titanium alloy is usually kept below 7% to prevent the formation of ordered phase Ti_3Al (α_2), which reduces the plasticity, toughness and the resistance to stress corrosion of the alloy. Additionally, Al has smaller density and atomic radius than Ti, so it can enhance the atomic bonding force of the β-titanium alloy solid solution, improving the alloy's specific strength while minimally reducing the plasticity. Furthermore, Al can also improve the alloy's oxidation resistance and significantly increase the recrystallization temperature, thereby enhancing its hot strength. When preparing the alloy, the desired Al element is typically added to the titanium alloy in the form of aluminium shot, and intermediate alloy V-Al or Mo-Al, to reduce the alloy cost and avoid the high density inclusions during melting. Interstitial elements such as O, C and N also stabilize the α phase and improve the alloy's strength. However, they may reduce alloy's plasticity. Therefore, the content of these elements must be strictly controlled during actual production. B is known as the "vitamin of metal materials", and the addition of a small amount of B in titanium alloy can refine the grain and improve the properties. Generally, other α-stable elements, such as Ga, Ge and rare elements, are not used as alloying elements due to their low solid solubility.

β-stable elements can be solidly dissolved into β-phase to expand the β-phase region and reduce the phase transformation temperature, which is subdivided into β-eutectic elements and β-isomorphous elements. β-eutectic elements can have eutectoid reaction with Ti to produce intermetallic compound, mainly including Fe, Mn, Cr, Co, Ni and Cu. Fe belongs to strong β-stable element, but it is prone to segregation during melting, which may affect the thermal stability. Therefore, Fe is rarely added to alloys and it is added more frequently in some low-cost titanium alloys to replace the expensive V element. Adding Cr element will improve the hardenability of alloys, but it is easy to precipitate compounds that may reduce the plasticity when the content is high. Mn can improve the strength and plasticity of alloy, but it is prone to eutectic decomposition, and is commonly added in the early design of titanium alloy. Si can improve the thermal stability and heat resistance of alloy, so it is commonly used in most high-temperature titanium alloys, but the content is generally less than 0.5%. Other elements such as Cu, Ag, and Ni are rarely used.

β-isomorphic elements including Mo, V, Nb, Ta ect., have the same lattice structure and similar atomic radius as Ti, so they are solidly soluted into β phase to expand β phase region, and enhance the stability of β phase, among the elements, Mo has the best stabilization effect. Mo and V are common elements in β-Ti alloy and can be added to the matrix in the form of substitutes. The addition of Mo and V can improve the strength and abitily for quenching (retaining β phase to the room temperature in case of fast cooling) while maintaining high plasticity. In addition, Mo and V are able to inhibit the peritectoid or eutectoid reaction that may occur after adding Al, Fe and Cr, and improve the microstructure stability of β-titanium alloy. Nb has weak strengthening effect, but can improve the plasticity and toughness of titanium alloy. Therefore, Nb is also added in titanium alloy as a common element. Ta has the weakest strengthening effect and high density, so it is used less in titanium alloy.

The neutral elements, which have little effect on β transformation temperature, mainly include Zr, Hf and Sn. With similar size and properties of atoms, Zr, Hf and Ti all have high solid solubility in α and β phases to exert strong high-temperature strengthening effect. Therefore, such elements are usually used in high-temperature strength titanium alloys. Sn element has weak strengthening effect at room temperature, but it can improve the high-temperature strength of titanium alloy.

The type and content of alloying elements have a significant impact on the phase composition of titanium alloys. Figure 1.5 shows the binary phase diagram of titanium alloy. According to its phase composition at room temperature, the titanium alloys are classified as α, α + β and β alloys. The typical grade of titanium alloys and their β transformation temperature are listed in Table 1.1. There is non-equilibrium microstructure of titanium alloy in the actual production and application. The titanium alloy can be further subdivided into near-α titanium alloy and metastable β titanium alloy according to the phase composition and β-stable element content in metastable state.

α titanium alloy comprise commercially pure titanium and alloys exclusively alloyed with a-stabilizing and/or other elements. Commercially pure titanium is divided into different levels according to oxygen content. Though with low strength,

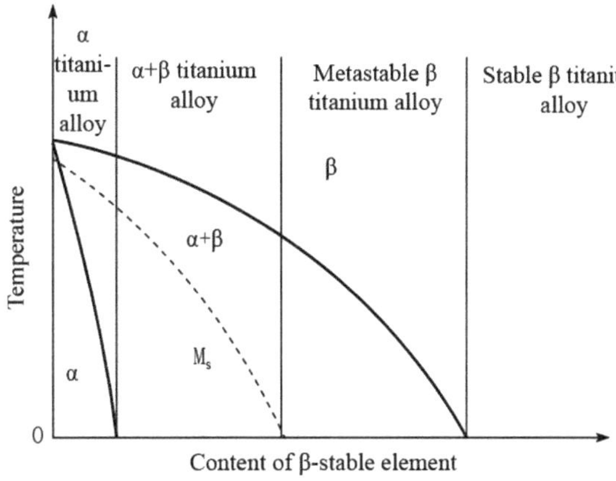

Fig. 1.5 Binary phase diagram of titanium alloy (M_S indicates the martensite starting temperature)

Table 1.1 Common α, α + β and β titanium alloys and their β transformation temperatures $T_β$

Name	Alloy component /% (by mass fraction)	$T_β$/°C
α titanium alloy		
Grade1	CP-Ti(0.2Fe, 0.180)	890
Grade2	CP-Ti(0.3Fe, 0.250)	915
Grade3	CP-Ti(0.3Fe, 0.350)	920
Grade4	CP-Ti(0.5Fe, 0.400)	950
Grade7	Ti-0.2Pd	915
Grade12	Ti-0.3Mo-0.8Ni	880
Ti-5–2.5	Ti-5Al-2.5Sn	1040
Ti-3–2.5	Ti-3Al-2.5V	935
α + β titanium alloy		
Ti-811	Ti-8Al-1V-1Mo	1040
IMI685	Ti-6Al-5Zr-0.5Mo-0.25Si	1020
IMI834	Ti-5.8Al-4Sn-3.5Zr-0.5Mo-0.7Nb-0.35Si-0.06C	1045
Ti-6242	Ti-6Ab2Sn-4Zr-2Mo-0.1Si	995
Ti-6–4	Ti-6Al-4V(0.200)	995
Ti-6-4ELI	Ti-6Al-4V(0.130)	975
Ti-662	Ti-6Al-6V-2Sn	945
IMI550	Ti-4Al-2Sn-4Mo-0.5Si	975
β titanium alloy		
Ti-6246	Ti-6Al-2Sn-4Zr-6Mo	940
Ti-17	Ti-5Al-2Sn-2Zr-4Mo-4Cr	890
SP-700	Ti-4.5Al-3V-2Mo-2Fe	900
β-CEZ	Ti-5Al-2Sn-2Cr-4Mo-4Zr-lFe	890
Ti-10–2-3	Ti-10V-2Fe-3Al	800
β-21S	Ti-15Mo-2.7Nb-3Al-0.2Si	810
Ti-LCB	Ti-4.5Fe-6.8Mo-1.5Al	810
Ti-15–3	Ti-15V-3Cr-3Al-3Sn	760
β-C	Ti-3Al-8V-6Cr-4Mo-4Zr	730
B120VCA	Ti-13V-11Cr-3Al	700
Ti-5553	Ti-5Al-5Mo-5V-3Cr-0.5Fe	855

it has good plasticity and weldability, so it is generally used in the case of high corrosion resistance requirements, and the long-term service temperature can reach 300 °C. α titanium alloy has high β-transformation temperature, and its equilibrium phase is α phase at room temperature. With stable microstructure and excellent high-temperature behavior, it is deemed as the basis of the development of high temperature titanium alloy. α titanium alloy cannot be strengthened by heat treatment due to lack of sensitivity to the microstructure type and heat treatment, so it only has moderate strength.

When a small amount of β-stable elements (< 2%) is added to α titanium alloy, nearl-α titanium alloy is formed and there is a small amount of β-phase or intermetallic compounds in the annealed microstructure. The near-α titanium alloy has good weldability, high-temperature strength and thermal stability, and its maximum service temperature can reach 600 °C. According to the different aluminium equivalents, it can be further divided into near-α titanium alloy with low aluminium equivalent and near-α titanium alloy with high aluminium equivalent. The former has the main characteristics of low tensile strength at room temperature, high plasticity and thermal stability, good weldability and forming performance, suitable for the production of complex-shape sheet stamping parts and welding parts, with the typical grades of TC1 and TC2, etc. The latter has relatively high aluminum content and is mainly used as high-temperature titanium alloys in the form of forgings. Currently, high-temperature titanium alloys such as IMI834, Ti-1100, Ti-6242S, BT36, Ti60 are the most widely used in the aviation field.

α + β titanium alloy contains both α-stable elements and β-stable elements, and it presents α + β microstructure at equilibrium state, which is generally based on Ti–Al system. As the microstructure of industrial α + β titanium alloys is dominated by α phase, a small amount of neutral elements Sn and Zr may be added to further strengthen the α-phase. Although the solid solution strengthening effect of β-phase stable elements is not obvious, it is possible to obtain quenched β-phase from low content of such element and further significantly improve the strength of the alloy by annealing or aging treatment. In addition to the content of stable elements in β phase, material processing and heat treatment process are also the factors affecting the properties of α + β titanium alloys. Regulation of microstructure is achieved by controlling the alloy component and processing technology, thus to meet different property requirements. Compared with α titanium alloy, α + β titanium alloy shows slightly lower high-temperature stability, and its upper limit of the service temperature is about 500 °C. The commonly used α + β alloys include Ti-6Al-4V, Ti-6242, Ti-62222S, TC11, TC17 and TC21, among which Ti-6Al-4V are most popular in aerospace applications.

β titanium alloy is divided into metastable β titanium alloy and stable β titanium alloy. The metastable β titanium alloy has excellent plasticity and good weldability after quenching. The strength of the alloy can be greatly improved by the fine α-crystal lamellae precipitated in β-phase matrix after aging treatment. The size, morphology, distribution and volume fraction of αlamellae can be adjusted by using different solid solution and aging temperatures as well as different cooling rates, thereby regulating the mechanical properties of the alloys. The tensile strength of some alloys can reach over 1500MPa. Metastable β titanium alloy has been used to manufacture large structural components. The typical class includes but not limited to Ti-5553, Ti-10–2-3, β-21S, β-C, Ti-15–3. When the β-phase stable elements exceed a certain critical content, the β transformation temperature drops below the room temperature, and when subject to annealing, the alloy will be composed entirely of stable β-phase, and is called stable β titanium alloy. As with high content of β-phase stable elements, the β titanium alloys freatures high corrosion resistance, with the example of TB7, Ti40 and Alloy C (burn-resistant titanium alloy). However, excessive content of β-phase stable elements contributes to increased density of the alloy, making it difficult to melt and form cast ingot. Hence, the workability of stable β titanium alloy is reduced.

1.1.3 Microstructure

The thermomechanical treatments of titanium alloys includes a complex sequence of deformation, solid solution treatment, aging, annealing for stress relief, etc. According to the content of alloying elements and cooling rate, the solid phase transformation of titanium alloy can be achieved by martensite transformation or nucleation-diffusion-growth. The microstructure depends on the composition of titanium alloy and parameters of thermomechanical treatment. There are four typical microstructure, namely, Widmanstatten, basket-weave, duplex and equiaxed microstructure (Fig. 1.6).

(1) Widmannstatten structure: it can be obtained by thermomechanical treatment in βphase region, consisting of parallel α lamellae and interphase original β.

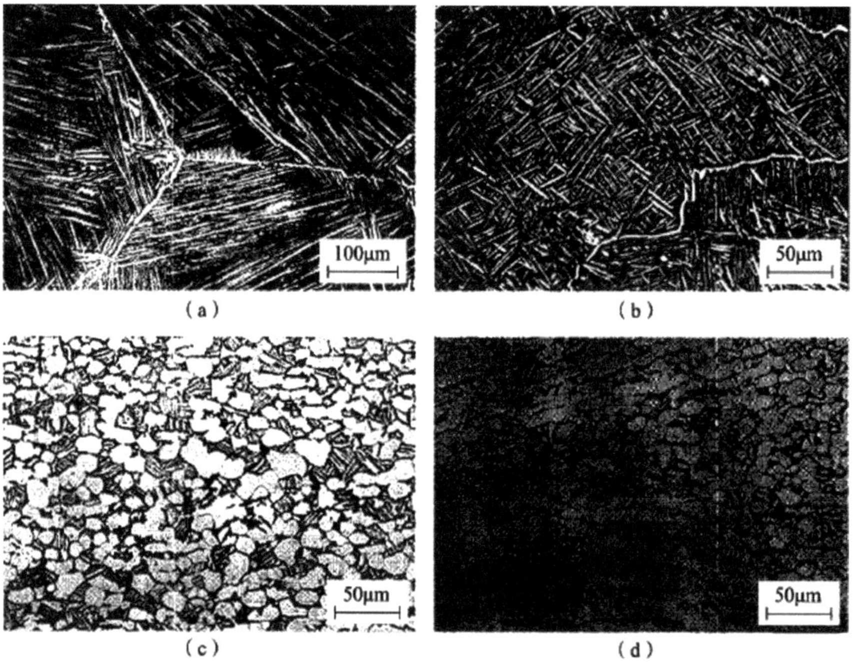

Fig. 1.6 Typical microstructure of titanium alloy. **a** Widmannstatten structure; **b** Basket-weave microstructure; **c** Duplex microstructure; **d** Equiaxed microstructure

In industrial production, the titanium alloy is deformed first in β phase region with low flow stress and then in α + β phase region, and finally annealed at a temperature higher than the β transformation temperature and slowly cooled to obtain Widmannstatten microstructure. During the cooling process, α phase nucleates at the original β grain boundary, and then grows along the specific crystal plane with low interfacial energy to form α lamellae, with adjacent α lamellae spaced by the original β. The α phase and the original β phase strictly follow the Burgers relationship, and the close-packed plane and densely packed direction of the two phases are parallel to each other. Titanium alloys with Widmannstatten microstructure show excellent fracture toughness, crack growth resistance and creep strength, but their plasticity, fatigue strength, notch sensitivity and thermal salt stress corrosion resistance are poor.

(2) Basket-weave microstructure: the basket-weave microstructure is formed after the alloy is properly deformed in the α + β phase region and cooled, with the grain boundary α phase outlining the profile of the original β grains. The original β grains are elongated, and the αlamellae inside are interleaved in a basket-weave pattern. Cooling rate and deformation are two factors affecting the formation of basket-weave microstructure. Faster cooling rate and higher degree of undercooling lead to more nucleation sites of α phase in β matrix and thereforeeasier formation of basket-weave microstructure. After deformation in the β phase region, the dislocation substructure formed in β grain provied nucleation site for α phase and facilitate the formation of basket-weave microstructure. Titanium alloys with basket-weave microstructure are characterized by high creep resistance, creep rupture strength, impact toughness and fracture toughness, but low plasticity.

(3) Duplex microstructure: duplex microstructure is obtained by heating or deformation in the upper part of α + β phase region, consisting of equiaxed primary αand transformed β. The thermomechanical treatment is implemented in four stages: homogenization in β phase region, deformation in α + β phase region, recrystallization and low temperature annealing in α + β phase region. Compared with Widmannstatten structure, its grain size is generally smaller. The microstructure characteristics are primarily determined by the cooling rate and recrystallization temperature after homogenization and recrystallization. For the duplex microstructure, the most important characteristics are the size and volume fraction of the primary α phase, which decrease with increase of recrystallization temperature. Inheriting the characteristics of both the basket-weave and equiaxed microstructures, the duplex microstructure is endowed with excellent comprehensive mechanical properties including high plasticity and fatigue strength, but its fracture toughness and high-temperature properties are lower than those of basket-weave microstructure.

(4) Equiaxed microstructure: the morphology of equiaxial structure is affected by heating temperature, deformation mode and deformation. The lower temperature, higher amount of deformation, and larger proportion of equiaxed α phase, results in smaller grain size. Repeated heating and deformation is necessary to acquire equiaxed microstructure. Insufficient deformation would result in

fiber-like microstructure, and even with original β grain boundary. Generally, equiaxed and duplex microstructures can transform mutually through certain thermomechanical treatment, but the grain size will change. Titanium alloy with the equiaxed microstructure generally exhibits excellent strength, plasticity and fatigue properties, but its crack growth resistance is poor.

1.2 Deformation Mechanism and Mechanical Properties of Titanium Alloy

Plastic deformation is closely related to the crystal structure, and the deformation mechanism of titanium alloy is controlled by the characteristics of α and β phases. When the alloy composition is fixed, controlling the proportion, morphology and distribution of the two phases through processing and heat treatment are the basis for regulating the mechanical properties of titanium alloy. There are complex factors that affect the strength, plasticity, toughness, fatigue, creep and other properties of titanium alloy, it is rather hard to ensure that all properties are optimized, so choice must be made for different components based on their requirements for specific properties.

1.2.1 Deformation Mechanism

The difficulty of metal plastic deformation gradually increases in the order of face-centered cubic (fcc) structure, body-centered cubic (bcc) structure, and hexagonal close packed (hcp) structure. For titanium alloy, the deformability of α phase (hcp structure) is lower than that of β phase (bcc structure). The slip plane of α phase mainly include basal plane (0002), cylindrical plane $\{10\bar{1}0\}$, cone plane $\{10\bar{1}1\}$ and cone plane $\{11\bar{2}2\}$. The slip direction may be $<11\bar{2}0>$, [0001], or $<11\bar{2}3>$, as shown in Fig. 1.7. The most common slip systems in α phase are basal a slip system $\{0002\}<11\bar{2}0>$, cylindrical a slip system $\{10\bar{1}0\}<11\bar{2}0>$, and conical a slip system $\{10\bar{1}0\}<11\bar{2}0>$. In addition, conical $<c+a>$ slip is also an important slip system, which normally occurs on the first conical plane $\{10\bar{1}0\}$.

The difficulty for activation of a slip system can be expressed by the critical resolved shear stress (CRSS). The smaller the CRSS, the easier the plastic deformation will occur. Close-packed plane of atoms is normally known as the slip plane in the crystal that is most easily to be initiated, and the direction where slip is most easily to occur is the close-packed direction. At room temperature, the cylindrical *a* slip system of titanium alloy α phase is most easily initiated, followed by basal *a* slip system, and the conical $<c+a>$ slip system is least easily initiated. The CRSS of the conical slip system decreases faster than that of the basal and cylindrical slip systems as the temperature rises. Therefore, the conical $<c+a>$ slip is easier to be initiated at elevated temperature.

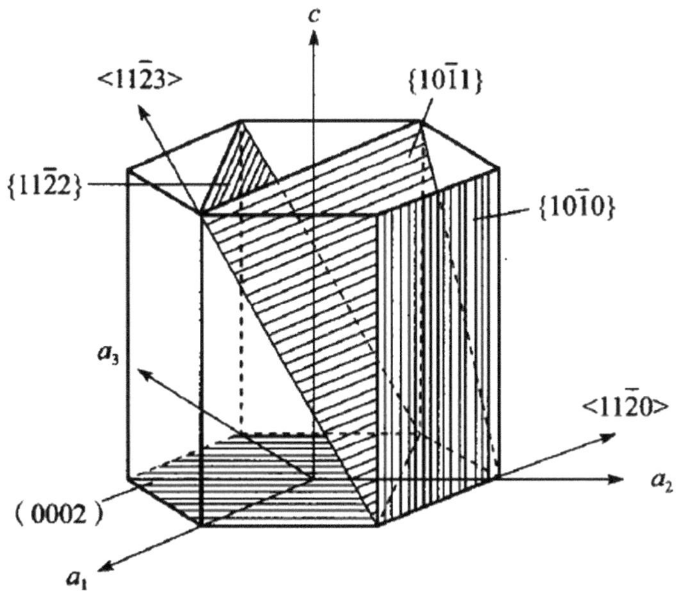

Fig. 1.7 Slip system of hexagonal crystal system

Twinning is also one of the important deformation mechanisms of titanium, mainly consisting of {1122} < 1123 > twinning system and {1012} < 1011 > twinning system. Twinning is formed to coordinate the deformation of α phase along the *c*-axis. However, with the increase of aluminum content, the probability of twinning deformation decreases dramatically and it can only occur at very low temperature. As the temperature increases, the dislocation slip becomes easier and the twinning deformation mechanism is further suppressed. Therefore, within the common range of strain rate and temperature, the deformation of titanium alloy in the *c*-axis direction occurs in the manner of conical $< c + a >$ slip.

1.2.2 *Texture and Anisotropy*

Texture is the preferential distribution of grain orientation in polycrystalline metals, where the microstructure and properties of materials show anisotropy from macroscopical view, and the grain orientation shows a significant deviation from random distribution from a microscopical view. Texture is a common phenomenon in materials, including casting texture, deformation texture, phase transformation texture and recrystallization texture. The presence of texture will lead to anisotropy in mechanical, optical, electromagnetic and other properties of polycrystalline materials. Crystallographic texture is greatly affected by the factors such as deformation mode, deformation amount, deformation temperature and recrystallization annealing. The texture intensity enhances with the increase of deformation.

The hcp α-phase shows inherent anisotropy. The texture exhibits different characteristics with the change of deformation temperature and deformation mode. When subject to rolling or upsetting below 900 °C, the material produced is dominated by basal texture, and the rolled material also shows cylindrical texture with certain strength. The texture is weak when hot worked at the temperature of 900 °C-930 °C. When the hot working is done at a temperature slightly lower than the β transformation temperature, the (0002) pole figure of rolled material shows the transverse texture, but shows the radial distributed texture under the upsetting,. When the processing is carried out above the β transformation temperature, cubic texture of the material is formed. Figure 1.8 shows the effect of deformation temperature and deformation mode on the texture of Ti-6Al-4V alloy.

Upon annealing treatment, the deformed titanium alloy exhibits the recrystallized texture. Formation of the recrystallization texture is affected by the factors such as crystal structure, original grain size, deformation temperature, deformation amount, heating temperature and cooling rate, pre-recovery and alloying elements. Recrystallization texture is formed on the basis of deformation texture, and is available in two types:① the recrystallization texture maintains the original deformation texture; ② the new texture replaces the original texture after recrystallization. According to the principle of thermodynamics, the formation of recrystallization texture is the process of changing towards the direction of system free energy reduction. There is a controversy on the formation mechanism of recrystallization texture. At present, the

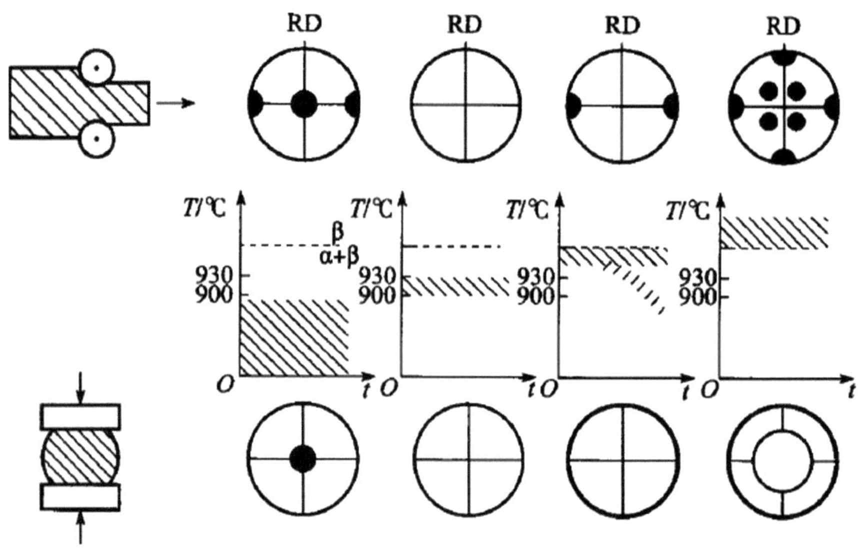

Fig. 1.8 Effect of deformation temperature and deformation mode on the texture of Ti-6Al-4V alloy

oriented nucleation and oriented growth theory is more acceptable. Later, the integrated theory of oriented nucleation and oriented growth was put forward, making the two theories complementary to each other.

1.2.3 Mechanical Properties

The microstructure charateristics of titanium alloy include the grain size of the original β phase, size of the α-phase lamellae, thickness of the α-phase lamellae, volume of the primary α-phase and the size of the primary α-phase. Following the change of these parameters, the mechanical properties of titanium alloy may be varied in a wide range. Based on the accumulation of massive data on microstructure and mechanical properties, the academic community has established property prediction models for various types of titanium alloys by way of regression analysis, crystal plastic finite element, artificial neural network and others. However, a majority of the models cannot accurately reflect the complex relationship between the mechanical properties and microstructure of titanium alloys, leading to a large deviation between the calculated values and the measured values, demonstrating less universality.

However, from a qualitative point of view, there is a basic consensus in the academic community regarding the relationship between microstructure characteristics and mechanical properties of titanium alloys (Table 1.2). For example, for α + β titanium alloy with fully lamellar microstructure, the size of αlamella is the most influencing factor as it determines the effective slip length. As the size of α phase lamellae decreases, the strength, plasticity and high cycle fatigue of titanium alloy improve. For duplex and equiaxed microstructures, the size of primary α phase on mechanical properties has the similar influence as the α lamellae. The qualitative relationship between the microstructure and properties of β titanium alloy is shown in Table 1.3.

Fatigue is the most critical mechanical property of titanium alloy, and also the main cause of failure in aviation service environment. Fatigue property of titanium alloys is affected by a number of factors, including chemical composition, microstructure, crystal texture, environment, test temperature and load bearing conditions (such as load amplitude, frequency, load sequence or average stress). For α + β titanium alloys, the fatigue properties are not only affected by the microstructure factors as listed in Tables 1.2 and 1.3, but also related to the microstructure type.

Figure 1.9 shows the effects of seven different microstructure types on the fatigue properties of Ti-6Al-4V alloy. For a smooth specimen, the α + aged Widmannstatten structure shows the highest high-cycle fatigue strength, followed by β processed fully lamellar microstructure, duplex microstructure and fully equiaxed microstructure in sequence. Among them, the fatigue strength of titanium alloy with duplex microstructure containing 15% α_P (α_P represents the primary α phase) is higher than that containing 50% α_P. This indicates that the duplex microstructure demonstrates higher fatigue strength when the volume fraction of α_P is small. Due to the coarsening of microstructure caused by over-aging, the fatigue strength of α + over-aged

Table 1.2 Qualitative relationship between important microstructure parameters and some mechanical properties of $\alpha + \beta$ titanium alloys

Characteristics of microstructure	Yield strength $\sigma 0.2$	Elongation ε_F	High-cycle fatigue (HCF)	Micro crack (ΔKth)	Macro crack			Creep strength 0.2%
					ΔKth (R = 0.7)	ΔK_{IC}	ΔKth (R = 0.1)	
Small α crystal cluster α sheet	+	+	+	+	−		−	±
Duplex microstructure	+	+	−	−	−		−	−
Small α grain size	+	+	+	+	−		−	−
Aging, oxygen	+	−	+	−	−		+	+
Secondary α phase in β phase	+	−	+	+	0		0	+
Texture: Stress axis is parallel to C-axis	+	0	+ (Vacuum) − (Air)	0 (Vacuum) −(Air)	0 (Vacuum) − (Air)		0 (Vacuum) − (Air)	+

Note Symbols + , −, and 0 indicate the change direction of mechanical properties with the change of the microstructure parameters, wherein + indicates the property is improved,—indicates the property is reduced and 0 indicates there is no effect. $\boldsymbol{\Delta K_{th}}$ indicates the threshold of fatigue crack growth, $\boldsymbol{K_{IC}}$ indicates the fracture toughness, and R indicates the stress ratio

Table 1.3 Qualitative relationship between the microstructure parameters and mechanical properties of β titanium alloys

Characteristics of microstructure	Yield strength σ0.2	Elongation ε_F	High-cycle fatigue (HCF)	Micro crack (ΔKth)	Macro crack			Creep strength 0. 2%
					ΔKth (R = 0.7)	ΔK_{IC}	ΔKth (R = 0.1)	
Grain boundary α lamelle in β annealed structure	0	–	–	–	0	+	0	0
Duplex structure	0	+	+	+	–	–	–	0
L-direction necklace-shaped structure	0	+	+	+	–	+	–	0
L-direction β processing structure	0	–	–	–	+	+	+	0
Reduced aging enhancement	–	+	–	+	+	+	+	–
Small β grain in a β-transformation microstructure	0	+	+	+	–	–	–	0

Widmannstatten structure is lower than that of α + aged Widmannstatten structure. For the notched specimens, titanium alloy with β-processed fully lamellar microstructure shows the highest fatigue strength, followed by titanium alloys with duplex microstructure and Widmannstatten structure, and fully equiaxed microstructure. However, concerning the influence of the microstructure on the high-cycle fatigue properties of titanium alloys, the conclusions obtained by other researchers are not consistent with the above. Zhao Yongqing et al. believe that the titanium ally with fully equiaxed microstructure shows higher fatigue strength. Zuo and Niinomi conclude that the fatigue strength of titanium alloy with duplex microstructure is higher than that of titanium alloy with lamellar microstructure. G. Q. Wu collected the data about the high-cycle fatigue properties of Ti-6Al-4V alloy published between 1972 and 2013, and analyzed the influence of microstructure on the high-cycle fatigue properties of Ti-6Al-4V alloy using statistical method. As a result, he found that the high-cycle fatigue strength reduces successively in the order of duplex microstructure, lamellar microstructure and equiaxed microstructure. In addition, it is pointed out that the fatigue properties of Ti-6Al-4V alloy with the same microstructure may be different due to the different test methods and hot working processes by different researchers.

The fatigue properties of Ti-6Al-4V alloy are also associated with the specific microstructure characteristics (Fig. 1.10). For titanium alloys with the lamellar microstructure, the major microstructure characteristics that affect the high-cycle

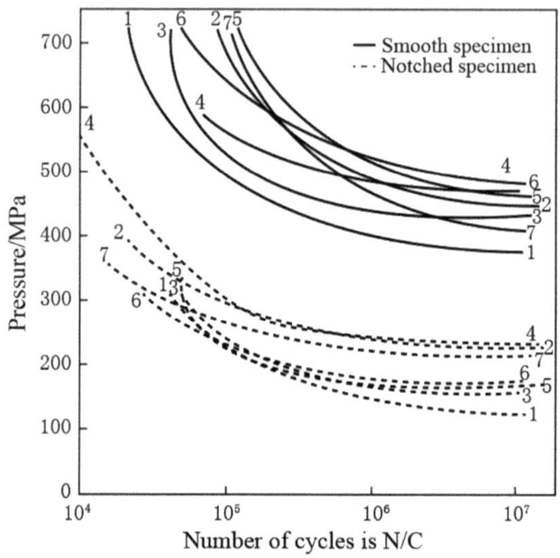

Fig. 1.9 Impact of different microstructure types on the fatigue properties of Ti-6Al-4V alloy

fatigue performance are the grain size of the original β phase, size of α and β cluster, and width of α phase lamella. The size of α cluster determines the effective slip length in lamellar microstructure. It decreases with the increase of cooling rate from β phase region, and the high-cycle fatigue strength increases with the decrease of α cluster size. Peters et. al. found that the high-cycle fatigue strength could be increased from 480MPa to 675Mpa by reducing the width of the α phase lamellar from 10μm to 0.5μm. Since the dislocation slip length is determined by the size of α-phase grains in equiaxed microstructure, titanium alloys with the fine equiaxed microstructure shows higher high-cycle fatigue strength. Trojahn found that the fatigue strength of materials increased from 560 to 720MPa when the α-phase grain of equiaxed microstructure was reduced from 12μm to 2μm. For titanium alloys with the duplex microstructure, the microstructure characteristics that affect the high-cycle fatigue strength mainly include the size and content of primary α phase grains, width of secondary α phase lamella and the size of original β phase grains. According to the work by Lütjerings, with the same content of the primary α phase content, the small secondary α lamellae could significantly improve the high-cycle fatigue strength of duplex microstructure. It is possible to increase the fatigue strength from 480 to 575MPa by reducing the width of α lamellae from 1μm to 0.5μm. The width of the secondary α lamellae is affected by the cooling rate. Hall used two different cooling modes, quenching and air cooling and found that the faster cooling rate can result in great improvement of the fatigue properties of duplex microstructure. In addition, the high-cycle fatigue strength usually decreases with increasing content of primary α phase. The reason lies in that alloying elements are reallocated during the separation of the primary α phase from β phase, in which case the basic strength of lamellar region in the duplex microstructure becomes lower than that of the fully lamellar microstructure, thereby causing impact on the overall high-cycle fatigue strength of

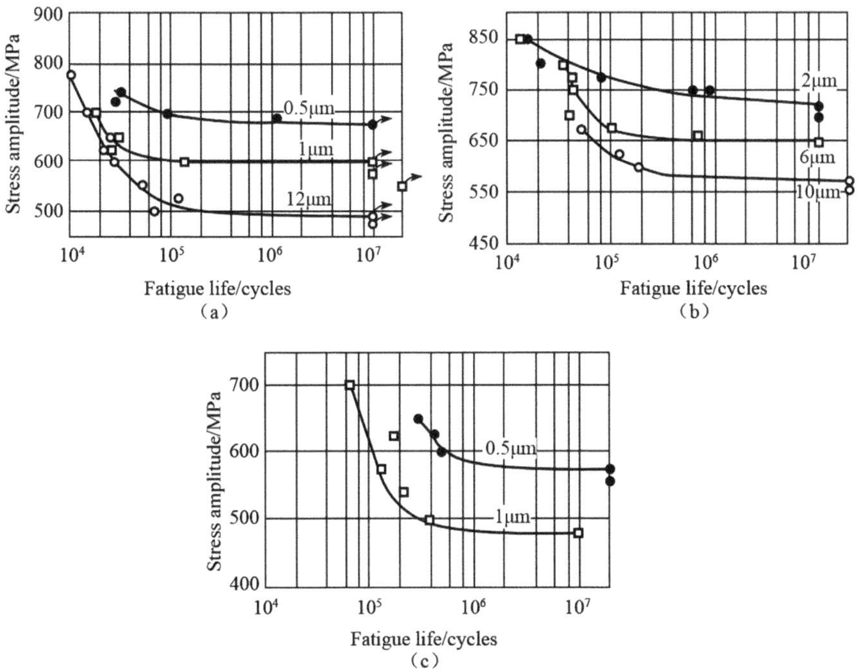

Fig. 1.10 Effect of microstructure characteristics on the high-cycle fatigue properties (R = −1) of Ti-6Al-4V alloy **a** Effect of lamellae width (lamellar microstructure); **b** Impact of α-phase grain size (equiaxed microstructure); **c** Effect of lamellae width (duplex microstructure)

the material. There are adverse effects from reallocation of alloying elements, which can be reduced and eliminated by adding the process of intermediate annealing.

Wagner et. al. presented the typical crack initiation location of α + β titanium alloy. In the lamellar microstructure, cracks are initiated in the slip band within the α-phase lamellae or along the α-phase of the original β-phase grain boundary. The width of α-phase lamellae determines the ability to resist the dislocation movement and fatigue crack initiation, so there is a direct correlation between the fatigue strength and yield strength. For equiaxed microstructure, cracks usually nucleate/initiate along the slip band in α-phase grains, so the fatigue strength has a close relationship with the grain size. In the duplex microstructure, cracks can initiate in the primary α phase, or in the lamellar matrix, or at the interface between the primary α phase and the lamellar matrix. The exact crack initiation location depends on the cooling rate, volume fraction and size of primary α phase.

The microstructure of titanium alloy affects the fatigue crack initiation and growth behaviors. The macroscopic crack growth resistance of microstructures is gradually reduced in the order of coarse lamellar microstructure, fine lamellar microstructure, coarse equiaxed microstructure, duplex microstructure and fine equiaxed microstructure, and this matches with the fact that lamellar microstructure showed higher fracture toughness than the equiaxed microstructure. However, the impact of microstructures on microscopic crack growth follows the opposite rules. The microscopic crack growth resistance shows a gradual increase trend for the coarse lamellar microstructure, fine lamellar microstructure, duplex microstructure and equiaxed microstructure. Figure 1.11 shows the growth behavior of microscopic crack in Ti-6Al-4V alloy with coarse lamellar microstructure and equiaxed microstructure.

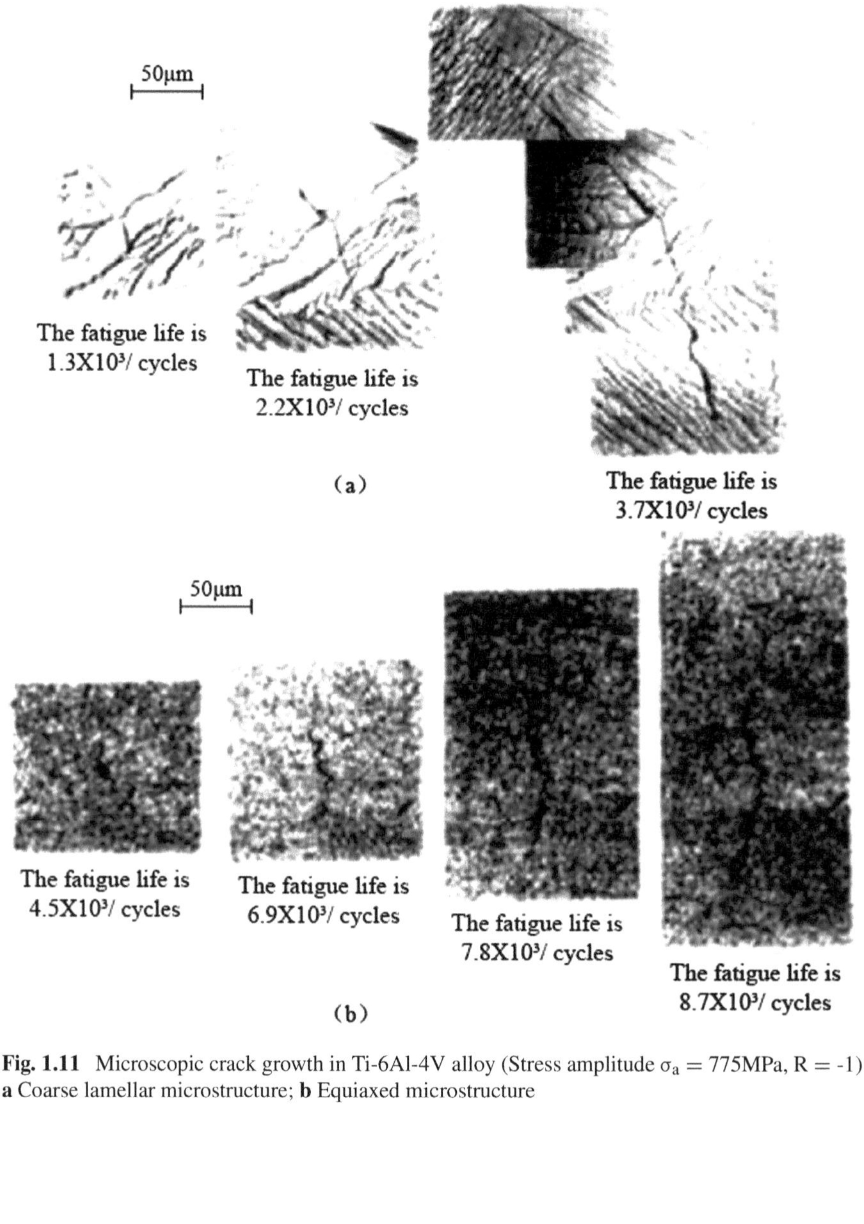

Fig. 1.11 Microscopic crack growth in Ti-6Al-4V alloy (Stress amplitude $\sigma_a = 775$MPa, R = -1) **a** Coarse lamellar microstructure; **b** Equiaxed microstructure

Comparison of the growth behavior of microscopic and macroscopic cracks in Ti-6Al-4V alloy is shown in Fig. 1.12. The coarse lamellar microstructure has lower microscopic crack growth resistance than the equiaxed microstructure, which is associated with the low interfacial density. However, the coarse lamellar microstructure is superior to the equiaxed microstructure in terms of the resistance to macroscopic crack growth. This is because the additional crack growth resistance caused by the crack front geometrical characteristics and the crack closure effect can delay the macroscopic crack growth. The crack growth test of Ti-6Al-4V alloy with fine lamellar microstructure and duplex microstructure showed that the crack growth curve of Ti-6AL-4V alloy is between that of coarse lamellar microstructure and equiaxed microstructure. Zhang Qingling further stated that the difference in fatigue properties between the titanium alloys with lamellar microstructure and the ones with duplex microstructure was caused by the different resistance to crack initiation and growth of the two microstructures. The crack initiation depends on the lattice strength to resist dislocation moment and the length of dislocation slip. Under low-cycle fatigue, the dislocation is prone to slip due to the high stress, and length of dislocation slip of the lamellar microstructure is much larger than that of the duplex microstructure, so the crack initiation is prone to occur in the lamellar microstructure; under high-cycle fatigue with low stress, the lattice strength represents the resistance to dislocation movement, playing a key role in crack initiation. With the increase of lattice strength, the crack initiation is less prone to occur.

Fig. 1.12 Comparison of microscopic crack and macroscopic crack growth behaviors of Ti-6Al-4V alloy

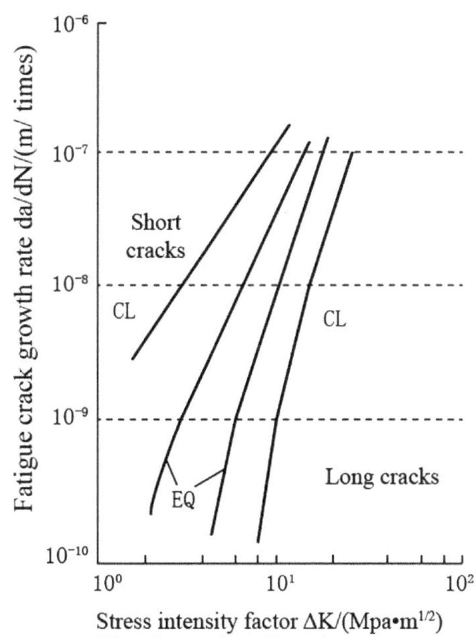

CL—Coarse lamellar microstructure;
EQ—Equiaxed microstructure

1.3 Application and Development Trend of Titanium Alloy in Aviation Field

Application of titanium alloy in aircraft structure contributes to excellent weight reduction of aircraft and meets the requirments of aircraft design for high maneuverability, high reliability and long life. Therefore, the usage of titanium alloy has become an important sign to measure the advancement of aircraft material selection. Thanks to the introduction of the design concept of damage tolerance in modern aircraft structures, the single pursuit for high-strength titanium alloys has gradually transformed to seeking comprehensive properties of strength, toughness, fatigue life, crack growth rate and other properties.

In the early 1950s, industrial pure titanium has been already ultilized in the manufacture of secondary bearing structural components, such as the rear fuselage insulation panel, tail cone fairing, and speed brake on some aircraft. In the 1960s, the application of titanium alloy in aircraft structural components expanded to major load-bearing structures such as flap rail tracks, load-bearing frames, central wing-box beams, landing gear beams, and helicopter propeller hub. Later, the application of titanium alloy expanded to military transport aircrafts and civilian aircraft. In various military aircrafts designed after the 1980s, the usage of titanium alloy has exceeded 20%, with the F-22 aircraft reaching an 41%. Among the 7 major titanium alloys, medium-strength titanium alloy Ti-6Al-4V, high-strength and high-toughness titanium alloys Ti-6242, Ti-1023, Ti-15–3, damage-tolerance titanium alloys Ti-6Al-4V ELI and Ti-6–22-22S, and tube-specialized titanium alloy Ti-3Al-2.5V, all satisfy the different design requirements for titanium alloys in various aircraft components. In addition, the usage of titanium in transport aircraft has increased from 6% in the early aircraft C-5 to 10.3% in the C-17 aircraft. In the civil aviation field, titanium usage has grown from 0.5% in the initial Boeing B707 to 7% in the B777, and up to 15% in the B787 aircraft. The Airbus company's new generation of wide-body passenger aircraft, the A350-XWB, also increased its titanium usage to 14%, mainly in areas such as fuselage joints, landing gear, wing suspension, pipelines, and others. The usage of titanium alloy in the composite materials is also expected to further increase due to its good chemical compatibility. It is predicted that the usage of titanium alloy will be increased to 20% in civil aircraft and 50% in military aircraft in the future. Figure 1.13 shows the comparison of titanium alloy usage in typical aircraft fuselages.

Currently, the medium-strength Ti-6Al-4V alloy is still the most widly used titanium alloy in the aviation field. Among the titanium alloys with higher strength, one type is the $\alpha + \beta$ titanium alloys represented by Ti-6–22-22S alloy (Ti-6Al-2Sn-2Zr-2Mo-2Cr-0.2Si) and TC21 alloy (Ti-6Al-2Sn-2Zr-3Mo-2Nb-1Cr-0.2Si), with a strength of about 1100MPa and a fracture toughness of about 70MPa·m$^{1/2}$. Another type is the β titanium alloy represented by Ti-10–2-3 alloy (Ti-10V-2Fe-3Al) alloy and Ti-5553 alloy (Ti-5Al-5V-5Mo-3Cr-0.5Fe), which are commonly referred to as high-strength titanium alloy. The strength of these alloys is in the range of 1100MPa-1250MPa, and the fracture toughness is 50MPa-80MPa m$^{1/2}$. Ti-10–2-3

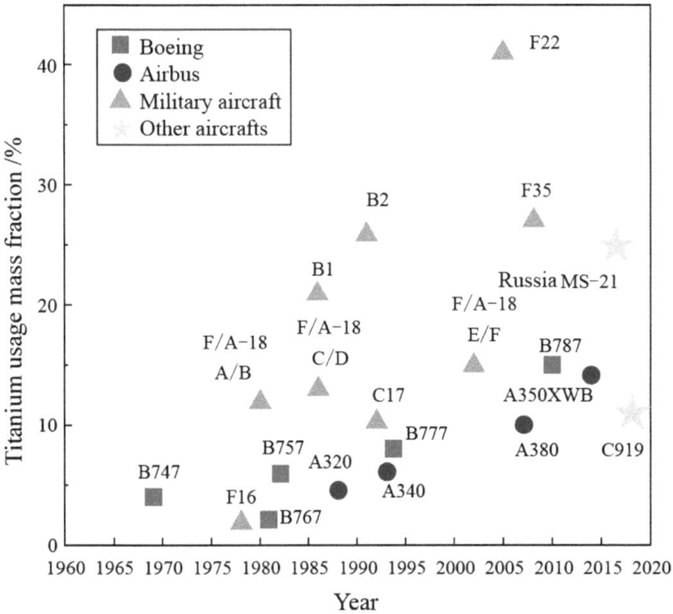

Fig. 1.13 Comparison of titanium usage in typical aircraft fuselage

alloy has advantages such as high specific strength and low forging temperature, making it particularly suitable for hot die forging or isothermal forging. It has been applied in the short cabin joint of the B737 aircraft, the landing gear of the large C-17 Globemaster and B777 passenger aircraft, and the main landing gear support of the A380. Ti-5553 and Ti-55531 alloys are similar to Ti-10–2-3 alloy but have better abitily for quenching, as well as good comprehensive properties of strength-plasticity-toughness, and are mainly used in high-strength and high-toughness parts such as aircraft landing gear, wing/hanger joints, and landing gear/wing joints. The β-21S alloy not only has high strength but also excellent creep resistance, and is used in the skin and various longitudinal beam structures of the B777 aircraft. BT22 alloy is easy to melt and cast, less prone to segregation for its low Fe content and has sound comprehensive mechanical properties and abitility for quenching, which is especially suitable for manufacturing large forgings. It has been used in fuselage wing, landing gear and other high-load aerospace components of IL-series transport aircrafts. On the basis of BT22 alloy, it is possible to improve the strength and high-temperature creep properties by adding Sn and Zr, thus forming BT37 alloy. Therefore, the strength and toughness of the high-strength titanium alloys need to be simultaneously enhanced to improve the structural efficiency of aircraft components and achieve greater weight reduction effects in application.

Apart from aircraft structure, titanium alloy are widely used in various parts of aerospace engine, such as fan blades, casing, compressor blades and disk. Good

Table 1.5 Titanium usage on typical aero-engines

Model of engine	TF36	TF39	JT90	F100	F101	CF6	V2500	F119	GE90	Trent900
Year of service	1965	1968	1969	1973	1976	1985	1989	1986	1995	2005
Equipment model	C-5A	C-5AB	B747/ 767	F-15/ 16	B1	A330 B747/ 767	A320 A321	F-22	B777	A380
Usage of titanium / %	32	33	25	25	20	27	31	39	40	41

comprehensive properties in strength, plasticity, toughness, creep and fatigue properties is necessary to meet the requirements of harsh working environments such as high temperature, vibration and airflow erosion. Ti-6Al-4V alloy has a maximum operating temperature of 350 °C, making it suitable for fan blades, low-pressure compressors blades and discs, while, near-α titanium alloys are generally used for high-pressure compressor parts. Currently, the highest operating temperature for titanium alloy is 600 °C. The Ti–Al-Sn-Zr-Mo-Si series titanium alloys, such as IMI-834, Ti-1100, BT18y, and BT36, have been used in Trent-700, EJ200, PW350, AL-31 and other engines as well as centrifugal impeller of turbo engines. The optimal microstructure is duplex microstructure containing 15% primary α phase. The α-phase lamellae may be formed on the surface of high-temperature titanium alloy during service, which can result in significant decrease of plasticity and fatigue properties. Therefore, in addition to the high strength and creep properties, the oxidation resistance is also one of the critical consideration for high-temperature titanium alloys.

Concerning the application history of titanium alloy in engines, Pratt & Whitney of the United States and Rolls-Royce of the United Kingdom had used titanium alloy in jet engines as early as in the 1950s. From then on, the usage of titanium alloy increased. In the third-generation aero-engines such as F100, the usage of titanium alloy was recorded to be 25%, while the value was increased to 40% for the fourth-generation aero-engines such as F119 (Table 1.5). The demand for aero-engines has driven the development of high-temperature titanium alloys, which in turn contributed to improved their thrust-to-weight ratio and fuel economy.

1.4 Forming Method of Titanium Alloy Structure

Due to the increasing demand for aircraft and engine components that require lightweight structures, integration, and high reliability, manufacturing technology now faces new challenges in producing titanium alloy parts. The forming methods for these parts include casting, powder metallurgy, additive manufacturing, and plastic forming. Selecting the appropriate forming method according to the different characteristics of the parts can achieve the desired shape while also improving the material microstructure and properties at a lower cost.

1.4.1 Casting

Investment casting of titanium alloy is a near net shape technique developed to meet the manufacturing requirements of thin-walled components in aviation. Currently, over 98% of titanium alloy castings used in the aerospace are produced by investment casting. The development of hot isostatic pressing technology and heat treatment techniques enables to improve the quality of titanium alloy castings to a level close to β annealed titanium alloy forgings. Casting techniques can produce parts with complex shapes, saving a significant amount of machining costs and improving the manufacturing efficiency.

Due to its high chemical activity, titanium in the molten state may react with commonly used casting materials to form a hard and brittle reaction layer on the casting surface when it reacts with oxygen, carbon or nitrogen. Investment graphite shell and tungsten surface ceramic shell were once used in titanium alloy precision casting. Currently, the most widely used is the oxide surface ceramic shell, and commonly used shell materials include yttrium oxide and zirconium oxide. Vacuum arc melting (VAR) and induction skull melting (ISM) are mainly used for melting and casting titanium alloy. Depending on the shape, size, and weight of the castings, either gravity casting or centrifugal casting may be selected. Centrifugal casting can be used to speed up the casting filling and improve fluidity and casting density for small and medium-sized castings with simple shapes, whereas gravity casting is typically used for complex and large components. Hot isostatic pressing (HIP) can be applied to eliminate casting defects such as shrinkage cavities and porosity in titanium alloy castings, refine the microstructure of castings, and improve their properties. Castings are now partially replacing forgings.

Early developments in titanium alloy precision casting took place in the United States, Japan, and Germany. The United States began studying the precision casting technique of titanium alloy in the 1960s, but it was not until the early 1980s that mass production of titanium alloy castings started. In the 1980s, Howmet and Tital became capable of producing complex, thin-walled integral titanium alloy precision castings with sizes over 400mm. Later, the number of titanium alloy castings achieved a gradual increase of 20% per year, benefiting from the investment casting technique, which allows the direct manufacturing of complex-shaped components, improving material utilization, and significantly shortening production cycles.

The new generation of American military aircraft is designed to reduce the overall weight of the aircraft by 50%, the number of fasteners by 80%, the production cost by 25% and the cycle by one third. Apart from extensive use of new materials, it is also necessary to use large integral precision castings to replace massive small components. Therefore, the investment casting technique of large integral structural components of titanium alloy is firstly promoted in the manufacturing of military aircraft structures such as F/A-22 and V-22. For example, Ti-6AL-4V alloy investment casting is used for the adapters on Boeing Bell V-22. Howmet Airlines and Bell Helicopter replaced the previous assembly consisting of 43 aluminum alloy forgings and 536 fasteners with 3 titanium alloy castings and 32 fasteners, resulting in a significant reduction in overall weight, a 62% reduction in production time and a 30% reduction in manufacturing costs. Ti-6Al-4V alloy castings were used in the side fuselage joints, vertical tail rudder actuator support and other key bearing parts on the wings of the American F/A-22 fighter, accounting for about 7.1% of the overall structural mass. Thin-walled Ti-6AL-4V castings with a thickness of 1.27mm were used for the complex parts of C-17 Globemaster, such as the cooling device. In 1999, titanium alloy investment casting was first used to manufacture the rear engine frame of Boeing 777, demonstrating the first successful application of titanium alloy investment casting in civil aircraft. The Airbus A380 also used a brake torsion hose made of titanium alloy investment casting technique to replace the previous forgings. Currently, the surface roughness R_a of titanium alloy castings can reach 6.3μm. With only a small amount of treatment, it can meet the application requirements. Investment casting is suitable for mass production of small titanium alloy parts with complex shapes and high surface requirements, as well as the production of large thin-walled castings. The smallest thickness of the wall can be up to 0.5mm.

1.4.2 Additive Manufacturing

As a new direction of near-net-shape manufacturing, additive manufacturing enables the direct manufacture of parts by 3D digital modeling based on layer-by-layer deposition. It can quickly and accurately manufacture the parts with complex shape on one set of equipment, without needing a mould, greatly reducing the process and shortens the cycle. Hence, it is particularly well-suited for manufacturing titanium alloys, high-temperature alloys and other materials difficult to process. Additive manufacturing offers significant advantages in cost and efficiency compared with traditional manufacturing methods, especially for parts with more complex structures. In the development of aircraft, the additive manufacturing is indispensable.

Additive manufacturing techniques of titanium alloy materials can be divided into two categories: directed energy deposition and powder bed fusion technologies. Depending on the heat source used, there are 5 techniques available: laser powder deposition, electron beam direct manufacturing, wire arc additive manufacturing, selective laser melting and electron beam selective melting. Directed energy deposition techniques offer the advantage of high efficiency, good platform flexibility, and the ability to produce large parts. Additionally, this technique enables the repair and secondary manufacturing of original parts. However, it is suitable for manufacturing complex hollow structures, and the formed blanks require complex mechanical processing to obtain the final parts. In contrast, selective laser melting offers the greatest advantage of forming complex and precise structures, such as blades with internal flow channels or spatial lattice structures. However, the size of the parts formed is limited by the forming cavity space. Each of these five techniques has its respective advantages, and it is important to choose the appropriate process based on the characteristics of the parts to be formed. Directed energy deposition techniques, such as laser powder deposition, electron beam direct manufacturing, wire arc additive manufacturing, are suitable for the rapid forming of large parts. while powder bed fusion techniques, such as selective laser melting and electron beam selective melting, are suitable for the precision forming of small complex parts.

The development of laser powder deposition technique started early. In 1995, Sandia National Laboratories in the United States developed a technique that uses a laser beam to melt the metal powder layer by layer to produce dense metal component. Since then, extensive research has been conducted on various materials such as titanium alloys, high-temperature alloys, and stainless steel. Since 1995, under the co-funding of Defense Advanced Research Projects Agency (DARPA) and United States Naval Research Laboratory (NRL), Johns Hopkins University, Penn State University, and MTS company have jointly developed the technique of using high-power CO_2 lasers to make large-size titanium alloy parts, and later AeroMet company was founded, accomplishing the 1-2kg/h deposition rate of Ti-6Al-4V alloy. AeroMet was sponsored by the U.S. military to conduct research on laser powder deposition technique of titanium structural components used on the aircraft fuselage, completed performance evaluation and standard formation, and realized the installation application of Ti-6Al-4V alloy secondary bearing components on F/A-18 and other aircrafts.

In the field of electron beam melting deposition forming, the American company Sciaky cooperates with Lockheed Martin, Boeing, and other companies to carry out research on electron beam melting deposition technology for large aerospace titanium alloy parts. Lockheed Martin has selected the flaperon beam of the F-35 aircraft as the test piece for electron beam melting deposition forming, which has reduced the part cost by 30% to 60%. In addition, a research group led by the American company CTC has developed the "Unmanned Aerial Vehicle Metal Manufacturing Technology Enhancement Program" for the Navy's unmanned combat drone program. The electron beam melting deposition technology is seen as a way to achieve low-cost and efficient manufacturing of large structures in the future, with the goal of reducing the weight and cost of unmanned aerial vehicle titanium alloy structures by 35%. Since 2010, Norway's Norsk Titanium has developed arc melting deposition equipment and produced titanium alloy parts up to 1 m in length. The company's titanium alloy wire arc additive manufacturing (WAAM) technology was certified as Technology Readiness Level 8 by the US Federal Aviation Administration in 2016. In addition, Cranfield University has developed plasma arc-based titanium alloy WAAM technology, which has higher deposition efficiency and is easier to control. These three direct energy deposition technologies can significantly reduce costs and shorten iteration cycles in the development and validation stages of aerospace titanium alloy structures. In the field of aviation, the most promising technologies are selective laser melting and electron beam selective melting, which is highly proved by acquisition of German Concept Laser company and Swedish Arcam company by General Electric (GE) in 2017. Only selective melting technology has the ability to shape complex and precise structures, which is highly compatible with the needs of structural optimization of aviation parts. The first part made by selective laser melting is a high-temperature alloy fuel nozzle in GE's LEAP engine, with the original 20 components now replaced by one component. As a result, the weight is reduced by 25% and the service life is prolonged by 5 times. Till now, there are more than 100,000 fuel nozzles fabricated by the selective laser melting technology installed in the LEAP engines, powering the B737 MAX and A320 neo aircraft. Instead of the original casting technology, GE has achieved first application of electron beam processing technology in manufacturing TiAl low pressure turbine blades, which has entered the production stage now.

However, random distribution of defects and special microstructures are formed during the layer-by-layer processing. Thus, the components exhibit location-dependent mechanical properties. Moreover, the mechanical properties of specimens fabricated in different batches varies from each other.

The properties of different parts of varied, so the batch properties are relatively dispersive. Therefore, currently, the titanium alloy parts manufactured by additive manufacturing technology are mainly used in functional structure or secondary bearing structure. How to detect small defects and establish the design criteria based on the impact of defect size, number and distribution on dynamic mechanical properties are the key to expand the application range of additive manufactured titanium alloys in aeronautical structures.

1.4.3 Powder Metallurgy

Powder metallurgy is a technology that realizes near-net forming of parts by pressing and sintering powder as the raw material. By controlling the quality and forming process parameters of titanium alloy powder, parts with uniform microstructure and excellent mechanical properties can be obtained, which greatly improves the material utilization rate and significantly reduces the amount of subsequent mechanical processing. Therefore, compared with traditional forging process, powder metallurgy has a cost advantage in the preparing of parts with more complex shape.

Hot isostatic pressing is an indispensable process for preparing high-quality parts. Casting or sheath filled with metal powder are placed in a hot isostatic pressing furnace. Taking the inert gas as the pressure transfer medium, the sheath and casting are subjected to uniform pressure (100-2000MPa) from all directions in the high temperature environment (normally 0.6–0.7 of the metal powder melting point). The metal powder inside the sheath softens and deforms under the action of high temperature and pressure, and the pores between the powders gradually close and finally form the densified parts. There are two typical processing t for titanium alloy powder metallurgy: the first is to densify the power by hot isostatic pressing, and then fabricate components by means of isothermal forging or superplastic forging; the second is direct forming of powder by hot isostatic pressing.

In 1956, GE corporation employed the hot-pressing method to produce GET73 turbojet engine bearing seat blanks made of sponge titanium. Due to its high forming accuracy and minimal cutting processing, the cost was reduced by over 25% compared to forging. Grumman Aerospace Corporation of the United States used ceramic film to produce the titanium alloy internal support rods and fuselage pillars for the F-14 fighter jet, increasing the material utilization from about 20% in forging to over 50%. The hot isostatic pressing of titanium alloy powder was also applied in producing the nose keel of F-15 fighter through the engine mounting bracket of F-18 fighter. The companies such as Bodycote, Crucible, ADMA, Pratt & Whitney in the United States have produced various titanium alloy parts, including the F-14A cockpit frame, Sidewind missile powder titanium alloy hood, F107 cruise missile engine powder titanium alloy impeller, etc.

With the increasing improvement of various types of software, simulation of hot isostatic pressing (HIP) process by finite element modeling has become a research hotspot in recent years. By combining the functions of 3D design software such as CATIA with finite element simulation software such as ABAQUS, researchers have investigated researchers have investigated the shrinkage laws of key dimensions, assisted in the design and prediction of critical dimensions of the HIP can, and integrated the design of the can, the densification process of titanium alloy during HIP, and simulation of powder metallurgy products. This approach has shortened the cycles and reduced cost, providing strong support for the preparation of various titanium alloy parts using HIP. Further study is needed to explore how to combine the related software with actual production to further reduce the cost of titanium alloy powder metallurgy parts.

1.4.4 Plastic Forming

Plastic forming uses the plasticity of materials to make parts under the action of external forces. Plastic forming can not only process the alloy into the desired shape, but also tailor microstrucure and improve mechanical properties. Titanium alloy is characterized by low plasticity and high deformation resistance at room temperature. For most titanium alloys, the traditional cold-working plastic forming process is difficult to achieve good forming results. When the titanium alloy is under heating, especially at a temperature above 500 °C, its plasticity significantly improves, the deformation resistance significantly decreases, and the cracking tendency reduces. Meanwhile, the thermal effect at high temperature enables to eliminate the internal stress generated from the material deformation, reduce the springback and improve the forming accuracy of the parts. Therefore, for titanium alloy, the widely used plastic forming process mainly forging process and sheet forming process represented by hot forming process.

Forging is one of the key plastic forming processes for manufacturing titanium alloy parts. Forging can not only obtain qualified forging shape, and more importantly, improve the mechanical properties of the alloy, thus to meet the engineering requirements. Due to the inherent crystal structure and sensitivity to process parameters, titanium alloy belongs to hard deformation material, which exhibits higher difficulties in processing as compared to other alloys. The forging process parameters of titanium alloy, including forging temperature, deformation amount, deformation rate and post-forging cooling rate, etc., have a significant effect on the the microstructure and properties of titanium alloy. Titanium alloy involves a number of forging methods such as free forging, mould forging, rolling, ring rolling and extruding. Taking β phase transformation temperature as the temperature boundary, titanium alloy forging is divided into two types of $\alpha + \beta$ forging and β forging. With the search on the technological process in recent years, many new forging processes have been developed, such as near-β forging, quasi-β forging, hot mould forging, isothermal forging and superplastic forging, which have been applied in production.

Titanium alloy forging has been widely used in the field of aerospace, including manufacture of aircraft. It mainly used in manufacturing the critical parts and important parts that bear alternating and concentrated loads in the aircraft and engines, such as the frames, beams, landing gear and joints in the aircraft body, engine disc, shaft, blade, ring, etc. For example, the four bearing frames of the F-22 fighter were made of Ti-6Al-4V ELI alloy large integral frame mould forging, and the projected area of the forging is 4.06–5.67m^2; Ti-6Al-4V titanium alloy forging was used to manufacture the main landing gear transmission beam of B747 aircraft, which was 6.20m long and 0.95m wide, with a projection area of 4.06m^2, a weight of 1545kg; Ti-10V-2Fe-3Al titanium alloy forging was used to manufacture the main landing gear load beam of B777 aircraft, with a mass of 3175kg and a projected area of 1.23 m^2, which is one of the largest β titanium alloy forgings till now. Further, Ti-10V-2Fe-3Al titanium alloys have been used by Bell Helicopter, Sikorsky, and Westland to manufacture the structural components of the helicopter rotor systems, showing excellent high-cycle fatigue properties. Thanks to the development and application of large forging equipment featuring large tonnage, high accuracy and high efficiency, the forging technology of titanium alloy large integral structure is able to change the traditional multi-piece combined components into integral structural components. This greatly reduces the structural weight of aircraft, and improves the structure beneficial results and the safety and reliability of parts.

Hot forming technology is one of the important manufacturing techniques for aerospace sheet metal parts. By utilizing the property of metal material softening under heating, the ductility of the material is improved and springback is reduced, thus achieving precise forming of aerospace sheet metal parts. The main influencing factors of hot forming are temperature, pressure, and time. As the temperature increases, the strength of titanium alloy decreases and the plasticity increases. For the commonly used Ti-6Al-4V alloys, the hot forming temperature is in the range of 680–750 °C. Higher pressure has significant benefits in eliminating wrinkles, but has little effect oninhibiting the springback after unloading. In case of forming at low temperature, the loading speed has little effect on deformation resistance and plasticity, but the effect becomes increasingly obvious at moderate and high temperature. However, longer hot forming time will increase the thickness of the oxide layer on the surface of the titanium alloy. Hence, it is advised to shorten the forming time accordingly in case of high hot forming temperature, which should not be more than 1h for Ti-6Al-4V alloy.

Since the 1950s, there was an increasing demand for titanium alloy in the aerospace field, hot forming technology has rapidly entered the engineering stage, and has been widely used in the manufacture of aircraft skin, heat insulation frame, engine cold-end parts, missile shells, wings and other structures. The United States, Russia, Japan, Britain, France, Italy and other countries have established production and research bases for hot forming technology. Starting from the SR-71 Blackbird (high-speed air early warning aircraft), the parts manufactured by hot forming technology have been applied in almost all aircrafts and missiles. The manufacturing precision and quality of aviation sheet metal parts will directly affect the aircraft shape, structural life, assembly quality and aircraft performance. From the view of overall development, sheet metal parts account for a large proportion in advanced aircraft manufacturing. Even though more composite materials are used innew fighter aircrafts and wide-body aircrafts, sheet metal parts still play an irreplaceable role.. For example, sheet metal parts on F-22 aircraft account for approximately 45%, but the labor required for the production of sheet metal parts only accounts for about 15% of the total, and the tooling required for the sheet metal parts accounts for about 70% of the total. The sheet metal parts also show great importance for large transport aircrafts. For example, IL-86 has approximately 70,000 sheet metal parts, with sheet metal parts accounting for about 18%, profiled parts accounting for about 78%, and pipe parts accounting for about 4% by quantity. The metal plate structural components for aviation purpose, play a very key role in improving the performance of aircraft and engines. However, the demand for lightweight and integrated aircraft fuselage and engine is becoming increasingly urgent, and conventional hot forming technology has difficulty fully meeting the manufacturing requirements of such structures.

References

1. Lutjering, G., and J.C. Williams. 2007. *Titanium*, 2nd ed. Berlin: Springer, Berlin Heidelberg.
2. Leyens, C., and M. Peters. 2005. *Titanium and titanium alloys*. Translated by Chen Zhenhua Beijing: Chemical Industry Press.
3. Xiangming, Wang, and Liu Wenting. 2010. *Structural design and application of aircraft titanium alloy*. Beijing: National Defense Industry Press.
4. Yongqing, Zhao, Chen Yongnan, Zhang Xuemin, et al. 2012. *Phase transformation and heat treatment of titanium alloys*. Changsha: Central South University Press.
5. Zhishou, Zhu. 2013. *Research and development of new-brand titanium alloys of high performance for aeronautical applications*. Beijing: Aviation Industry Press.
6. Boyer, R.R. 1996. An overview on the use of titanium in the aerospace industry. *Materials Science and Engineer-ing A* 213 (1–2): 103–114.
7. Ting, Lei. 2018. *Titanium and titanium alloys*. Beijing: Metallurgical Industry Press.

8. Huang, Xu., Zhu Zhishou, and Wang Honghong. 2012. *Advanced aeronautical titanium alloys and application*. Beijing: National Defense Industry Press.
9. Huang, Xu. 2012. General development situation of titanium alloys for aviation. *Dual Use Technologies and Products* 7: 12–14.
10. Ling, Guo. 2011. *Foring techniques of advanced aeronautical materials and components*. Beijing: National Defense Industry Press.
11. Yang Baoxiang, Hongfei Hu, Jinyong He, et al. 2015. Titanium based material manufacture. Beijing: Metallurgical Industry Press.
12. Zhu, Zhang, Xie Shuisheng, and Zhao Yunhao. 2010. *Plastic working techniques of titanium material*. Beijing: Metallurgical Industry Press.
13. Beijing Aeronautical Manufacturing Technology Research Institute. 2013. *Aeronautical manufacturing technology*. Beijing: Aviation Industry Press.
14. Boyer, R.R., and R.D. Briggs. 2005. The use of β titanium alloys in the aerospace industry. *Journal of Materials Engineering and Performance* 14 (6): 41–43.
15. Zhiqiang, Li., and Guo Heping. 2010. Application progress and development tendency of superplastic forming/diffusion bonding technology. *Aeronautical Manufacturing Technology* 8: 32–35.
16. Xiuquan, Han. 2013. Titanium alloy forming technology: Challenges and opportunities. *Aeronautical Manufacturing Technology* 438 (18): 68–69.
17. Zhishou, Zhu. 2014. Research status and development of aviation titanium alloy technology in China. *Journal of Aeronautical Materials* 34 (4): 44–50.
18. Huang, Z., H. Qu, C. Deng, et al. 2011. Development and application of aerial titanium and its alloys. *Materials Reports* 25 (1): 102–107.
19. Yongqing, Zhao. 2010. Development status and trend of titanium alloy research at home and abroad. *Materials China* 10: 1–8.
20. Jian, Yang. 2006. Application of titanium alloy on aircraft. *Aeronautical Manufacturing Technology* 11: 41–43.
21. Beijing Institute of Aeronautical Materials. 2013. *Material technology of aeronautics*. Beijing: Aviation Industry Press.
22. Xiyan, Zhang, Zhao Yongqing, and Bai Chenguang. 2005. *Titanium alloy and its application*. Beijing: Chemical Industry Press.
23. Cotton, J.D., R.D. Briggs, R. Boyerr, et al. 2015. State of the art in beta titanium alloys for airframe appli-cations. *JOM Journal of the Minerals Metals and Materials Society* 67 (6): 1281–1303.
24. Banerjee, D., and J.C. Williams. 2013. Perspectives on titanium science and technology. *Acta Materialia* 61 (3): 844–879.

Chapter 2
Principles of Superplastic Forming/Diffusion Bonding

Superplastic forming (SPF) is a branch of thermoforming technology. From the perspective of materials science, it belongs to the creep deformation process under specific conditions. Diffusion bonding (DB) is a process in which materials are bonded by atomic diffusion at the interface under a certain temperature and pressure. Titanium alloy has good superplasticity and diffusion bonding properties. The superplastic forming/diffusion bonding (SPF/DB) technology can be used to manufacture thin-walls with multi-layer hollow structures, which realizes lightweight, shortens the production cycle and reduces the manufacturing costs. Titanium alloy SPF/DB technology has been widely used in aviation, aerospace and other fields, resulting in huge economic value and social benefits.

2.1 Principle of Superplasticity

Plasticity refers to the ability of materials to undergo permanent deformation under external force without damage. Elongation is the main index to measure the plasticity. The elongation of metallic materials at room temperature usually does not exceed 80%, and it is difficult to reach 100% even at high temperatures. In 1934, when C. E. Pearson, a British scholar, conducted a tensile test at a slow speed on the Sn-37%Pb and Bi-44%Sn eutectic alloy, the elongation reached 1950%. In 1945, the Soviet scholar A. A. Sogeap et al. also obtained abnormally high elongation in Zn-Al alloys, and proposed the term "superplasticity". In 1964, the American scholar Backofen put forward the significant parameter—strain rate sensitivity exponent (m value) and its measurement method, which opened a new chapter in the study of superplasticity theory. Scoloars from various countries have successively carried out a lot of research work on the superplasticity mechanism, mechanical properties, application technologies, and etc.

Z. Li, *Superplastic Forming/Diffusion Bonding Technology of Titanium Alloys*,
https://doi.org/10.1007/978-981-99-3909-1_2

2.1.1 Connotation of Superplasticity

When a metal is elongated under certain conditions such as microstructure, deformation temperature and strain rate, it shows an abnormally high elongation (more than 100%) without necking and fracture, which is called superplasticity. The superplastic deformation of materials is sensitive to the strain rate. The high strain rate sensitivity can make the necking in the deformation process of materials diffuse and transfer. The materials in the superplastic state has the characteristics of strong deformation ability, good fluidity, small deformation resistance and low springback, so superplastic forming/diffusion bonding technology is especially suitable for forming low plastic materials under conventional states.

At present, superplasticity is mainly divided into three categories: microstructural superplasticity, phase transformation superplasticity and other superplasticities. Among them, the microstructural superplasticity is generally called static superplasticity, fine grain superplasticity or constant temperature superplasticity, and the commonly mentioned superplasticity also refers to microstructural superplasticity. There are three conditions to be fulfilled in order to achieve microstructural superplasticity: ① Uniform and fine equiaxed grains. The grain size is usually less than 10 μm, and it should be stable at high temperature; ② Deformation temperature $T > 0.5T_m$ (T_m represents the melting point temperature of the material, expressed in thermodynamic temperature), and the temperature remains constant during deformation; ③ The strain rate is low, and the optimal strain rate range is 10^{-4}–$10^{-2}s^{-1}$.

The evaluation method of superplastic performance index of materials generally adopts the tensile test method in which the specimen of standard size is subjected to axial tension. Figure 2.1 shows the relationship between tensile force F and displacement ΔL of TC4 titanium alloy at constant tensile speed. Figure 2.2 shows the morphology of the TC4 titanium alloy superplastic tensile specimen. At the beginning of tension, the tensile force rises rapidly to the highest, then the necking instability occurs, and the tensile force decrease as the displacement increases. From the Backofen relationship between superplastic flow stress and strain rate (constitutive equation $\sigma = K\dot{\varepsilon}^m$, where σ represents the flow stress, $\dot{\varepsilon}$ represents the strain rate; m represents the strain rate sensitivity exponent; K is a constant depending on the material and deformation conditions), it can be seen that when the necking occurs locally in the tensile specimen, the strain rate of the necking increases and the flow stress increases, which suppresses the deformation of the thin neck position, so that necking occurs on the cross-section with low flow stress, and necking can be diffused and transferred. Therefore, we can say that there exists necking phenomena in the superplastic forming process, and the necking is constantly diffusing and transferring. The deformation gradient of the whole specimen is slow and uniform, and it can be clearly seen from the fractured specimen that there is no obvious necking. When the m value of the material is larger, the change of the flow stress with the strain rate is more severe, the diffusion and transfer ability of necking is stronger, and the elongation of the material is larger. Generally, the m value of superplastic materials is between 0.3 to 0.9, and mostly between 0.4 to 0.8. Therefore, superplasticity is

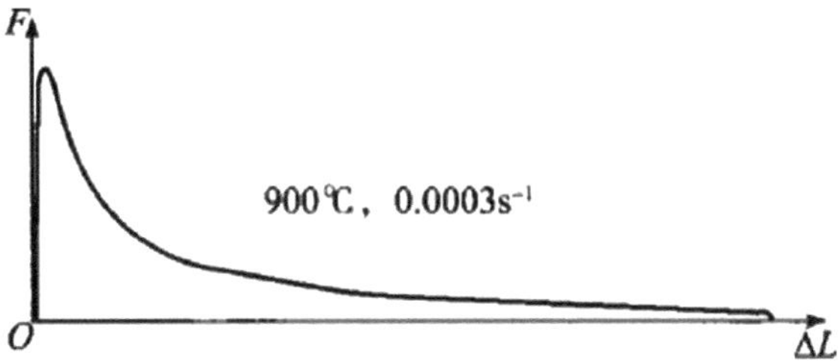

Fig. 2.1 Superplastic tensile curve of TC4 titanium alloy

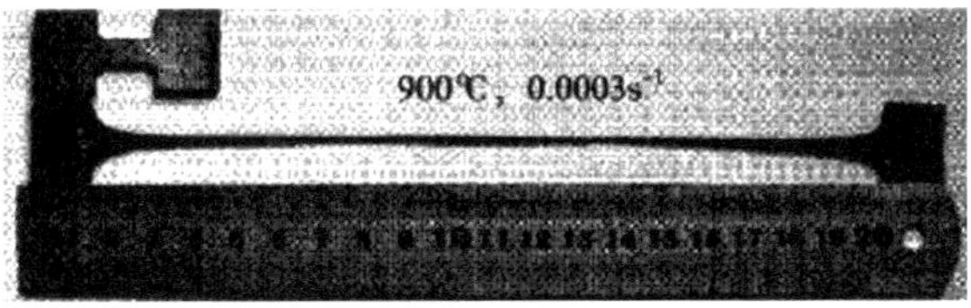

Fig. 2.2 Morphology of TC4 titanium alloy superplastic tensile specimen 900 °C, 0.0003 s^{-1}

also defined by the value (greater than 0.3) of the strain rate sensitivity exponent m (or the ability to resist necking).

The flow stress curve of the material can also be obtained according to the tensile test. As shown in Fig. 2.3, the curve is expressed in double-logarithmic coordinates, and the slope of the curve is the strain rate sensitivity exponent m:

$$m = \frac{d[\ln(\sigma)]}{d[\ln(\dot{\varepsilon})]} \tag{2.1}$$

The curve is divided into three areas according to the strain rate: Area 1 is the low strain rate zone; Area 2 is a medium strain rate zone; Area 3 is a high strain rate zone. The value of $\dot{\varepsilon}$ is $10^{-1} \sim 10^{-2}s^{-1}$. Area 2 corresponds to the superplastic deformation zone. The shape of the flow stress curve in Fig. 2.3 is similar to the English letter "S", so this curve is also called S curve. The maximum value of m is at the inflection point of the curve, as shown in Fig. 2.4, which is the optimum strain rate for superplastic deformation. It is possible to obtain the maximum elongation when the material undergoes superplastic deformation at this optimum strain rate.

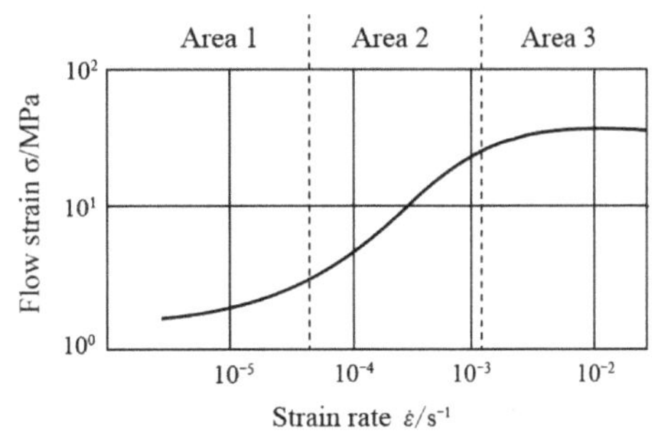

Fig. 2.3 Flow stress curve

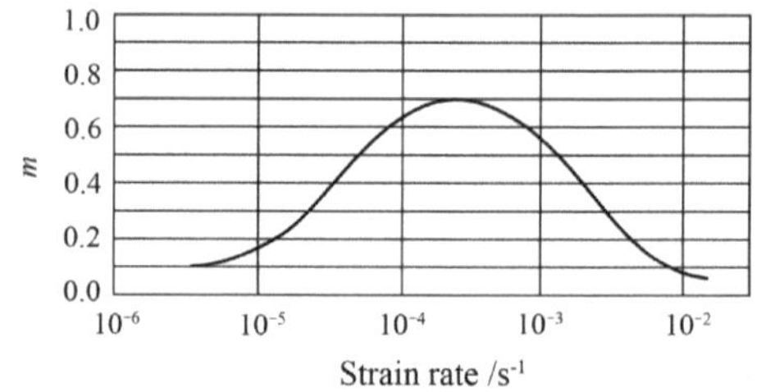

Fig. 2.4 The relationship between *m* and strain rate

2.1.2 Superplastic Deformation Mechanism

The superplastic deformation process of materials is affected by stress, deformation temperature, strain rate, strain, microstructure, time and other factors. Currently, there are two widely accepted theoretical models of superplastic deformation: the diffusion creep model, coordinated deformation model of diffusion creep and grain boundary sliding. According to the diffusion creep theory, the stress field leads to the directional migration of cavities and the deformation of grains. Based on different diffusion pathways, diffusion creep mechanisms include the bulk diffusion mechanism proposed by Nabarro and Herring and the grain boundary diffusion mechanism proposed by Coble. According to the corrdinated deformation model of diffusion creep and grain boundary sliding, the main way of superplastic deformation is grain boundary sliding, and diffusion creep can adjust grain boundary sliding, among which the most famous model is proposed by Ashby and Vdrall (abbreviated as A-V model). The A-V model is more accepted in the academic world.

Figure 2.5 shows the deformation process of a group of two-dimensional hexagonal grains rearranging from the initial state a ($\varepsilon = 0$) to the intermediate state b ($\varepsilon = 0.275$) under vertical tensile stress, and finally to the final state c ($\varepsilon = 0.55$). During the deformation process, the relative position of grains has changed, but the shape of grains remains unchanged. The work done by the external force on this grain group is consumed in the following four irreversible processes:

(1) Diffusion process: temporary change of grain shape caused by grain boundary diffusion or bulk diffusion to achieve self-adaptation.
(2) Interface reaction: the diffusion of cavities into and out of the grain boundary consumes energy to overcome the interface barrier.
(3) Grain boundary sliding: cosume energy to overcome the grain boundary viscosity before sliding.
(4) Increase and decrease of interface area: the area increase (a, b) and decrease (b, c) of the grain group also consumes energy.

Ignoring the work done by interface reaction and grain boundary sliding, the A-V theory suggests that grain variation is mainly controlled by grain boundary diffusion, accompanied by a small amount of bulk diffusion. The main mechanism of superplastic deformation is grain boundary sliding deformation, which is accompanied by grain rotation and grain boundary migration. The grain boundary sliding leads to

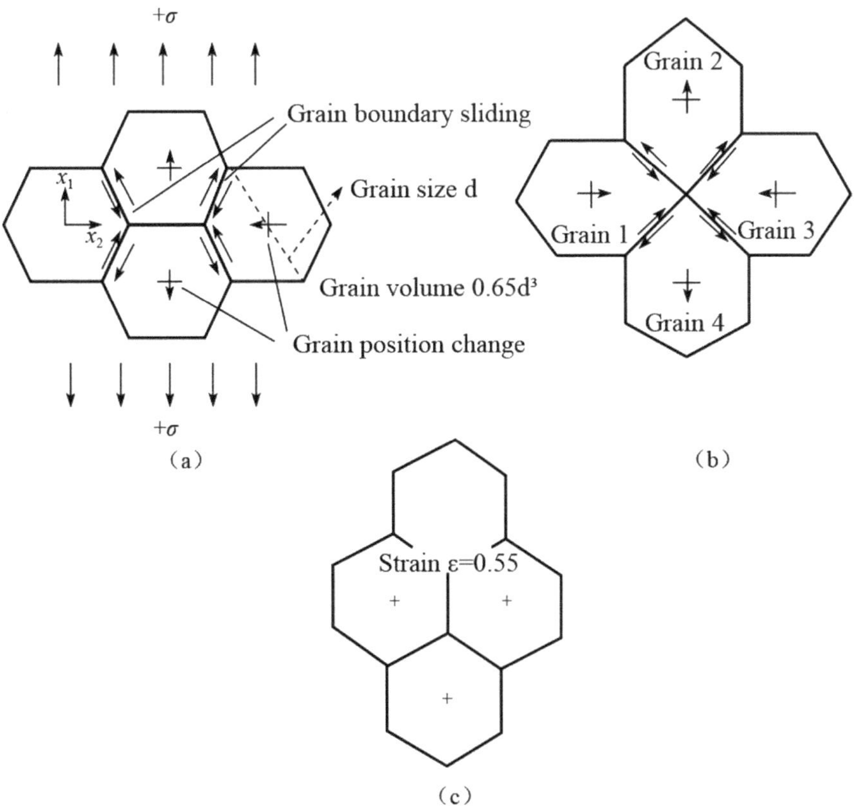

Fig. 2.5 Diagram of A-V theoretical grain rearranging process. **a** Initial state; **b** Intermediate state; **c** Final state

form cativies at the triple line boundary junction, and these cavities can only be eliminated mainly by strong diffusion creep. The coordinated adaptation of diffusion creep and grain boundary sliding changes the grain position and structure, which finally achieves the grain transposition. Lots of experimental observations have basically confirmed the main contents of the A-V theory, such as the grain boundary sliding, grain rotation, grain transposition, cavities, and the grain maintenance equiaxation. There are also some unreasonable situations in the A-V model. The results calculated by the A-V model are only in agreement with the low strain rate part of the S curve. The m value predicted by the A-V model is equal to 1, while the actual value is between 0.4 and 1. The A-V model is only a two-dimensional model that considers the change of grain position, and does not consider the three-dimensional movement process of grains, such as the process of the grain under the surface exposed to the surface and surface area growth.

Because of the complexity of superplastic deformation, one mechanism can only explain the deformation of one or several alloys, or can only explain part of the phenomenon in superplastic deformation. However, the essence of the evolution of various processes in nature is energy dissipation. The formation of various structures and various changing processes in materials are directly related to energy. Energy controls structures and changing processes. For many physical processes in the superplastic deformation process, such as grain boundary sliding, dislocation slip, diffusion control, grain stability, pore nucleation and growth, deformation instability and other microscopic phenomena, the scientific essence is caused by energy conversion inside the system, which is the relaxation of energy along the path of minimum resistance under the premise of following the law of minimum resistance in nature, leading to energy dissipation and redistribution. Therefore, it is more reasonable, scientific and accurate to comprehensively analyze the superplastic deformation process from the perspective of energy.

The numerical simulation of superplastic deformation process based on crystal plasticity theory can analyze the superplastic deformation mechanism from the perspective of energy. The superplastic deformation mechanisms of titanium alloys involves: grain boundary sliding and diffusion, phase boundary (α phase and β phase boundary) sliding, grain rotation, intragranular dislocation slip and diffusion. It is also affected by factors such as grain orientation, grain size, temperature and strain rate during the deformation process. Therefore, the study of the superplastic deformation mechanism of titanium alloys is a multiscale problem.

The mathematical models of grain boundary sliding, intragranular sliding and grain rotation during superplastic deformation of TC4 titanium alloy were established by] crystal plasticity theory. At the same time, based on the stress–strain data of superplastic tensile deformation of titanium alloy and the electron backscatter diffraction (EBSD) data of original microstructure, fitting the parameters of the constitutive model of titanium alloys (such as grain boundary diffusion energy, grain boundary sliding rate, dislocation slip activation energy, critical resolved shear stress of intragranular dislocation slip, etc.), and finally embedding these models into the theoretical framework of crystal plasticity, the crystal plastic finite element simulation of superplastic deformation can be realized.

Figure 2.6 shows the crystal plastic finite element simulation results of the superplastic deformation process of TC4 titanium alloy. During the deformation process, the per unit volume plastic work of the intergranular phase is slightly greater than that of intragranular phase, the stress of the intergranular phase is smaller than that of the intragranular phase, and the strain of intergranular phase is greater than that of intragranular phase. The results show that the intergranular phase of titanium alloy undergoes obvious plastic deformation during the superplastic deformation process, while the plastic deformation in the grain is relatively small, and the superplastic deformation is dominated by the sliding and rotation of the grain boundary. The mechanism of superplastic deformation of titanium alloys is better explained from the perspective of energy.

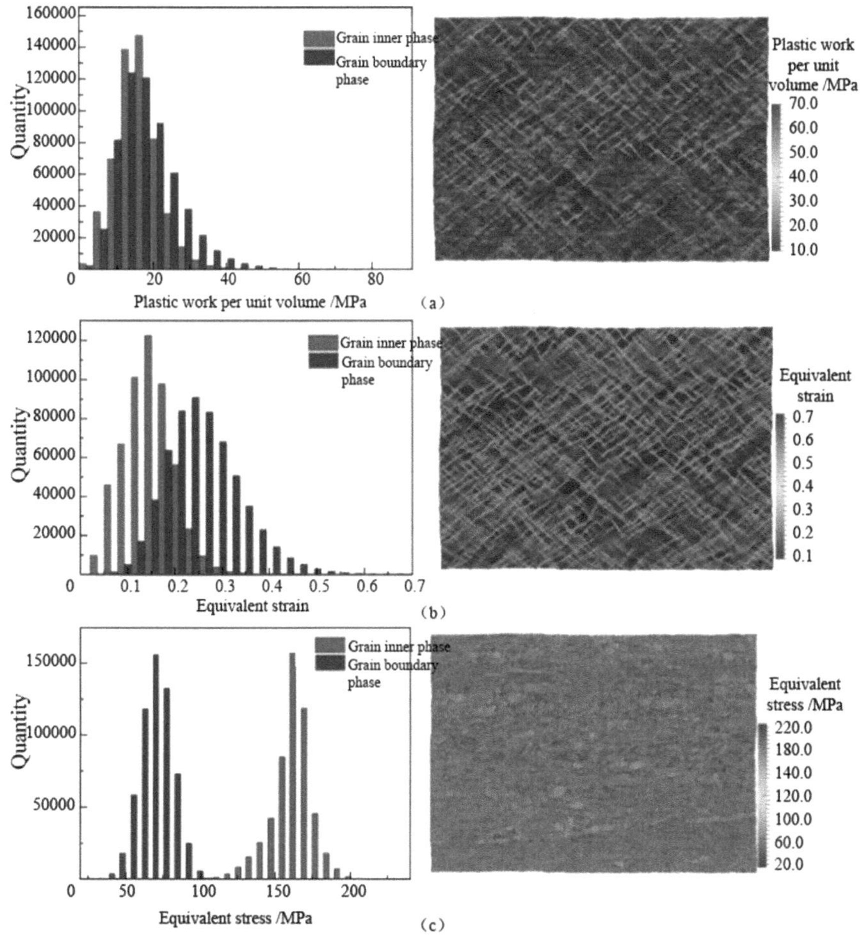

Fig. 2.6 Distribution of intragranular phase and intergranular phase **a** Distribution of plastic work per unit volume; **b** Distribution of equivalent strain; **c** Equivalent stress distribution

2.1.3 Influencing Factors of Superplasticity

The study on the influencing factors of superplastic deformation can seek the optimal combination of superplastic conditions, so as to give full play to the superplastic performances of materials. However, superplastic deformation of metals is a complex physical and chemical process with many influencing factors. This book mainly introduces several important influencing factors:

1. **Deformation temperature**

Superplastic deformation is a deformation behavior in a certain temperature range. Generally, the deformation temperature is required to be higher than half of the melting temperature ($T > 0.5T_m$). The superplasticity temperature of most materials is a specific temperature range lower than the critical temperature. In the case of other conditions certained, when the deformation temperature is lower than the critical temperature, the flow stress decreases as the temperature increases, and the elongation and m value increase slowly; when the deformation temperature reaches the critical temperature, the elongation and m value increase to the maximum; when the deformation temperature exceeds the critical temperature, the elongation and m value decrease gradually.

The critical temperature of superplastic deformation is not fixed, but is related to factors such as strain rate, grain size, phase composition and distribution. Only when other influencing factors are constant, the critical temperature can be regarded as the best deformation temperature under a certain condition.

2. **Grain size**

For microstructuresuperplasticity, it is generally required that materials possess equiaxed and fine grains. The grain diameter is less than 10 μm, and the grain grows slowly during the deformation process. Within a certain range, the grain size decreases, the flow stress decreases, and the m value and elongation increase.

3. **Strain rate**

There exists a strain rate range in superplastic deformation of materials. Generally, increasing the temperature or reducing the grain size will increase the strain rate range of superplastic deformation. Outside this strain rate range, the material exhibits little superplasticity. When other conditions remain unchanged, the superplastic deformation of the material has an optimal strain rate, which corresponds to the peak of m value. When the strain rate is less than or greater than the optimal value, the m value and elongation decrease.

The above three are the main influencing factors of superplastic deformation. In addition, the phase composition and distribution, stress state, deformation degree, loading mode and environments will have a certain impact on superplastic deformation.

2.2 Superplasticity of Typical Materials

In order to realize superplastic deformation, three basic conditions are required: fine grain, higher deformation temperature and lower strain rate. According to the superplastic deformation characteristics of typical materials, the analysis of the influence of temperature and strain rate on deformation is the basis for engineering applications. However, harsh superplastic deformation conditions lead to lower production efficiency. Therefore, how to reduce the superplastic deformation temperature and increase the superplastic deformation rate has become the focus of domestic and foreign scholars in recent years.

2.2.1 Conventional Superplasticity

1. **SP700 titanium alloy**

SP700 is an α + β type titanium alloy by addingβ-stablizing elements Mo and Fe on the basis of TC4 titanium alloy composition, which is rich in β phase. The composition of SP700 titanium alloy is: 4.5%Al, 3%V, 2%Mo and 2%Fe (i.e., Ti-4.5Al-3V-2Mo-2Fe). It is the first titanium alloy with SP (the abbreviation for "superplasticity") as the brand, which has excellent superplasticity at 700°C. Compared with TC4 alloy, SP700 alloy possesses better cold and hot working formability, higher strength, plasticity, fracture toughness and fatigue strength.

The maximum *m* value method and the constant strain rate method were used to conduct superplastic tensile tests on SP700 titanium alloy. The deformation temperature range was 755–785 °C. The tensile axis direction of the specimen was 0°, 45° and 90° with the rolling direction of the plate. The strain rate is 0.005–0.1s^{-1}. The results of the elongation after fracture by the maximum *m* value method are shown in Figs. 2.7 and 2.8. The elongation of the specimen at 775 °C and 45° direction is the highest, which is 3110%; the elongation at 785 °C and 90° direction is the lowest, which is 1127%.

Figure 2.9 shows the elongation obtained by superplastic tensile with constant strain rate method under different conditions. Under the same tensile direction and different strain rates, the elongation increases with the decrease of the strain rate, and the highest elongation is obtained at 775 °C and 0.005s^{-1}. The maximum elongation at 0°, 45° and 90° direction is 960%, 750% and 1121%, respectively. Comparing the elongation in different tensile specimen directions under the same deformation temperature and strain rate, it can be seen that the elongation in three directions is basically the same at the strain rate of 0.1s^{-1} and 0.01s^{-1}. However, when the strain rate decreases to 0.005s^{-1}, the elongation at 45° direction is significantly lower than that at the other two directions, and the elongation at 90° direction is higher than that

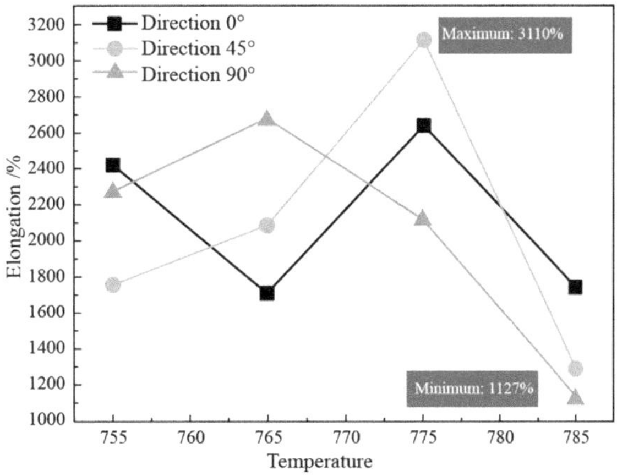

Fig. 2.7 Superplastic elongation by the maximum m method

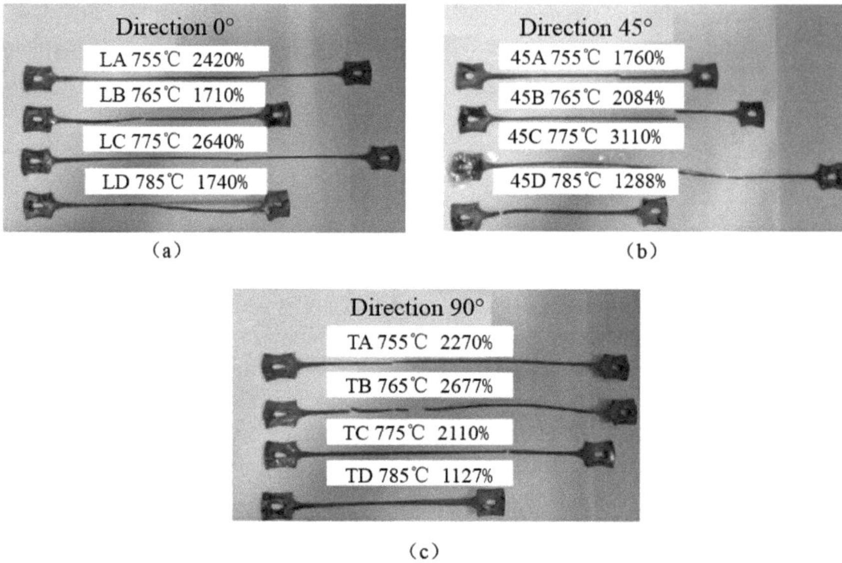

Fig. 2.8 Morphology of specimens after tensile fracture in three directions. **a** Direction 0°; **b** Direction 45°; **c** Direction 90°

at 0° direction, which indicates that the material will exhibit anisotropy only when the strain rate is low.

2. **TA32 titanium alloy**

TA32 (TA12A) is a new type of near-α titanium alloy obtained by optimizing the composition on the basis of TA12 (Ti55) alloy. TA12 is a representative 550 °C high temperature titanium alloy. This alloy possesses good comprehensive properties at 550 °C, and can be used for a long time at a temperature lower than 550 °C, and the short-term service temperature can reach 600 °C.

The stress–strain curve of TA32 alloy after superplastic tensile under the condition of deformation temperature of 880 ~ 940 °C and the strain rate of 0.0005 ~ $0.01s^{-1}$ is shown in Fig. 2.10. It can be seen that at the initial stage of tension, the flow stress increases rapidly with the increase of strain. When the strain is about 0.1, the flow stress curve tends to be horizontal, and then changes differently with the increase of strain. At 880 ~ 920 °C temperature and 0.005–$0.01s^{-1}$ initial strain rate, the material softens continuously under the dynamic recovery and dynamic recrystallization, and the flow stress decreases; at 880 ~ 900 °C temperature and $0.001s^{-1}$ initial strain rate, and at 940 °C temperature and 0.005–$0.01s^{-1}$ initial strain rate, the strain hardening induced by dislocation pile-up, grain growth and the strain softening induced by dislocation annihilation and dynamic recovery in the material reach a dynamic equilibrium, and the flow stress curve tends to be horizontal. At 880–940 °C temperature and $0.0005s^{-1}$ initial strain rate, and at 920–940 °C temperature and $0.001s^{-1}$ initial strain rate, the initial strain rate is low, and the material is kept

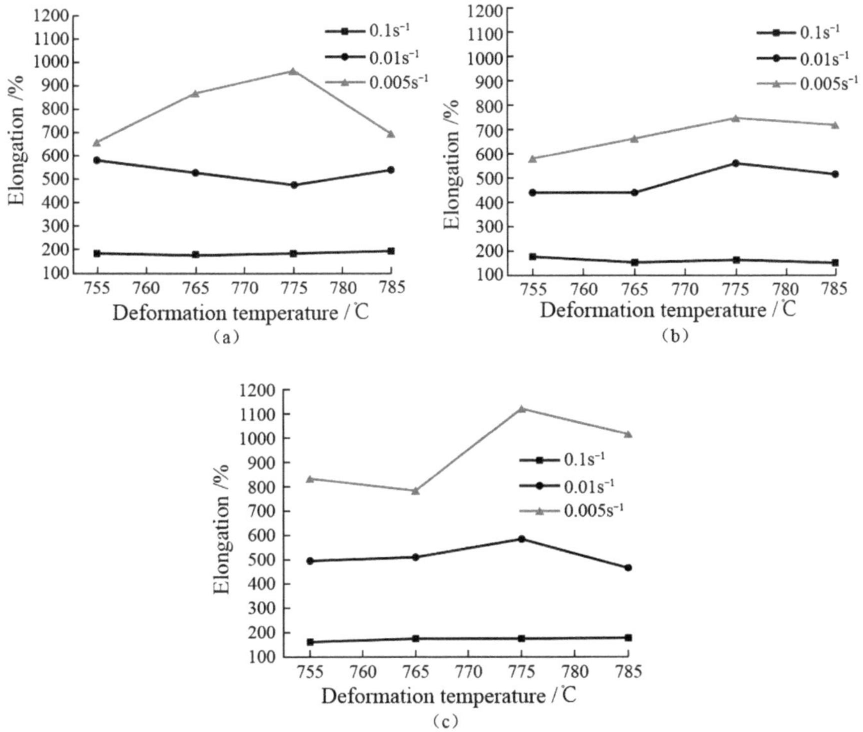

Fig. 2.9 Elongation of superplastic tensile under different conditions by constant strain rate method. **a** Direction 0°; **b** Direction 45°; **c** Direction 90°

in a high-temperature environment for too long, which will cause the grain growth rate to be higher than the rate of dynamic recrystallization to refine grains. Then the two-phase grains are fully coarsened, grain boundary sliding and grain rotation are reduced, so that the alloy continues to harden, and the flow stress increases with the increase of strain until the peak stress is reached before t the specimen fractures.

The elongation after fracture of TA32 titanium alloy under different deformation conditions is shown in Table 2.1. The elongation after fracture increases with temperature increases. This is because the atomic kinetic energy increases at higher temperatures, the vacancy diffusion and atomic diffusion effects are enhanced, and the grain boundary sliding increases, so the elongation after fracture is improved. The elongation after fracture shows a complex trend with the change of strain rate. At 940 °C deformation temperature and 0.005 s^{-1} initial strain rate, the maximum elongation after fracture of 949% is obtained.

3. **Ti65 titanium alloy**

Ti65 is a kind of Ti–Al-Sn-Zr-Mo-Si-Nb–Ta-W–C 10-element near α-type high-temperature titanium alloy with the characteristics of low density, high specific

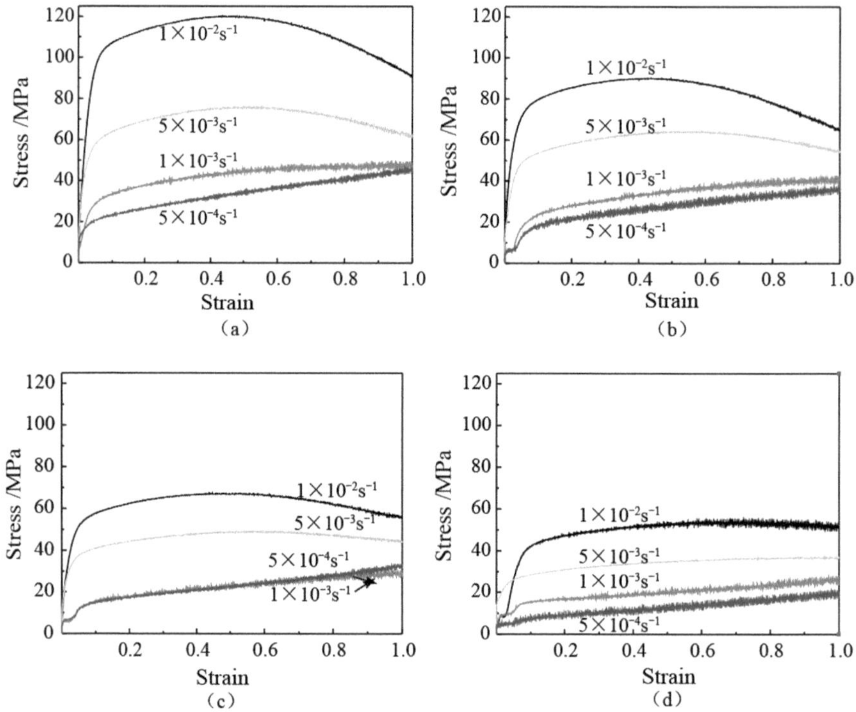

Fig. 2.10 Stress–strain curves of TA32 at different temperatures and initial strain rate **a** 880 °C; **b** 900 °C; **c** 920 °C; **d** 940 °C

Table 2.1 Elongation after fracture of TA32 under different deformation conditions

Strain rate ε	Elongation after fracture			
	880 °C (%)	900 °C (%)	920 °C (%)	940 °C (%)
$5 \times 10^{-4}\ s^{-1}$	610	654	882	894
$1 \times 10^{-3}\ s^{-1}$	585	676	774	769
$5 \times 10^{-3}\ s^{-1}$	322	331	331	949
$1 \times 10^{-2}\ s^{-1}$	328	309	352	662

strength and good high temperature performance. The long-term service temperature is 650 °C, and the service temperature under short-term high stress conditions is 650 ~ 750 °C.

Figure 2.11 shows the macroscopic morphology of Ti65 titanium alloy after superplastic tensile at 900 ~ 960 °C deformation temperature and 0.001 ~ 0.03s^{-1} strain rate. The deformation of Ti65 alloy is relatively uniform during superplastic deformation, no obvious necking occurs, and the superplasticity of the material is good.

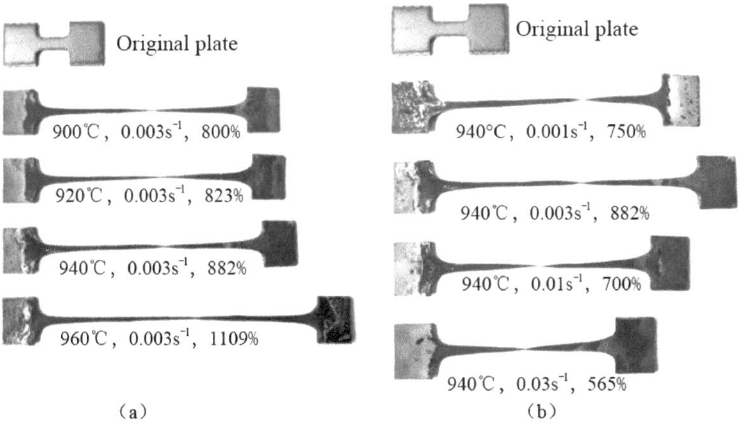

Fig. 2.11 Macro morphology of superplastic tensile. **a** Different deformation temperature; **b** Different strain rates

Figure 2.12 shows the relationship between ture stress and ture strain of Ti65 superplastic tensile. It can be seen that the shape of true stress-true strain curve is basically the same under different deformation conditions, showing obvious superplastic deformation characteristics: at the initial stage of deformation, the stress increases rapidly with the increase of strain, showing an obvious hardening effect; as deformation proceeds, the stress increases slowly and enters the steady flow stage due to the strain softening; finally, when necking or fracture occurs, the stress drops sharply.

Under the condition of $0.003s^{-1}$ strain rate, the elongation increases and the peak stress decreases with the increase of deformation temperature. At 940 °C deformation temperature, with the increase of strain rate, the elongation first increases and then

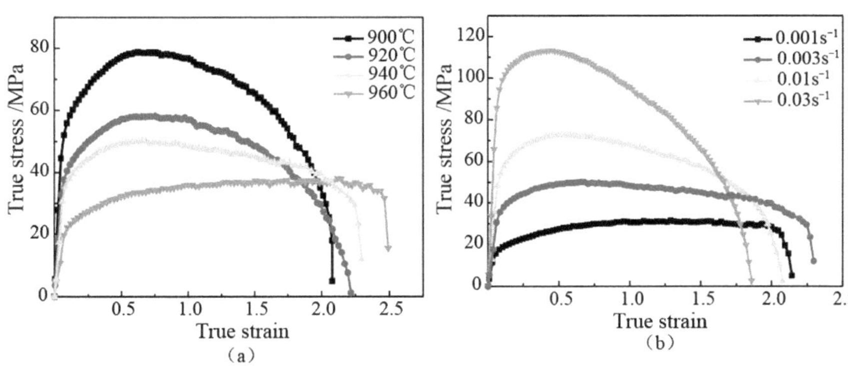

Fig. 2.12 Ti65 superplastic tensile stress–strain curve. **a** Different deformation temperature; **b** Different strain rates

decreases, and the peak stress increases. At 960 °C temperature and $0.003s^{-1}$ strain rate, the maximum elongation of Ti65 alloy is 1109%.

4. **TNW700 titanium alloy**

TNW700 is a nearly α-type multi-component strengthened high-temperature titanium alloy with the nominal chemical composition of Ti-5.86Al-3.4Sn-5.56Zr-1.15Nb-1.6W-0.19Si. This alloy is based on the Ti–Al-Sn-Zr-Mo-Si alloy system, using Nb element to replace Mo element, and adding W, C and other elements. The thermal stability and thermal strength of the alloy can be improved by the addition of Nb and W, that can help form high-temperature stable phases. TNW700 can be used as a load-bearing material for short-term use at 600 ~ 750 °C, and its oxidation resistance is better than that of TC4 alloy.

Figure 2.13 shows the appearance of the TNW700 specimen after superplastic tensile at the deformation temperature of 890–950 °C and the strain rate of 5×10^{-4} ~ $1 \times 10^{-2}s^{-1}$. When the high strain rate is $1 \times 10^{-2}s^{-1}$, the tensile section width of TNW700 changes linearly and the fracture is sharp. However, under the condition of low strain rate ($5 \times 10^{-3}s^{-1}$ and $5 \times$ 10-4s^{-1}), the width of some specimens along the length direction changes in a "wave pattern", and the fracture surface is straight or 45° inclined. While at higher temperature and lower strain rate, the uniformity of the specimen width is improved, which may be the case that due to high temperature and low strain rate, the coordinated deformation of the microstructure can occur with enough time and energy, so as to obtain a uniform width.

Figure 2.14 shows the superplastic tensile stress–strain curve of TNW700 under different deformation conditions. The common feature of the data is that the peak stress decreases with the increase of deformation temperature, but the peak strain increases with the increase of temperature, and the work hardening stage increases with the decrease of strain rate. Especially when the strain rate is $5 \times 10^{-4}s^{-1}$, once

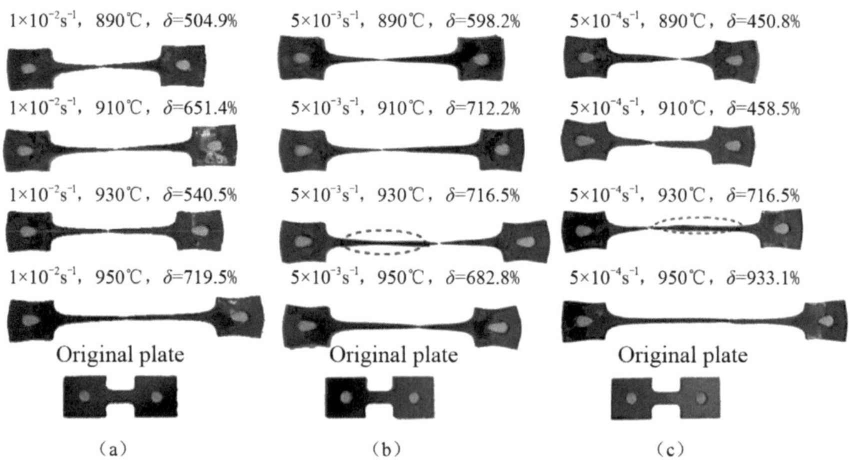

Fig. 2.13 Appearance of TNW700 superplastic tensile specimen. **a** The strain rate is $1 \times 10^{-2}s^{-1}$; **b** The strain rate is $5 \times 10^{-3}s^{-1}$; **c** The strain rate is $5 \times 10^{-4}s^{-1}$

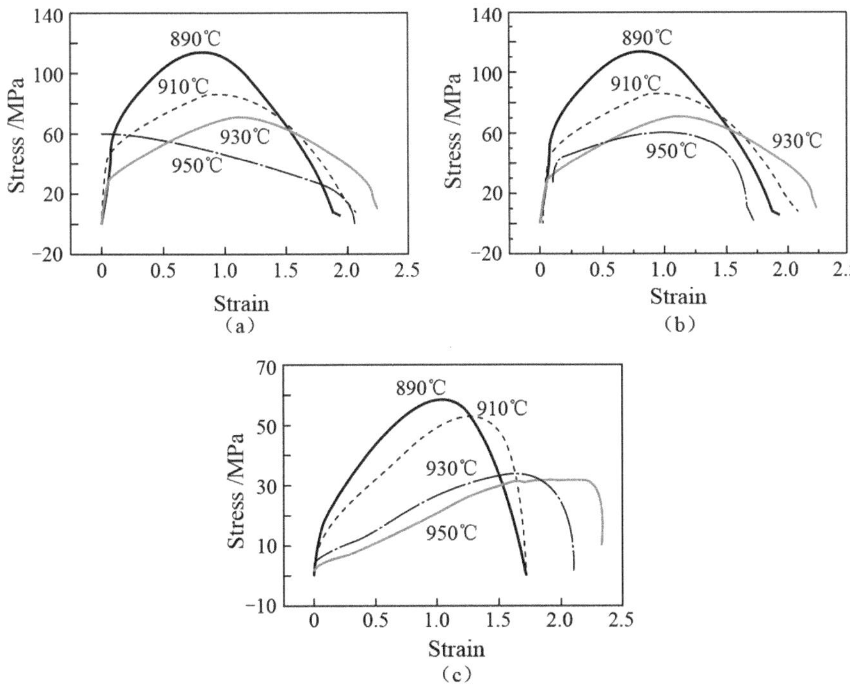

Fig. 2.14 Superplastic tensile stress–strain curve of TNW700. **a** The strain rate is $1 \times 10^{-2}s^{-1}$; **b** The strain rate is $5 \times 10^{-3}s^{-1}$; **c** The strain rate is $5 \times 10^{-4}s^{-1}$

the deformation exceeds the peak strain, the stress drops rapidly until the specimen breaks. The "work hardening" deformation behavior of TNW700 is different from that of ordinary titanium alloys, which may be caused by the microstructure evolution during superplastic deformation.

5. **Ti2AlNb alloy**

The outstanding advantages of titanium-aluminum intermetallic compounds are low density, excellent high-temperature specific strength and specific elastic modulus. There are three main types of Ti–Al intermetallic compounds: α_2-Ti_3Al alloy, γ-TiAl alloy and δ-$TiAl_3$-based alloy. Among them, Ti2AlNb has higher service temperature and better high-temperature performance than Ti3Al. The long-term service temperature of Ti2AlNb is 700–800 °C, and the short-term service temperature of Ti2AlNb can be higher than 1100 °C.

In this book, the superplastic properties of Ti2AlNb by using three superplastic tensile test methods is studied. The superplastic tensile temperature range is 940–980 °C, and the strain rate is $10^{-5} \sim 10^{-3}s^{-1}$. Figure 2.15 shows the elongation obtained by using three unidirectional superplastic tensile methods, and the tensile axis direction is parallel to the rolling direction of the plate. The maximum elongation obtained by the maximum m value method, the constant strain rate method and the constant

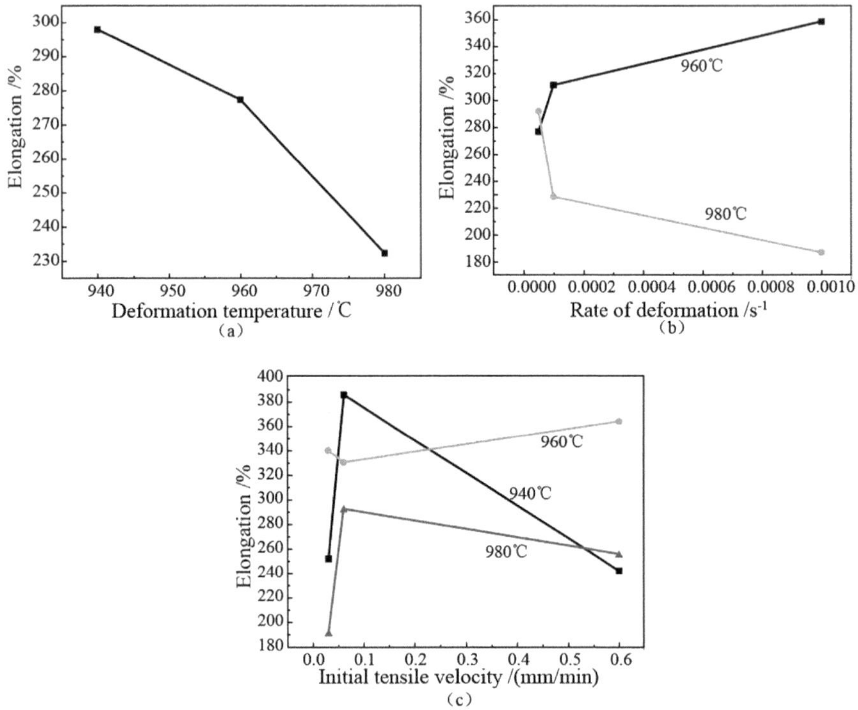

Fig. 2.15 Elongation obtained by three superplastic tensile methods. **a** Tensile results of maximum m value method; **b** Tensile results of constant strain rate method; **c** Tensile results by constant velocity method

velocity method are: 298% (940 °C), 358.9% (960 °C, $5 \times 10^{-5}s^{-1}$), 385.8% (940 °C, 0.06mm/min), respectively. It can be seen that the elongation obtained in this test is not high, which may be related to the strong texture of the plate used and the microstructure with three-phases.

2.2.2 *Low Temperature High Strain Rate Superplasticity*

The superplasticity of fine-grained material (with grain size less than 10μm) can only be achieved at higher temperatures and lower strain rates. However, high-temperature and long-term superplastic deformation will not only lead to coarsen microstructure and deteriorate the performance of materials, but also reduce the production efficiency of parts with the increase of forming time. Therefore, how to reduce the deformation temperature and increase the strain rate of titanium alloys is an important direction in the study of titanium alloy superplasticity.

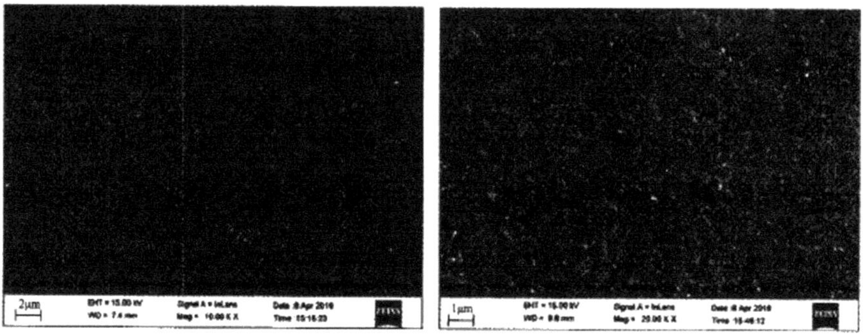

(a) (b)

Fig. 2.16 Scanning image of TC4 titanium alloy **a** Quenching state; **b** Processed by FSP after quenching state

1. **Superplasticity of ultrafine-grained materials**

According to the relationship between superplastic grain size and strain rate ($\dot{\varepsilon} \propto 1/d^n$), the strain rate during superplastic deformation can be increased by refining grains. Therefore, ultrafine grains are a necessary condition for achieving low temperature high strain rate superplasticity. The studies show that when a material has submicron or nanoscale grains microstructure, superplasticity can be obtained at a lower temperature or at a higher rate than a material with a micron grains microstructure.

TC4 titanium alloy is heat treated by 1010 °C thermal insulation for 1h then water cooling, and 550 °C thermal insulation for 3h then air cooling, and then the friction stir processing is carried out at a rotation speed of the stirring head of 120r/min and a forward speed of the stirring head of 30mm/min, thus obtaining the ultrafine-grained microstructure. Figure 2.16 shows the microstructure morphology of the material after friction stir processing. After friction stir processing, equiaxed α phases and intergranular β phases were formed in TC4 titanium alloy, with the grain size of 0.51μm.

The superplastic tensile deformation experiments of ultrafine-grained quenched TC4 titanium alloy under conditions of deformation temperature of 550 °C, 600 °C, 650 °C and strain rate of $1 \times 10^{-4}s^{-1}$, $3 \times 10^{-4}s^{-1}$, $1 \times 10^{-3}s^{-1}$, $3 \times 10^{-3}s^{-1}$ were carried out for. The macroscopic morphology of the specimen after fracture is shown in Fig. 2.17. Under the test conditions, the materials show uniform superplastic flow characteristics, without obvious necking, and the elongation is greater than 400%. The maximum elongation of the material is 1130% at 600 °C temperature and $3 \times 10^{-4}s^{-1}$ strain rate.

Table 2.2 summarizes the superplastic tensile elongations obtained by other researchers and this book after severe plastic deformation of TC4 titanium alloy. Compared with other results, the ultrafine-grained titanium alloy prepared by friction stir processing in this book obtained the highest elongation of 1130% at a lower deformation temperature (600 °C).

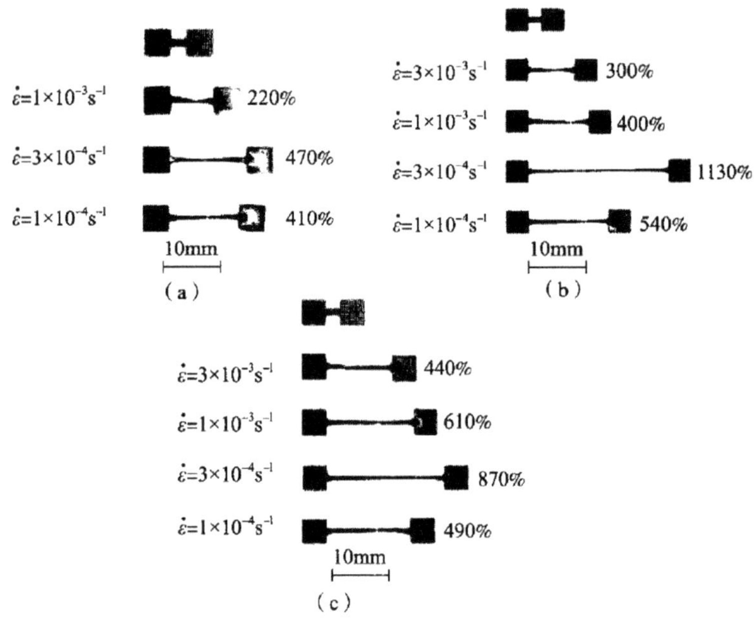

Fig. 2.17 Macroscopic morphology of specimen after fracture under different superplastic tensile conditions **a** The deformation temperature is 550 °C; **b** The deformation temperature is 600 °C; **c** The deformation temperature is 650 °C

Table 2.2 Elongation of fine-grained TC4 under superplastic tensile

Processing method	Grain size d/μm	Temperature T/°C	Strain rate ε/ $(10^{-4}s^{-1})$	Elongation δ/%
Equal channel angular extrusion	0.3	600–700	1×10^{-4}- 5 $\times 10^{-4}$	296–700
High pressure torsional machining	0.075–0.3	600–650	5×10^{-4}-1 $\times 10^{-3}$	575–780
Multi-directional forging	0.135–0.4	550–700	5×10^{-4}–7 $\times 10^{-4}$	640–910
Hot rolling	0.1–0.3	650–700	1×10^{-4}- 1 $\times 10^{-2}$	220–516
Friction stir processing	0.51	600	3×10^{-4}	1130

2. **Superplasticity of hydrogenated materials**

Hydrogen treatment of titanium alloys is a new method to reconstruct the microstructure by hydrogen deduced plasticity, hydrogen deduced phase transformation and

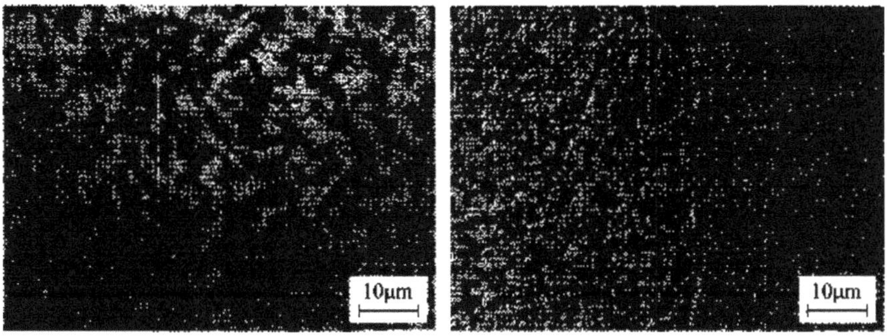

Fig. 2.18 Original microstructure and after 0.11% hydrogen treatment microstructure of original plate

reversible alloying of hydrogen to achieve the optimal microstructure of titanium-hydrogen system, so as to improve the processing performance. By hydrogen treatment, reducing the deformation temperature of titanium alloys, and increasing the deformation rate and production efficiency can be achieved from two aspects of hydrogen deduced superplasticity and hydrogen deduced grain refinement.

Figure 2.18 shows the original microstructure and 0.11% hydrogen treatment microstructure of fine-grained TC4 titanium alloy plates. The original microstructure is an incompletely recrystallized $\alpha + \beta$ equiaxed microstructure, in which the grain boundaries of α phases are connected, presenting an with equiaxed or strip morphology, while the β phases are distributed on the α phase grain boundaries with a small quantity. In the microstructure of 0.11% hydrogen treatment at 750 °C, the α phases and β phases do not grow significantly; the color of β phases becomes lighter and closer to that of α phase, and it is not easy to be corroded. But the proportion of two phases has not changed significantly, and the overall morphology is not much different from the original microstructure.

The superplastic tensile deformation is conducted for titanium alloy subjected to hydrogen treatment at the deformation temperature of 800 ~ 900 °C and the strain rate of $3 \times 10^{-4}s^{-1} \sim 1 \times 10^{-2}s^{-1}$, and the macroscopic morphology of the specimen after fracture is shown in Fig. 2.19.

The deformation uniformity of the hydrogenated and non-hydrogenated materials is basically the same. The maximum elongation of non-hydrogenated titanium alloy is more than 1700% at 900 °C and $0.003s^{-1}$, while the maximum elongation of titanium alloy with 0.11% hydrogenated is 1530% at 860 °C and $0.0003s^{-1}$. The optimum temperature range for superplastic deformation of the alloys subjected to hydrogen treatment is narrowed and moves to the low temperature range. After hydrogenation, the elongation of the material reaches 1190% at 840 °C, which is 75% higher than 680% elongation at the same temperature withouthydrogenation, and is only 20% lower than 1480% maximum elongation at 900 °C without hydrogenation.

The optimum temperature of superplastic deformation after hydrogenation is 40 ~ 60 °C lower than the value when without hydrogenation, while the optimum strain

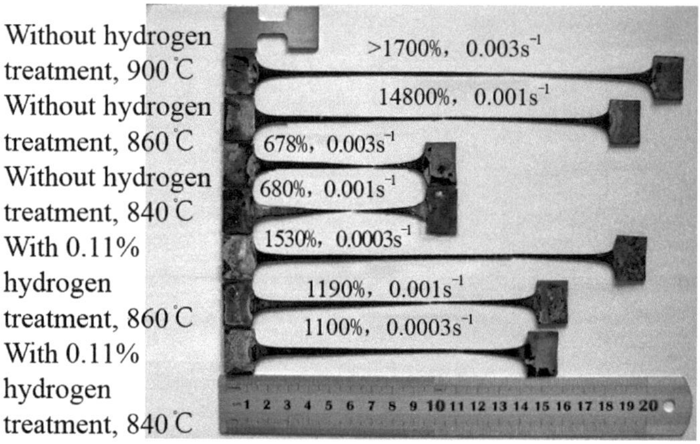

Fig. 2.19 Macroscopic morphology after deformation under different tensile conditions

rate moves to the low strain rate region.This is because the addition of hydrogen, as a β-stabilizing element, the optimum ratio of α phases and β phases can be achieved at lower temperatures. At the same time, the addition of a small amount of hydrogen can reduce the dislocation density in the microstructure, facilitate the sliding of grain boundaries, and improve the fluidity of the β phases between α-α grain boundaries, thereby improving the ability of diffusion and coordinated deformation during superplastic deformation, so the optimal superplastic temperature is reduced. On the other hand, hydrogen has strong diffusivity at high temperature, and it is easy to be accumulated at the stress concentration under tensile stress, resulting in excessive hydrogen concentration at defects (e.g. grain boundaries) and dislocation pinning. Therefore, when severe high strain rate deformation occurs, the higher stress concentration will cause the reduction of superplasticity, while the effect of stress concentration on superplasticity is weaker during low strain rate deformation, so it is easy to obtain a higher elongation.

Figure 2.20 shows the true stress—true strain curves of TC4 without hydrogenation and with 0.11% hydrogenation. Under the same deformation conditions, the stress of 0.11% hydrogenation is significantly lower than that of non-hydrogenation. With the increase of the strain capacity, the change trend of the stress under the two conditions is similar, that is, after the short period of hardening reaches the peak value,, both show continuous softening behaviors.

When the material with 0.11% hydrogenation and 840 °C temperature is deformed under the conditions of $0.001s^{-1}$ and $0.0003s^{-1}$ strain rates, the softening of the true stress—true strain curve is relatively gentle, which belongs to quasi-steady state deformation. The steady state deformation does not appear. Therefore, a higher strain can be obtained. The addition of hydrogen promotes dynamic recrystallization and dynamic recovery. On the one hand, it reduces the flow stress of the material, and on the other hand, it can be coordinated with the strain hardening rate of the material to promote the diffusion and transfer of the necking, which is beneficial to start the grain boundary sliding, grain rotation, diffusion creep, etc., so as to achieve uniform deformation.

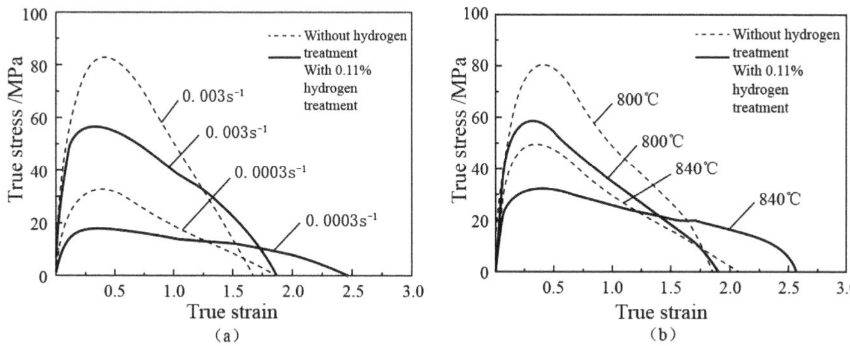

Fig. 2.20 the true stress - true strain curves of TC4 without hydrogenation and with 0.11% hydrogenation during superplastic tensile **a** 840 °C; **b** $0.001s^{-1}$

3. **Pulse current-assisted superplasticity**

In the process of superplastic deformation, the application of the pulse current (or additional electric field) can reduce the flow stress of materials, reduce the strain hardening exponent, and increase the strain rate sensitivity exponent m, so that high elongation at lower temperatures and higher strain rates can be obtained, and the harsh deformation conditions of materials that are difficult to be deformed can also be optimized.

1420 aluminum–lithium alloy (Al–Mg-Li-Zr) contains 1.8wt.% lithium and 4.8wt.% magnesium, and zirconium (Zr) alloy element is added to the alloy. The alloy does not contain copper and other heavy metal elements, so such alloy features low density, high elastic modulus and good weldability. Therefore, the sheets, forgings and profiles made of 1420 aluminum–lithium alloys possess broad application prospects in aerospace shuttles, fighters and civil aircrafts. The true stress - true strain curves of 1420 aluminum lithium alloy during superplastic tensile deformation under different deformation conditions are shown in Fig. 2.21 a–d. It can be seen that the flow stress increases with the increase of the strain rate and decreases with the increase of deformation temperature. Therefore, 1420 aluminum–lithium alloy belongs to both strain rate sensitive material and temperature sensitive material.

Figure 2.22 shows the relationship between elongation and temperature under different initial strain rates. With the increase of temperature, the elongation increases first and then decreases, and the maximum elongation is at 480 °C under all strain rate conditions. And under the same temperature conditions, the elongation increases with the decrease of the initial strain rate. The optimum superplastic deformation condition of the materials is 480 °C and $0.0003s^{-1}$, and the elongation reaches 332%.

Figure 2.23 shows the superplastic tensile test results of 1420 aluminum–lithium alloy with and without pulse current at a superplastic tensile temperature of 480 °C and a strain rate of $1 \times 10^{-3}s^{-1}$. The parameters of the specimen applied pulse current are: the pulse current density of $192A/mm^2$, the pulse current frequency of 150Hz, and the pulse current time of 30s. The application of pulse current reduced the peak

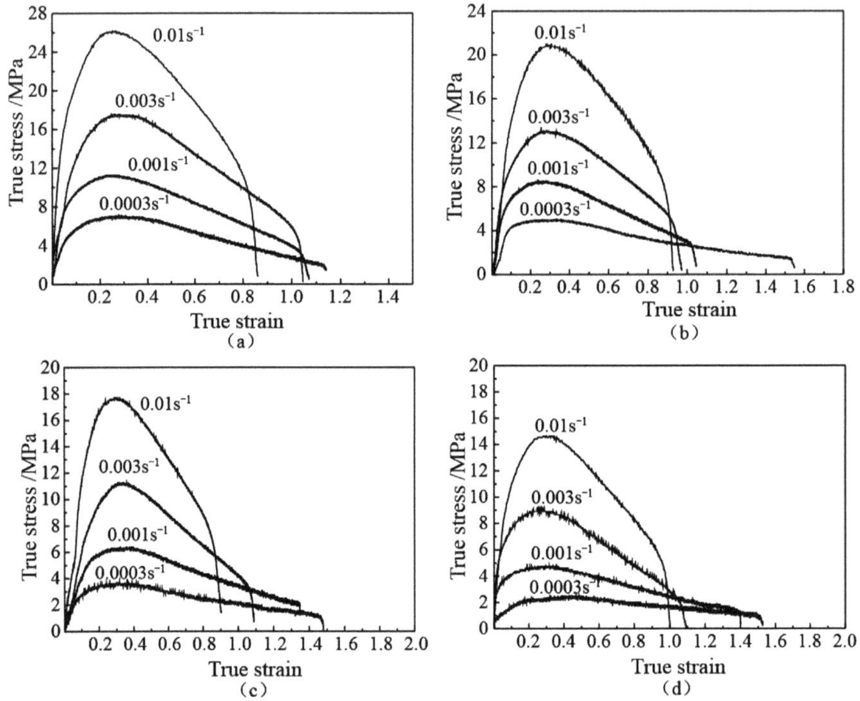

Fig. 2.21 True stress - true strain curves under different deformation conditions. **a** T = 440°C; **b** T = 460 °C; **c** T = 480 °C; **d** T = 500 °C

Fig. 2.22 Relationship between elongation and temperature under different strain rates

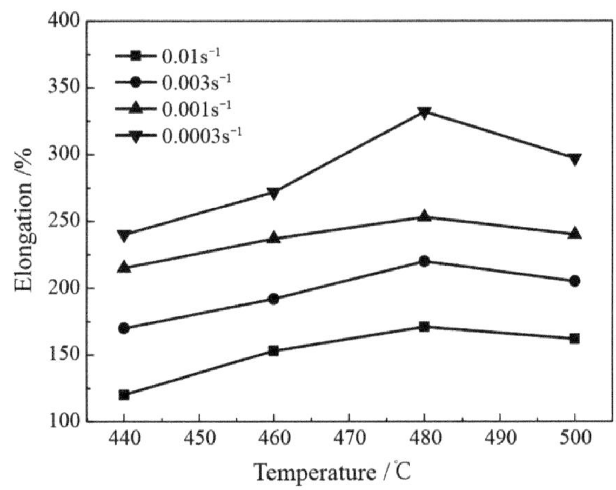

Fig. 2.23 True stress - true strain curve of superplastic tensile with and without pulse current at 480 °C and 1 × $10^{-3}s^{-1}$

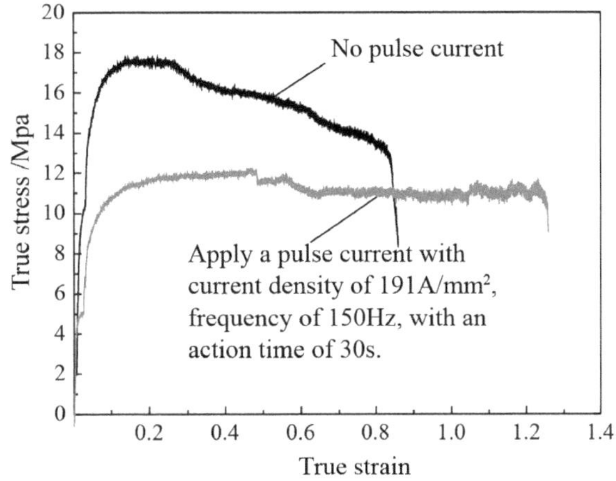

stress (from 17.8MPa to 12MPa) and increased the elongation of the material (from 160 to 270%).

The superplastic tensile test results of 1420 aluminum–lithium alloy with and without pulse current under different deformation temperatures and strain rates are shown in Table 2.3. The elongation (200%) of the material at the lower deformation temperature after applying a pulse current is higher than the enlongation (161%) at higher deformation temperature without applying a pulse current, and the elongation (270%) of the specimen at higher strain rate is higher than the elongation (207%) at lower strain rate without applying pulse current. Therefore, the application of the pulse current not only improves the superplastic deformation capability of 1420 aluminum–lithium alloy under the same deformation conditions, but also reduces the superplastic deformation temperature and increases the strain rate, so that the low-temperature and high-strain rate superplasticity of the material can be achieved.

Table 2.3 Effect of pulse current on superplastic deformation parameters

Deformation temperature/°C	Current density /(A/mm^2)	Strain rate /s^{-1}	Elongation /%
440	0	$1 \times 10^{-3}s^{-1}$	118
	192	$1 \times 10^{-3}s^{-1}$	200
480	0	$1 \times 10^{-3}s^{-1}$	161
	0	$3 \times 10^{-4}s^{-1}$	207
	192	$1 \times 10^{-3}s^{-1}$	270

2.3 Diffusion Bonding Principle

Diffusion bonding can achieve the reliable bonding of materials without melting them. It is widely used in the bonding of the same or different materials, and it has been applied and developed in fields that require high bonding quality, such as aviation, aerospace, nuclear and electronic industries.. The quality of diffusion bonding directly determines the mechanical properties of the structure. Therefore, scholars around the world have focused on exploring the effect of many factors such as temperature, pressure and time, providing a theoretical basis for achieving high-quality diffusion bonding.

2.3.1 Connotation of Diffusion Bonding

As a type of solid-phase welding process, diffusion bonding is a method to obtain an integral joint through mutual diffusion between the atoms, depending on the local plastic deformation of the material surface at high temperature to make the contact surfaces tighten in a certain temperature range ((0.5–0.75) T_{m}, where T_{m} is the melting point of the base metal) and pressure conditions.

Different from the traditional welding method, there is no metal melting in diffusion bonding, but achieves the bonding through the mutual diffusion between the atoms of the material. The bonding depth increases with the continuous migration and diffusion of atoms at the contact surface until the original contact interface disappears, and a new whole is formed.

The diffusion bonding technology possesses the following characteristics:

(1) Good quality of diffusion bonding joint: the microstructure and properties of diffusion bonding joints are similar to or the same as those of the base metal. There are no fusion welding defects or heat affected zone with overheated microstructures at the bonding interface, and the joint quality is excellent.
(2) High precision of diffusion bonding: due to the low pressure during diffusion bonding, the workpiece is heated as a whole and cooled by furnace cooling, so the overall deformation of diffusion bonding joints and parts is very limited and the residual stress is small, which can not only ensure the dimensional accuracy of diffusion bonding, but also realize the precision assembly and bonding after machining.
(3) Bondability of large-section joints: due to the low bonding pressure and the small-tonnage of equipments required for bonding, it is easy to achieve the bonding of large-section joints. When gas pressure diffusion bonding technique is adopted, it is easy to conduct laminating composite diffusion bonding of two plates.
(4) Wide types of bonding: due to the low bonding temperature and low thermal deformation, diffusion bonding can be used to bond parts with complex structure, large thickness difference and high precision requirements.

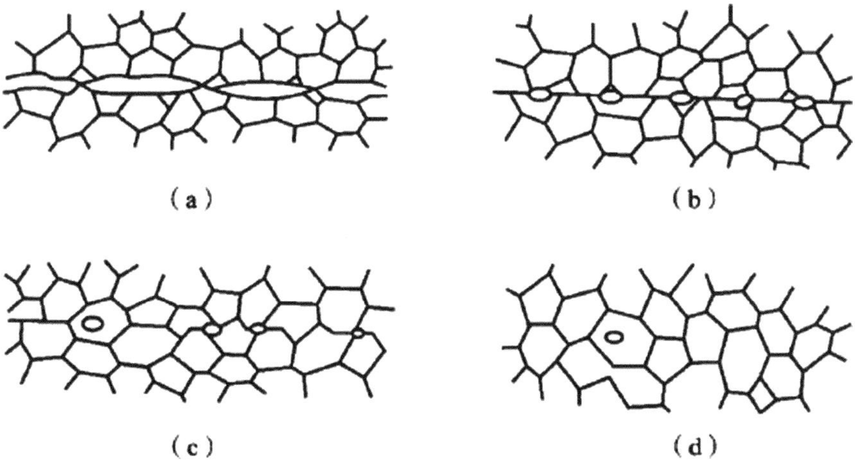

Fig. 2.24 Schematic diagram of solid diffusion bonding process. **a** Rough initial contact; **b** First stage: deformation and interface; **c** Second stage: grain boundary migration and micropore elimination; **d** Third stage: bulk diffusion micropore elimination

(5) Bonding the materials difficult to weld by other welding methods: for the same materials with poor plasticity or high melting point, or for different materials that do not dissolve each other or form brittle intermetallic compounds during fusion welding, including some metals and ceramics, diffusion bonding is the only reliable bonding method. The diffusion bonding process can be divided into three stages, as shown in Fig. 2.24.

The first stage: the physical contact stage of the material surface. Under the applied pressure at high temperature, some local contact points of the uneven microscopic surface first undergoes plastic deformation through the yield and creep; under continuous pressure, the contact area gradually increases, eventually achieving the reliable contact across the entire surface. At the end of this stage, there are still gaps between the interfaces, but the contact area has basically achieved the bonding between grains.

The second stage: interdiffusion between the atoms on the contact surfaces to form a firm bonding layer. Under high temperature and high pressure, the atoms at the grain boundaries of the contact surface continue to diffuse, leading to the disappearance of many micropores. At the same time, the grain boundary migration at the interface leaves the original interface of the joint and reaches an equilibrium state, but there are still many small micropores left in the grains.

The third stage: the formation of reliable joints. The bonding layer formed at the contact sites gradually diffuses towards the substrate, then the defects (pores, oxide inclusions, etc.) are eliminated. Common grains are formed at the contact surface, leading to internal stress relaxation, thereby forming a reliable joint. At this stage, the remaining micropores completely disappear.

The three processes of diffusion bonding are not completely separated, but are carried out alternately. Finally, metallurgical bonding will be formed in the joint bonding area through diffusion, recrystallization and other processes. The microstructures formed by joints include solid solutions, eutectics, intermetallic compounds, etc. The third stage of diffusion bonding usually stabilizes the strength and improves the plasticity of the joint. However, for some materials (such as dissimilar metals that may form brittle phases at the interface), the generation of brittle phases is extremely detrimental for the performance of diffusion bonding joints. In this case, this stage needs to be strictly controlled.

2.3.2 Diffusion Bonding Mechanism

For two substances that are in contact with each other, the thermal motion will cause the atoms to penetrate into each other. Diffusion always proceeds in the direction of decreasing material concentration, so that substances are evenly distributed in the space they occupy, which can be the diffusion between the same atoms, or the diffusion between different atoms. There are four different diffusion pathways for atoms in solid metals: bulk diffusion, surface diffusion, intergranular diffusion and dislocation diffusion.

In the actual diffusion process, these diffusion mechanisms exist simultaneously with similar laws, while bulk diffusion is the most basic diffusion process. The diffusion mechanism of atoms in solid crystals is generally related to the position of diffusion atoms in the crystal and the crystal structure of the diffusion medium. So far, there are four main diffusion mechanisms that have been found and proposed, as shown in Fig. 2.25:

(1) Transposition mechanism: in the substitutional solid solution, the diffusion of interstitial atoms in equilibrium is considered to achieved by the direct exchange of positions between two adjacent atoms due to the high energy required in formation of interstitial atoms, as shown in Fig. 2.25a. According to this mechanism, when two atoms exchange positions, they will inevitably require adjacent atoms to give up enough space, which will cause serious lattice distortion in the nearby space and consume a lot of energy. Therefore, it is generally believed that such a direct exchange mechanism is difficult to implement in practice.
(2) Vacancy mechanism: in the case of interdiffusion between lattice vacancies and adjacent atoms, position exchange between vacancies and atoms can be realized by vacancy transfer (i.e. atoms migrate by the movement of vacancies), as shown in Fig. 2.25b. It is well known that there is a certain concentration of vacancies at different temperatures. The existence of vacancies will cause the adjacent atoms to deviate from their equilibrium positions, resulting in an increase in potential energy, so that the barriers that the atoms need to overcome to jump into the vacancies will be reduced. In this way, it is easier for adjacent atoms to jump towards vacancies. The higher the temperature, the greater the concentration of the vacancies, and the easier the diffusion of atoms in the metals. The atoms adjacent to the vacancy are much easier to enter the vacancy position and make

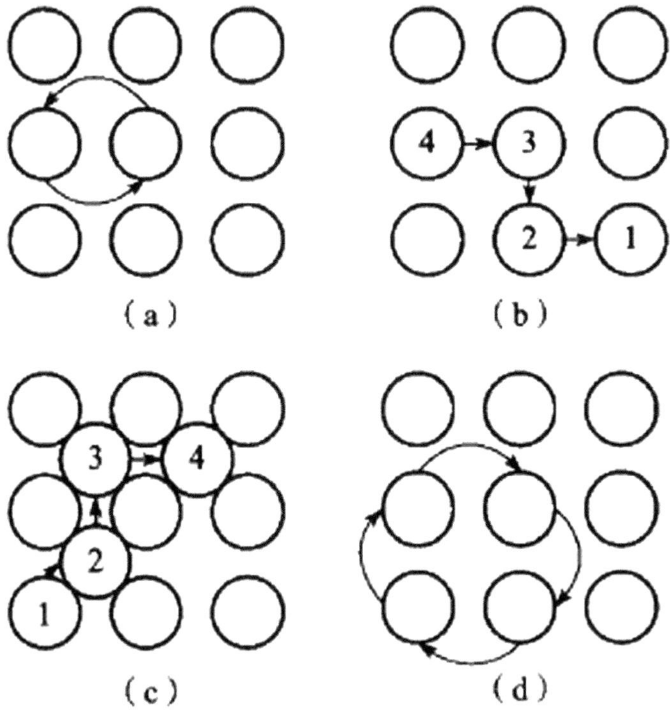

Fig. 2.25 Four basic mechanisms of diffusion **a** Transposition mechanism; **b** Vacancy mechanism; **c** Interstitial mechanism; **d** Annular mechanism

the original atomic position become a vacancy. With this process going on, the counter flow of diffused atoms and vacancies is formed, which is the vacancy diffusion mechanism. The basic principle of this model is that: diffused atoms migrate by transposition with adjacent vacancies.

(3) Interstitial mechanism: it refers to the diffusion of diffusing atoms through transitions between lattice interstitial positions, as shown in Fig. 2.25c. In interstitial solid solutions, since the radius of solute atoms is usually much smaller than that of matrix atoms, infinite lattice distortion will be formed during the transition, and the low energy is consumed, so the diffusivity is usually large. The main model is that interstitial atoms with small diameters transition from one interstitial position to another adjacent interstitial position. At this time, the diffusion coefficient of interstitial atoms is 10^4–10^5 times larger than the self-diffusion coefficient of the matrix metal atoms. When the diameter of the interstitial atom is larger, the interstitial atom occupies the original lattice position by pushing its adjacent atoms to the nearby void.

(4) Annular mechanism: it is considered that n atoms at equal intervals on the same crystal place can alternate positions at the same time to diffuse, which requires several atoms in the crystal to move regularly at the same time, as shown in

Fig. 2.25d. The probability of such behavior in solid metals and alloys is obviously very small. Therefore, it is also difficult to realize the annular exchange actually, and there is no experimental support for this mechanism up to now. In fact, a lot of metals during interdiffusion are often accompanied by the Kirkendall effect, which at least shows that the diffusion mechanism of these metals is not simply an exchange mechanism (including direct exchange mechanism and annular mechanism). The reason is that according to the exchange mechanism, the net value of the number of atoms passing through the plane perpendicular to the diffusion direction should be zero, that is, the diffusion coefficients of the two components should be equal. Essentially, the mechanism relies on the mutual transposition between lattice atoms and adjacent atoms for diffusion. Metals and binary alloy solid solutions with FCC lattice (when the concentration of solute is low) may diffuse according to this mechanism.

2.3.3 Influencing Factors of Diffusion Bonding

The main factors affecting the diffusion bonding are: temperature, pressure, time and surface state.

1. **Temperature**

Temperature is the most important process parameter of diffusion bonding, and even a small change in temperature can cause a large change in diffusion bonding speed. Within a certain temperature range, the higher the temperature, the better the material plasticity, the faster the diffusion process, and the higher the joint strength obtained. From this point of view, the diffusion bonding temperature should be selected as high as possible. However, the heating temperature is limited by the high temperature strength of the bonded workpiece and fixture, the phase transformation, recrystallization and other metallurgical characteristics, which will lead to the coarsening of the base metal structure, composition segregation, etc. Therefore, the temperature and the diffusion bonding quality are not simply positively correlated, but there is a better diffusion temperature range, usually (0.6 ~ 0.8) T_{m} (T_{m} is the melting point of the base metal).

2. **Pressure**

The function of pressure in the diffusion bonding is to make the joint undergo certain plastic deformation at the bonding temperature, eliminating interface pores, and increasing the actual contact area of the joint surface. Pressure is the main factor affecting the diffusion bonding of thick titanium alloy plates. If the applied pressure is too low, the surface plastic deformation will be insufficient, the process of forming physical contact on the surface may be incomplete, and the remaining pores on the interface are too large and there is a large number. Higher diffusion pressure can produce greater plastic deformation of the surface layer, and can also reduce the recrystallization temperature of the surface layer, thus accelerating the migration of grain boundaries. High pressure is conducive to the shrinkage and elimination of micropores in solid diffusion welding, and can also reduce or prevent diffusion pores in diffusion welding of dissimilar metals. When other parameters are fixed, higher pressures enable to produce better joints. When the grain size of the workpiece is larger or the surface is rougher, higher diffusion pressure is required. However, the pressure should not be too large, or serious plastic deformation will occur, which may lead to a decrease in the dimensional accuracy of joints, and this also depends on the limit of the overall deformation, equipment tonnage, etc.

3. **Time**

Diffusion bonding time refers to the time that the bonded workpiece is kept under diffusion temperature and diffusion bonding pressure. The diffusion process must be fully completed within the bonding time, so as to achieve the required strength. Diffusion time is not a completely independent parameter, and should be reasonably selected according to temperature and pressure. The average migration distance X of atoms in the diffusion process can be expressed as $X = C\sqrt{Dt}$ (where, C is a constant depending on the material, D is the diffusion coefficient, and t is the diffusion time). When other parameters remain unchanged, proper extension of the diffusion time can make the diffusion proceed more fully, which is conducive to achieve a more uniform distribution of the joint composition and microstructure. At the same time, the temperature of diffusion bonding is generally above the recrystallization temperature. Further extension of diffusion time will not further improve the joint quality, but will make the joint grains grow. For joints that may form brittle intermetallic compounds, the diffusion time should be governed to control the thickness of the diffusion layer to avoid affecting the the joint performance.

4. **Surface state**

The surface state includes surface roughness, cleanliness (including oxide film, organic film, water film and gas film on the surface) and flatness. The surface of parts to be diffused and bonded should be pickled or chemically etched before heating and pressure, and vacuum sintering for vacuum degassing to obtain a clean surface. The cleanliness of the bonding surface has a great impact on the mechanical properties of diffusion bonding joints. Certain straightness and roughness are to ensure that the interface can reach the required joint area without high deformation. The surface of the specimen for diffusion bonding has high machining accuracy, and the microscopic flatness of the contact surface is small. The reliable bonding of the whole bonded surface can be realized at a lower temperature and pressure. On the contrary, if the processing accuracy is low, the reliable bonding can be realized at higher temperatures and pressures.

2.4 Superplastic Forming/Diffusion Bonding Technology

Superplastic forming (SPF) refers to the method of forming materials by using the superplasticity of materials under specific conditions, as shown in Fig. 2.26. Superplastic forming includes pneumatic forming, hydraulic forming, bulk forming, sheet forming, tube forming, cupping forming, dieless forming, dieless drawing, etc. Among them, pneumatic forming, also known as blow moulding, is the most widely used superplastic forming method in the aerospace industry. Blow moulding is a sheet forming technology that can obtain high deformation with low energy and low pressure. mould is used to form a closed pressure space on the outside of the blank. After the thin plate/sheet is heated to the superplastic temperature,, the blank will produce superplastic deformation under the pressure of compressed gas and gradually approach the mould surface until it is completely fitted with the mould.

The superplastic forming/diffusion bonding (SPF/DB) technology is a near net shape integrated manufacturing technology that combines superplastic forming and diffusion bonding to manufacture high-precision large parts. This technology utilizes the superplasticity of materials and the diffusion bonding characteristics within the same thermal specification to complete superplastic forming and diffusion bonding in a thermal cycle, and to manufacture complex integral structural parts that are difficult to taking shape by conventional processes. The hollow structures manufactured by SPF/DB technology are usually classified according to the number of blank layers, and the most common are the following four types of structures (Fig. 2.27):

1. **Single-layer plate structure**

Single-layer plate structure includes two structural forms: one is the single-layer plate structure formed by a layer of sheet metal through SPF technology. which is mainly applied to components such as pressure vessels, shells, fairings; the other is the single-layer plate reinforced structure (Fig. 2.27a). Before or at the last stage of superplastic forming, a layer of reinforced plates is diffused and bonded at the local part of the sheet metal to increase the local thickness of the single-layer plate structure or serve as a stiffening rib, so as to improve the structural strength and

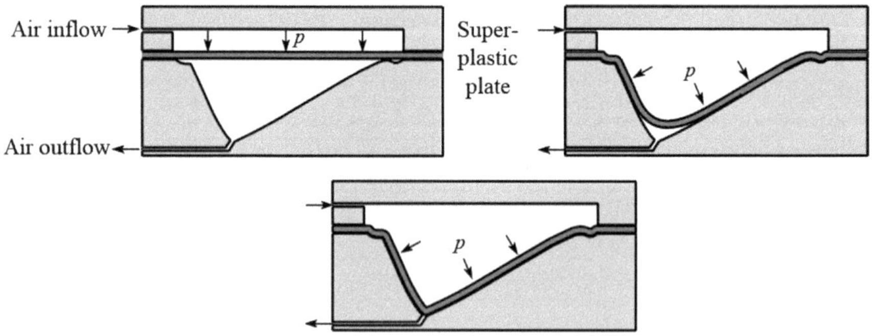

Fig. 2.26 Schematic diagram of superplastic forming

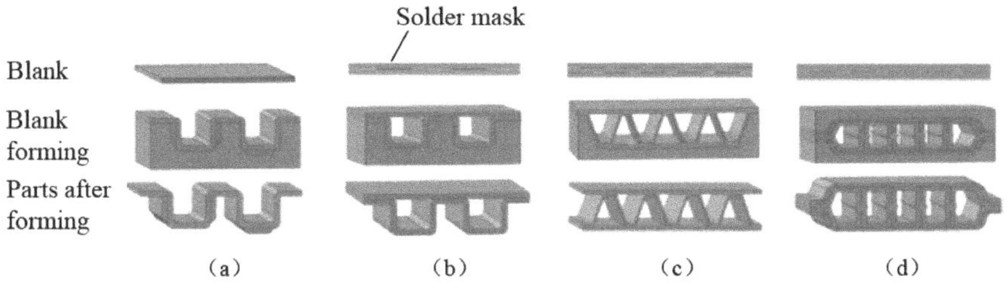

Fig. 2.27 Four basic forms of titanium alloy SPF/DB structural parts **a** Single-layer plate; **b** Two-layer plate; **c** Three-layer plate; **d** Four-layer plate

stiffness of the parts. The plates used for reinforcement can be ordinary titanium plates or the machined titanium alloy reinforcements. This reinforcement method is suitable for strengthening internal structures such as frames and ribs.

2. **Two-layer plate structure**

The original blank is a two-layer plate. Before forming, the areas that do not need to be bonded between the two layers of plates are coated with solder resists. After forming by SPF/DB technology, the areas without coating solder resists are diffused and bonded together, and the areas coated with solder resists are superplastic formed into hollows, finally forming a two-layer plate structure. The structure has significant advantages such as good integrity, strong stiffness and lightweight, which can replace or partially replace the skin-truss riveted welding structure and it can be used in non-load-bearing and secondary load-bearing components such as aircraft tegmen oris, cabin doors and wall panels.

3. **Three-layer plate structure**

The original blank is composed of three layers of plates: two layers of panels and one layer of a core. Before forming, solder resists are applied to the appropriate area between the plates. Subject to SPF/DB technology, the upper and lower panels form a part contour, and the middle core forms a corrugated plate with a strengthening effect, which plays a role in improving the structural strength and stiffness. This structure is mainly used in structures of inlet lips, missile airfoil surface, blades, etc.

4. **Four-layer plate structure**

The original blank is composed of four layers of plates: two layers of panels and two layers of cores. Before forming, solder resists are applied to the appropriate area between the two layers of cores. Subject to SPF/DB technology, the upper and lower two panels form a part contour, and the middle two cores form a vertical spacer plate to strengthen the structure. This structure has the obvious advantages of lightweight and integration while meeting the requirements of load-bearing capacity. It is mainly used in the structures of missile airfoil surfaces, rudder surfaces, aircraft slats, pelvic fins, engine straightener blades, adjust

blades, etc.

SPF and SPF/DB technologies have the following advantages due to their process characteristics:

(1) During the forming process, materials are able to withstand high deformation without cracking, and form structural parts with complex profiles, which can not be achieved by conventional thermoforming methods or can only be achieved by multi-pass forming.
(2) In the forming process, the material flow stress must be kept low. Large-size structural parts can be formed with small-tonnage equipment, and the processed structural parts have small springback, low residual stress and high forming precision.
(3) The structural parts manufactured by SPF and SPF/DB technology have good overall performance, higher strength and stiffness, and better fatigue and corrosion resistance. Compared with the traditional welded or mechanically-bonded structural parts, it greatly reduces the number of parts and toolings, shortens the manufacturing cycle, and reduces the manufacturing cost.
(4) The adoption of SPF and SPF/DB technology can improve the freedom of structural design, further improve the structural load-bearing efficiency and reduce the weight of structural parts while ensuring the structural strength and stiffness.

With its characteristics of high performance, high efficiency, low cost and near net shape, SPF/DB technology breaks through the limitations of traditional sheet forming methods, promotes the innovative development of aerospace structural design concepts, and has become an advanced manufacturing technology with great development prospects in the aerospace field.

References

1. Zhaorong, Lin. 1990. *Principle and application of metal superplastic forming*. Beijing: Aviation Industry Press.
2. Shidun, Wu. 1997. *Theory of superplastic deformation of metals*. Beijing: National Defense Industry Press.
3. Kayeboshev, O.A. 1982. *Plasticity and superplasticity of metals*. Translated by Wang Yanwen. Beijing: China Machine Press.
4. Nieh, T.G., J. Wadsworth, and O.D. Sherby. 1997. *Superplasticity in metals and ceramics*. New York: United States of America by Cambridge University Press.
5. Tikhonov, A.C. 1983. *Superplastic effect of metals and alloys*. Translated by Liu Chunlin. Beijing: Science and Technology of China Press.
6. Underwood, E.E. 1962. Review of superplasticity. *Journal of Metals* 14 (3): 414–419.
7. Backofen W. A., J. R. Turner, and D.H. Avery. 1964. Superplasticity in an Al-Zn alloy. Transactions on ASME 57(6): 980–990.
8. Artzy, A.B., and A. Shtechman. 2003. Plastic deformation of wrought magnesium alloys AZ31. *Proceedings of the Second Israeli International Conference on Magnesium Science and Technology* 2: 151–158.
9. Yingbing, Luo. 2007. *Study on superplastic deformation mechanism and forming process of light alloy [D]*. Shanghai: Shanghai Jiaotong University.

10. Hongbo, He. 2008. *Superplastic deformation mechanism and behavior characterization of fine-grain TC4 titanium alloy [D]*. Da Lian: Dalian University of Technology.
11. Liu Yu. 1998. Superplasticity and superplastic forming of metals. *Machine Building and Automation*, 2: 14–15.
12. Liu Weixian, and Liping Huang. 1982. Superplasticity and superplastic forming of titanium alloy. *Aeronautical Manufacturing Technology*, 8: 3–9.
13. Chen Puquan, Zhongqi Cui, and Min Zhao. 1990. Research on progress, direction and deformation mechanism of superplasticity. *Materials Science and Technology*, 2: 16–22.
14. Xiaoqi, Shi. 2007. *Study on combined superplastic forming/diffusion bonding process for titanium alloys [D]*. Nanjing: Nanjing University of Aeronautics and Astronautics.
15. Zhang Xuexue. 2012. Experimental study on superplastic forming/diffusion bonding of TC4 multilayer plate [D]. Nanjing University of Aeronautics and Astronautics.
16. Wang Dagang. 2013. Study on superplastic forming/diffusion bonding process of titanium alloy rudder body [D]. Nanjing University of Aeronautics and Astronautics.
17. Kim, J.S., Y.W. Chang, and C.S. Lee. 1998. Quantitative analysis on boundary sliding and its accommodation mode during superplastic deformation of two-phase Ti-6Al-4V alloy. *Metallurgical and Materials Transactions A* 29: 217–226.
18. Ko, Y.G., C.S. Lee, D.H. Shin, et al. 2006. Low-temperature superplasticity of ultra-fine-grained Ti-6Al-4V processed by equal-channel angular pressing. *Metallurgical and Materials Transactions A*. 37: 381–391.
19. Ko, Y.G., W.G. Kim, and C.S. Lee. 2005. Microstructural influence on low-temperature superplasticity of ultrafine-grained Ti-6Al-4V alloy. *Materials Science and Engineering A*. 410–411: 156–159.
20. Roy, S., and S. Suwas. 2014. Enhanced superplasticity for ($\alpha + \beta$)-hot rolled Ti–6Al–4V–0.1B alloy by means of dynamic globularization. *Materials and Design*, 58: 52–64.
21. Gao Fei, Weidong Li, and B. Meng, et al. 2017. Rheological law and constitutive model for superplastic deformation of Ti-6Al-4V. *Journal of Alloys and Compounds*, 701: 177–185.
22. Lutfullin, R.Y., O.A. Kaibyshev, and R.V. Safiullin, et al. 2000. Superplasticity and solid state bonding of titanium alloys. *Acta Metallurgica Sinica(English Letters)*, 2: 561–566.
23. Salishchev, G.A., R.M. Galeyev, O.R. Valiakhmetov, et al. 2001. Development of Ti–6Al–4V sheet with low temperature superplastic properties. *Journal of Materials Processing Technology* 116 (2–3): 265–268.
24. Wang, Y.C., and T.G. Langdon. 2013. Influence of phase volume fractions on the processing of a Ti–6Al–4V alloy by high-pressure torsion. *Materials Science and Engineering A* 559: 861–867.
25. Wang, Y.C., and T.G. Langdon. 2013. Effect of heat treatment on microstructure and microhardness evolution in a Ti–6Al–4V alloy processed by high-pressure torsion. *Journal of Materials Science*. 48: 4646–4652.
26. Fu, J., H. Ding, Y. Huang, et al. 2015. Influence of phase volume fraction on the grain refining of a Ti-6Al-4V alloy by high-pressure torsion. *Journal of Materials Research and Technology* 4 (1): 2–7.
27. Fu, J., H. Ding, Y. Huang, et al. 2015. Grain refining of a Ti-6Al-4V alloy by high-pressure torsion and low temperature superplasticity. *Letters on Materials* 5 (3): 281–286.
28. Salishchev, G.A., O.R. Valiakhmetov, V. Valitov, et al. 1994. Submicrocrystalline and nanocrystalline structure formation in materials and search for outstanding superplastic properties. *Materials Science Forum* 170–172: 121–130.
29. Zhang, T., Y. Liu, D.G. Sanders, et al. 2014. Development of fine-grain size titanium 6Al–4V alloy sheet material for low temperature superplastic forming. *Materials Science and Engineering A* 608: 265–272.
30. Matsumoto, H., K. Yoshida, S.H. Lee, et al. 2013. Ti–6Al–4V alloy with an ultrafine-grained microstructure exhibiting low-temperature–high-strain-rate superplasticity. *Materials Letters*. 98: 209–212.
31. Zhang, W.J., Dingh, and M.H. Cai, et al. 2018. Ultra-grain refinement and enhanced low-temperature superplasticity in a friction stir-processed Ti-6Al-4 Valloy. *Materials Science and Engineering A*, 727: 90~96.

32. Qinxin, Yang. 2017. *experimental study on superplastic forming/diffusion bonding of TAL5 two-layer plate [D]*. Nanjing: Nanjing University of Aeronautics and Astronautics.
33. Chen, Zhang. 2017. *Study on superplastic forming process of TC4 alloy four layer rudder structure [D]*. Beijing: China Academy of Machinery Science and Technology Group Co., Ltd.
34. Kui, Wang. 2016. *Study on superplastic deformation characteristics of Ti-55 high temperature titanium alloy with hydrogen treatment [D]*. Shanghai: Shanghai Jiaotong University.
35. Xiaojie, Chen. 2012. *Study on superplasticity of tal5 alloy with optimum deformation rate and its application [D]*. Nanchang: Nanchang Hangkong University.
36. Liangliang, Yan. 2015. *Experimental study on superplastic formation-diffusion bonding of Ti-55 two-layer plate [D]*. Nanjing: Nanjing University of Aeronautics and Astronautics.
37. Tsuzuku, T. 1991. Superplasticity in advanced materials. *Japan Society for Research on Superplasticity*, 611.
38. Tuoyang, Zhang. 2014. *Study on low temperature superplastic deformation of fine-grain TC4 alloy [D]*. Changsha: Central South University.
39. Tanaka, K., and R. Iwasaki. 1985. A phenomenological theory of transformation superplasticity. *Engineering Fracture Mechanics* 21 (4): 709–720.
40. Han, H.N., and J.K. Lee. 2002. A constitutive model for transformation superplasticity under external stress during phase transformation of steels. *ISIJ international.* 42 (2): 200–205.
41. Xiangfeng, Lin, and Yun Junbi. 1996. Study on phase transformation superplastic diffusion welding of steel. *Journal of Nanjing University of Aeronautics and Astronautics* 28 (2): 199–204.
42. Sun Chao. 2014. Diffusion bonding superplastic forming technology for TC4 titanium alloy hollow blades [D]. Harbin: Harbin Institute of Technology.
43. Shuixing, Zhu, and Yang Zhenheng. 1988. Stress-strain rate relation of superplastic deformation. *Materials Science and Technology* 4: 44–49.
44. Yuquan, Song, and Zhao Jun. 1984. Superplastic constitutive equation with variable M-value. *Materials Science and Technology* 3: 4–17.
45. Liqin, Tan, Wang Gaochao, Gan Wenqing, et al. 2014. Superplastic constitutive equation of TAL5 titanium alloy based on strain rate cycling method. *Journal of Aeronautical Materials* 34 (6): 21–27.
46. Tan, L.Q., G.C. Wang, and W.Q. Gan, et al. 2014. Superplastic constitutive equation of TAL5 titanium alloy based on strain rate cycling method. *Journal of Aeronautical Materials*, 6: 21–27.
47. Fan, X.G., H. Yang, and P.F. Gao. 2013. Prediction of constitutive behavior and microstructure evolution in hot deformation of TA15 titanium alloy. *Materials & Design* 51: 34–42.
48. Zhiping, Guan. 2008. *Quantitative mechanical analysis of superplastic tensile deformation [D]*. Changchun: Jilin University.
49. Zhiqiang, Li., and Guo Heping. 2004. Application and development of superplastic forming/ diffusion bonding technology. *Aeronautical Manufacturing Technology* 11: 50–52.

Chapter 3
Typical Structure Forming and Process Quality Control

The rational design and precise control are the keys to ensure the forming quality of typical SPF/DB titanium alloy structure. In this chapter, SPF/DB process principle and route of typical structures such as single-layer plate structure, two-layer plate structure, three-layer plate structure and four-layer plate structure are introduced. The structural parameter design principle, process parameter selection criteria and mould design method based on process feasibility are put forward. The control methods of quality defects such as wall thickness uniformity, surface wrinkle, local fracture, cooling deformation and surface step difference by using the method of numerical simulation and experimental verification are introduced in detail, typical examples of titanium alloy SPF/DB parts in the aerospace field are shown for demonstration.

3.1 Typical Structure Forming Process

The perfect combination of titanium alloy superplastic forming and diffusion bonding technology can achieve the forming and manufacturing of different types of SPF/DB structures. According to the number of layers of the blank material, typical SPF/DB structures can be divided into single-layer sheet structure, double-layer sheet structure, triple-layer sheet structure, and quadruple-layer sheet structure, which have different process requirements. This section mainly introduces the process principles and process routes corresponding to four typical structure forms.

Z. Li, *Superplastic Forming/Diffusion Bonding Technology of Titanium Alloys*,
https://doi.org/10.1007/978-981-99-3909-1_3

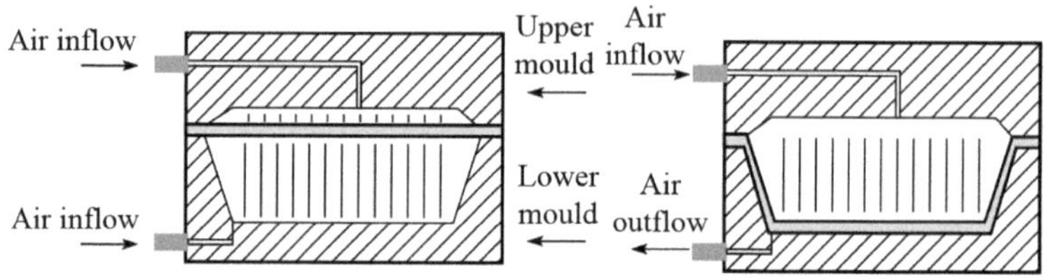

Fig. 3.1 Schematic diagram of superplastic forming process of single-layer sheet structure

3.1.1 Superplastic Forming/Diffusion Bonding Principle of Typical Structures

1. Single-Layer Sheet Structure

The original blank of single-layer sheet structure is a titanium alloy sheet with superplasticity, and the process is shown in Fig. 3.1. The titanium alloy sheet is placed into the forming mould, and the upper and lower moulds are closed to form a closed space around the sheet, which is pressed tightly against the mould edges. When the sheet is heated to the superplastic temperature, inert gas is injected into the mould cavity. Under the pressure difference of the gas between the upper and lower mould cavities, the sheet is slowly formed at a low strain rate and gradually approaches the mould surface until it is completely fitted to the mould surface.

2. Two-Layer Sheet Structure

Mould

The original blank of the SPF/DB two-layer sheet structure are two titanium alloy sheets. The process is shown in Fig. 3.2. Firstly, the sheets 1 and 2 with local coating of anti-welding coating are welded and placed in the heating furnace, and the space between the two sheets is under vacuum. Inert gas is injected from the lower mould to make the uncoated part of sheet 1 and sheet 2 diffuse-connect. Then, inert gas is injected between sheet 1 and sheet 2, causing the part coated with anti-welding coating in sheet 2 to undergo superplastic deformation, and at the same time, the gas in the mould cavity is discharged until sheet 2 is completely fitted to the lower mould cavity.

3. Three-Layer Sheet Structure

The three-layer sheet structure is shown in Fig. 3.3. The original blank consists of three sheets. Both sides of sheet 2 are coated with stop-off and sealed with sheet 1 and sheet 3 to form a sealed pocket. The end of the pocket is reserved with a gas

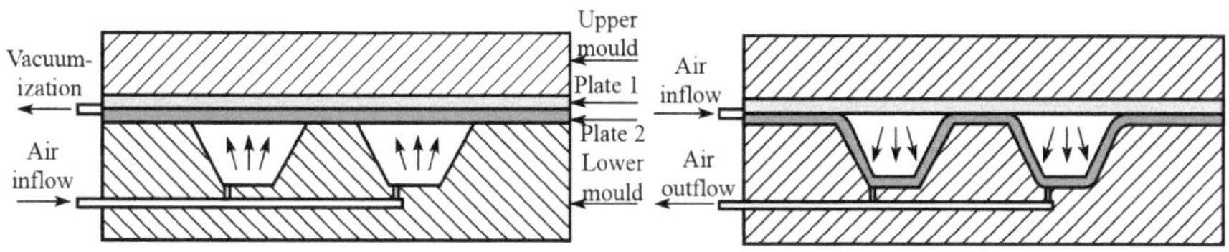

Fig. 3.2 Schematic diagram of SPF/DB process for two-layer sheet structure

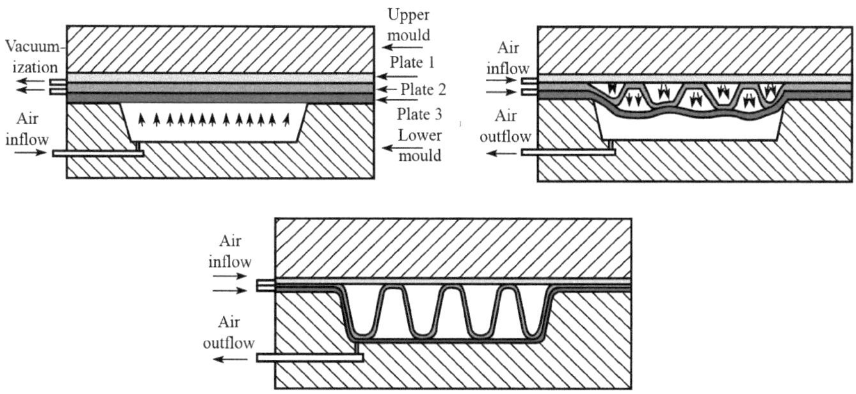

Fig. 3.3 Schematic diagram of SPF/DB process for three-layer sheet structure

tube. First, the interior of the three-layer sheet is vacuumized, and then inert gas is injected from the lower mould to achieve diffusion bonding between the parts of the three-layer sheet that are not coated with stop-off. Then, inert gas is injected into the air inlet tube of the sealed pocket of the three-layer sheet; the sheets 1 and 3 are driven by the air pressure to inflate to the mould on both sides until they are tightly attached to the mould to form a cavity with a corrugated truss structure.

4. Four-Layer Sheet Structure

The four-layer sheet structure is composed of four sheets. Firstly, a specific area between sheet 2 and sheet 3 is coated with anti-welding coating and formed by SPF/DB. Sheet 1 and sheet 4 form the panel, while sheet 2 and sheet 3 form vertical partitions to strengthen the structural. The main process principles for forming the four-layer sheet structure are shown in Fig. 3.4:

(1) Superplastic forming of outer layer: when the sheets are heated to the SPF/DB temperature, inert gas is injected between sheet 1 and sheet 2, and between sheet 3 and sheet 4. The strain rate of the sheets is controlled within the optimal range, which ensures that the wall thickness of the deformed sheet is as uniform as

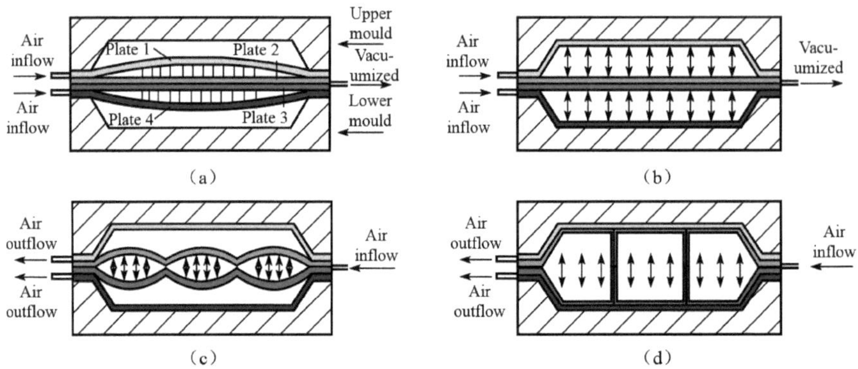

Fig. 3.4 Schematic diagram of SPF/DB process for four-layer sheet structure

possible until sheet 1 and sheet 4 are completely fitted to the mould, as shown in Fig. 3.4a.

(2) Diffusion bonding of the inner layer: during the superplastic forming process of sheet 1 and sheet 4, a vacuum state is maintained between sheet 2 and sheet 3 to prevent contamination of 'diffusion bonding interface. The outer layer is pressurized for a period of time, and sheet 2 and sheet 3 are diffusion bonded, as shown in Fig. 3.4b.
(3) Superplastic forming of the inner layer: the gas is discharged from between sheet 1 and sheet 2, and between sheet 3 and sheet 4, while inert gas is injected between sheet 2 and sheet 3. The sheet is slowly formed within the optimal strain rate range until sheet 2 is completely fitted to sheet 1 and sheet 3 is completely fitted to sheet 4, as shown in Fig. 3.4c.
(4) Diffusion bonding of inner and outer layer: the gas pressure between sheet 2 and sheet 3 is maintained to ensure that sheet 2 is fully diffusion bonded with sheet 1 and sheet 3 is fully bonded with sheet 4, as shown in Fig. 3.4d.

3.1.2 Superplastic Forming/Diffusion Bonding Process Route of Typical Structures

1. Forming Process Route of Single-Layer Sheet Structure

The superplastic forming process route for single-layer sheet structure forming is relatively simple, and the following aspects need to be focused on during the forming process: cleaning and oxidation protection of the mould and sheet surface, uniformity of furnace temperature during forming process, control of strain rate during forming process, and deformation during discharging. The specific process route is shown in Fig. 3.5.

(1) Sheet preparing: carried out using a shearing machine or by wire cutting.

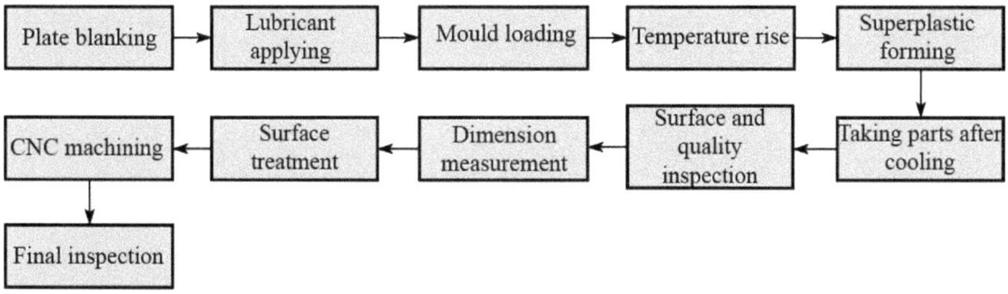

Fig. 3.5 Superplastic forming process route of single-layer sheet structure

(2) Lubricant coating: After drying the cleaned sheets, a lubricating anti-oxidation coating is applied.
(3) Furnace loading and heating: After loading the sheets into the forming mould, the mould is closed and heated to the forming temperature together with the sheets.
(4) Superplastic forming: Inert gas is filled into the mould cavity, and the sheet is slowly formed under low strain rate. If bi-directional forming is required, gas channels are set in the upper and lower mould cavities for stepwise forming.
(5) Furnace cooling and discharging: After the furnace chamber is cooled down, the parts are removed, and the surface quality and dimensional accuracy of the parts are checked.
(6) Inspection: Surface quality, material quality, and dimensional inspections are conducted.
(7) Surface treatment: The parts are alkaline washed and acid washed or sandblasted to remove oxide scale.
(8) CNC machining: Parts are CNC machined and surface polished according to product inspection requirements.
(9) Final Inspection: The parts are subject to final inspection according to relevant testing requirements.

2. Forming Process Route of Multi-Layer Sheet Structure

For multilayer sheet structure, the hollow structure should be manufactured through the combination of diffusion bonding and superplastic forming process, so its process flow is relatively complex. Compared with the single-layer sheet structure, the multi-layer sheet structure also applies t the process of pattern preparation, assembly welding and edge sealing, ultrasonic testing, etc. sheet. The specific process route is as shown in Fig. 3.6. The key processes different from the single-layer sheet structure are described as follows:

(1) Degreasing and acid washing: Metal cleaning agent is used to remove contaminants, oxides, acid traces, water stains, etc. from the surface of the sheet.
(2) Graphic preparation: The contour of the anti-welding coating should be marked on the sheet according to the simulation results, then the coating is applied

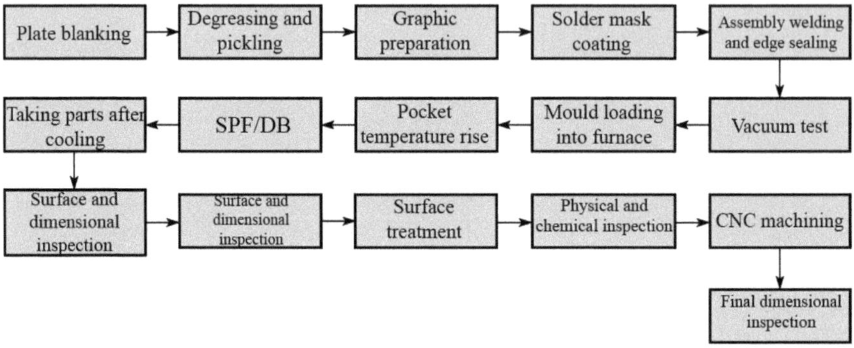

Fig. 3.6 Process route of SPF/DB with multilayer sheet structure

according to the contour using methods such as marking, laser projection, or laser engraving.

(3) Assembly welding and edge sealing: After the multi-layered sheets are positioned in order, they are fixed with special welding clamps, and the pockets are sealed with arc welding, leaving air inlets and exhaust pipes.
(4) Vacuum inspection: The closed cavity is vacuumed and the vacuum degree is inspected to ensure good sealing of the cavity.
(5) Diffusion bonding: Nitrogen is pressurized in the mould cavity to perform diffusion bonding between the two titanium alloy sheets where the anti-welding coating is not applied.
(6) Superplastic forming: The mould cavity is vacuumed, and the pocket is pressurized with inert gas. During pressurization, the strain rate is controlled according to the preset pressure–time curve to ensure that the material is formed at the best deformation rate. Finally, the part is kept under pressure for a period of time to ensure complete moulding.
(7) Ultrasonic testing: The part is subjected to ultrasonic non-destructive testing to detect the quality of the diffusion bonding.
(8) X-ray testing: The part is subjected to X-ray testing to detect the quality of the reinforcing bars.

3.2 Numerical Simulation of Forming Process of Typical Structures

The process of SPF/DB is complex and is carried out in a high-temperature, sealed environment, making it difficult to observe and control the intermediate process of forming. Additionally, due to the varied forms of SPF/DB structures, it is not possible to use a unified process to solve the forming quality issues of different types of structures. With the development of finite element numerical simulation technology, using nonlinear finite element methods to numerically simulate the SPF/

DB process can provide a convenient and accurate method for optimizing process parameters and controlling process quality. This section introduces the basic theory of numerical simulation for superplastic forming and describes the finite element modeling method and numerical simulation results for typical SPF/DB structural components.

3.2.1 Basic Theory of Numerical Simulation of Superplastic Forming

3.2.1.1 Basic Mechanical Equations of Rigid Plastic/Rigid-Viscoplastic Finite Element Method

Superplastic deformation is highly sensitive to strain rate, and the material exhibits minimal elastic recovery after deformation, which belongs to a typical problem of nonlinear large deformation. Therefore, rigid-plastic and rigid-viscoplastic models are generally used to calculate superplastic forming.

The rigid-plastic and rigid-viscoplastic finite element method is generally based on the variational principle or upper-bound theorem. It uses the nonlinear function of node velocity to represent the energy dissipation rate functional in the finite element mode. The optimal velocity field with the minimum total energy dissipation rate is obtained using mathematical optimization theory. Finally, the deformation velocity field, stress field, strain field, and other deformation parameters are obtained using the basic mechanical theory formulas for superplastic deformation. In the solution process, compared with the elastic–plastic or elastic-viscoplastic finite element method, the rigid-plastic and rigid-viscoplastic finite element method has no stress accumulation error problem and no gradual yielding of the elements. Therefore, it has high computational accuracy and low computational workload, and is an effective method for finite element numerical simulation of the superplastic forming process.

In order to reduce the computational workload and improve simulation efficiency, some assumptions need to be made about the deformation characteristics and physical properties of the material when using the rigid-plastic and rigid-viscoplastic finite element method to simulate superplastic deformation. These assumptions include:

(1) elastic deformation of the material is not considered during the deformation process;
(2) the volume of the material remains constant during the deformation process;
(3) the material is homogeneous and isotropic;
(4) volume force and inertia force are not considered;
(5) the material deformation flow conforms to the Levy–Mises yield criteria.

Based on these assumptions, it can be approximately assumed that the strain hardening index $n = 0$, and the flow state equation can be expressed as

$$\sigma = K\dot{\varepsilon}^{m} \tag{3.1}$$

where: σ is stress; $\dot{\varepsilon}$ is strain rate; K is material constant; m is strain rate sensitivity exponent. The coefficients in the constitutive equation are obtained from the uniaxial superplastic tensile test.

In the process of finite element numerical simulation, when the rigid viscoplastic material undergoes superplastic deformation, the following basic mechanical theoretical equations of rigid plasticity/rigid viscoplasticity must be satisfied:

(1) Equilibrium equation:

$$\sigma_{ij,j} = 0 \tag{3.2}$$

where: $\boldsymbol{\sigma}_{ij}$ is stress tensor; i,j = x,y,z; "," represents the partial differential of a coordinate variable.

(B) Geometric equation:

$$\dot{\varepsilon}_{ij} = \frac{1}{2}\left(u_{i,j} + u_{j.i}\right) \tag{3.3}$$

where: ε_{ij} is strain rate tensor; $\mathbf{u}_i$ and $\mathbf{u}_j$ are displacement velocity components.

(C) Incompressible volume equation

$$\dot{\varepsilon}_v = \dot{\varepsilon}_{ij}\delta_{ij} = 0 \tag{3.4}$$

where: $\dot{\varepsilon}_v$ is volumetric strain rate; δ_{ij} is Kronecker symbol, $\delta_{ij} = \begin{cases} 1\ (i = j) \\ 0\ (i \neq j) \end{cases}$.

(D) Yield criterion:

$$\overline{\sigma} = \sqrt{\frac{3\sigma_{ij}{'}\sigma_{ij}{'}}{2}} \tag{3.5}$$

where: $\overline{\sigma}$ is equivalent stress; σ_{ij}' is the stress deviation. For rigid-plastic materials,

$$\overline{\sigma} = \overline{\sigma}(\overline{\varepsilon}) \tag{3.6}$$

where: $\overline{\varepsilon}$ is the equivalent strain.

For rigid-viscoplastic $\overline{\varepsilon}$ materials,

$$\overline{\sigma} = \overline{\sigma}\left(\overline{\varepsilon}, \dot{\overline{\varepsilon}}\right) \tag{3.7}$$

where: is the equivalent strain rate.

(E) Levy–Mises relation:

$$\dot{\varepsilon}_{ij} = \dot{\lambda}\sigma_{ij}{'} \tag{3.8}$$

where: $\dot{\lambda}$ is a non-negative proportionality constant, and

$$\dot{\lambda} = \frac{3}{2}\frac{\dot{\bar{\varepsilon}}}{\bar{\varepsilon}} \tag{3.9}$$

where: $\dot{\bar{\varepsilon}} = \sqrt{\frac{2}{3}\dot{\varepsilon}_{ij}\dot{\varepsilon}_{ij}}$

(F) Boundary condition:

The boundary condition includes force boundary conditions and velocity boundary conditions. Suppose that the volume of the deformed body is V, the surface area is S, the surface force F_i is applied on the force surface S_F, the velocity $\overline{u}_i$ is given on the velocity surface S_u, and the stress boundary conditions meet:

$$\sigma_{ij}n_j = F_i \tag{3.10}$$

where: n_j is the component of the normal vector outside the unit at any point on the S_F surface.

The velocity boundary conditions meet:

$$u_i = \overline{u}_i \tag{3.11}$$

3.2.1.2. Variational Principle of Rigid Plastic/Rigid-Viscoplastic Finite Element Method

The variational principle is the theoretical foundation of the rigid-plastic and viscoplastic finite element method. It transforms the problem of solving partial differential equations, which are difficult to solve mathematically, into a problem of finding the extremum of a functional through energy integration. This provides a relatively easy method for solving various practical problems.

Based on the Markov variational principle, in a permissible velocity field that satisfies the boundary conditions, geometric equations, and the incompressibility condition of the volume, the true solution makes the functional:

$$\pi = \int E(\dot{\varepsilon}_{ij})dV - \int S_F F_i u_i dS \tag{3.12}$$

take the minimum value, where $E(\dot{\varepsilon}_{ij})$ is the work function, and

$$E(\dot{\varepsilon}_{ij}) = \int_0^{\dot{\bar{\varepsilon}}} \overline{\sigma} d\dot{\bar{\varepsilon}} \tag{3.13}$$

According to different constitutive relations of materials, the expression forms of work functions are different. For rigid plastic materials, the equivalent stress is independent of the strain rate, then: $E\left(\dot{\varepsilon}_{ij}\right) = \overline{\sigma}\dot{\overline{\varepsilon}}$.

According to different methods of introducing the volume-constancy condition into the functional expression, the rigid plastic/rigid-viscoplastic finite element solution methods and formulas are different, which mainly include the Lagrange multiplier method, the penalty function method and the material compressibility method.

The Lagrange multiplier method introduces additional unknowns, which increases the total number of unknowns in the equation system and changes the characteristic of the stiffness matrix by distributing non-zero elements along the diagonal band. This increases the required storage capacity and makes programming more difficult, and also increases computational time. The material compressibility method is more suitable for analyzing the forming process of porous compressible materials.

In comparison to the Lagrange multiplier method, the penalty method does not introduce additional unknowns, and the solution for the unknowns maintains the characteristic of the stiffness matrix with non-zero elements along the diagonal band. This allows for compressed storage using a half-bandwidth storage method, which improves computational efficiency.

The penalty method introduces a penalty factor α to the functional expression, which adds a volume conservation constraint $\frac{\alpha}{2}\int\dot{\varepsilon}_V dV$ to the functional expression, obtaining a new functional:

$$\pi = \int E\left(\dot{\varepsilon}_{ij}\right)dV - \int S_F F_i u_i dS + \frac{\alpha}{2}\int\dot{\varepsilon}_V dV \tag{3.14}$$

where,

$$\left.\begin{aligned} \pi_D &= \int E\left(\dot{\varepsilon}_{ij}\right)dV \\ \pi_P &= \tfrac{\alpha}{2}\int\dot{\varepsilon}_V dV \\ \pi_F &= -\int_{S_F} F_i u_i dS \end{aligned}\right\} \tag{3.15}$$

When the functional takes on its stationary value, the first-order variation is zero. Taking the variation of Eq. (3.14) based on Eqs. (3.12) and (3.13) yields the following variation equation:

$$\begin{aligned} \sigma\pi &= \sigma\pi_D + \delta\pi_P + \delta\pi_F \\ &= \int\overline{\sigma}\delta\dot{\overline{\varepsilon}}dV + \alpha\int\dot{\varepsilon}_V\delta\dot{\varepsilon}dV - \int_{S_F} F_i\delta u_i dS = 0 \end{aligned} \tag{3.16}$$

The penalty function method can only be used to solve for the stress deviation σ'_{ij}, and cannot directly calculate the average stress σ_m. However, it can be shown that the average stress can be calculated using the following equation when the solution converges to the true value.

$$\sigma_m = \alpha \dot{\varepsilon}_V \tag{3.17}$$

Mathematically speaking, the volume constraint can only be strictly satisfied when the penalty factor α approaches infinity, which is practically impossible to achieve. In general, the penalty factor α should be chosen so that the volume strain rate approaches zero. For fluid flow problems with stress units of MPa, α typically ranges from 10^5 to 10^6, while for problems with stress units of Pa, α typically ranges from 10^{11} to 10^{12}.

3.2.1.3. Solution Formulas of Rigid Plastic/Rigid-Viscoplastic Finite Element

The solution process of the rigid-plastic and rigid-viscoplastic finite element methods is similar to that of the general finite element method. Firstly, the type of element is selected to discretize the deformation body, and then the element stiffness matrix is established and assembled into the overall stiffness matrix. Finally, the equation system is solved, which is nonlinear in nature. Therefore, linearization must be performed when solving the element stiffness equations and the overall stiffness equations. This section introduces the main solution expressions of the rigid-plastic and rigid-viscoplastic finite element methods.

1. Rigid Plastic/Rigid-Viscoplastic Finite Element Discretization and Linearization

The discretization process involves the discretization of the spatial domain as well as the discretization of the energy functional. The total energy functional of the deformed body is the sum of the energy functionals of each element, and the variational Eq. (3.17) can be expressed using the nodal velocities u and their variations δu through discretization.

$$\delta\pi = \frac{\partial \pi}{\partial u}\delta u = 0 \tag{3.18}$$

Due to the arbitrariness of the variation, when the functional takes a stationary value, the following algebraic equations (i.e., the stiffness equations) can be obtained from Eq. (3.18):

$$\frac{\partial \pi}{\partial u_i} = \sum_e \frac{\partial \pi^e}{\partial u_i} = 0 \tag{3.19}$$

where π^e is the element functional and e is the unit.

For metal forming problems, Eq. (3.19) is nonlinear and needs to be linearized before solving. Usually, a Taylor series expansion method is used around the velocity field $\{u\}_n$ to achieve linearization, and then the equation system is solved iteratively. Assume that the velocity field for the (n + 1)th iteration $\{u\}_{n+1}$ is the sum of the velocity field for the nth iteration $\{u\}_n$ and the velocity field correction $\{\Delta u\}_{n+1}$ for the (n + 1)th iteration:

$$\{u\}_{n+1} = \{u\}_n + \{\Delta u\}_{n+1} \tag{3.20}$$

Expanding Eq. (3.20) using Taylor series and ignoring high-order terms of second order and above, we obtain the linearized equation system as follows:

$$\sum_e \frac{\partial \pi^e}{\partial \{u\}_{n+1}} = \sum_e \frac{\partial \pi^e}{\partial \{u\}_n} + \sum_e \frac{\partial^2 \pi^e}{\partial^2 \{u\}_n} \cdot \{\Delta u\}_{n+1} = 0 \tag{3.21}$$

Equation (3.21) can be simplified into the following matrix form:

$$[K]\{\Delta u\} = \{F\} \tag{3.22}$$

In the equation, K is the stiffness matrix and F is the residual nodal force vector. Solving the linear equation system (Eq. 3.22) can obtain the velocity increment $\{\Delta u\}$, which is then used to reconstruct a new stiffness equation. This process is iterated and corrected repeatedly until convergence is achieved, thereby obtaining the true velocity field $\{\mathbf{u}\}^e$.

In this book, quadrilateral isoparametric elements are used to discretize the spatial domain V of the deformed body into N nodes and M elements. The coordinates of the element nodes and the velocity at each node are represented by (x_i, y_j) and $\left(u_x^{(i)}, u_y^{(j)}\right)$ respectively, and the velocity of element nodes is expressed by the vector $\{\mathbf{u}\}^e$:

$$\{u\}^e = \{u_x^{(1)}, u_y^{(1)}, u_x^{(2)}, u_y^{(2)}, u_x^{(3)}, u_y^{(3)}, u_x^{(4)}, u_y^{(4)}\} \tag{3.23}$$

The shape functions of the four-node isoparametric element are bilinear functions of the natural coordinates of the element, i.e.,

$$N_i(\xi, n) = \frac{1}{4}(1 + \xi_i \xi)(1 + \eta_i \eta) \tag{3.24}$$

Here, (ξ_i, η_i) is the natural coordinates of the element nodes. According to the definition of isoparametric elements, for any point P inside the element with natural coordinates (ξ, η), the global coordinates and velocity at that point can be represented by:

$$\left.\begin{aligned} x(\xi, \eta) &= \sum_{i=1}^{4} N_i(\xi, \eta) x_i \\ y(\xi, \eta) &= \sum_{i=1}^{4} N_i(\xi, \eta) y_i \end{aligned}\right\} \tag{3.25}$$

$$\left.\begin{aligned} u_x(\xi, \eta) &= \sum_{i=1}^{4} N_i(\xi, \eta) u_x^{(i)} \\ u_y(\xi, \eta) &= \sum_{i=1}^{4} N_i(\xi, \eta) u_y^{(i)} \end{aligned}\right\} \tag{3.26}$$

The above equation can be expressed by the following matrix equation:

$$\{u\} = \{u_x(\xi, \eta), u_y(\xi, \eta)\}^T = [N]\{u\}^e \tag{3.27}$$

where: $\{\mathbf{u}\}$ is the velocity vector of point P; N is the shape function matrix.

$$[N] = \begin{bmatrix} N_1 & 0 & N_2 & 0 & N_3 & 0 & N_4 & 0 \\ 0 & N_1 & 0 & N_2 & 0 & N_3 & 0 & N_4 \end{bmatrix} \tag{3.28}$$

2. Rigid Plastic/Rigid-Viscoplastic Finite Element Strain Rate Matrix

For plane strain problems, the geometric Eq. (3.3) can be expressed as (3.29), and the symmetric figure is (3.30):

$$\{\dot{\varepsilon}\} = \begin{Bmatrix} \dot{\varepsilon}_x \\ \dot{\varepsilon}_y \\ \dot{\varepsilon}_z \\ \dot{\gamma}_{xy} \end{Bmatrix} = \begin{Bmatrix} \frac{\partial u_x}{\partial x} \\ \frac{\partial u_y}{\partial y} \\ 0 \\ \frac{\partial u_y}{\partial x} + \frac{\partial u_x}{\partial y} \end{Bmatrix} \tag{3.29}$$

$$\{\dot{\varepsilon}\} = \begin{Bmatrix} \dot{\varepsilon}_r \\ \dot{\varepsilon}_z \\ \dot{\varepsilon}_\theta \\ \dot{\gamma}_{rz} \end{Bmatrix} = \begin{Bmatrix} \frac{\partial u_r}{\partial r} \\ \frac{\partial u_z}{\partial z} \\ \frac{u_r}{r} \\ \frac{\partial u_z}{\partial x} + \frac{\partial u_r}{\partial z} \end{Bmatrix} \tag{3.30}$$

Substituting the expression (3.28) into Eqs. (3.29) and (3.30) yields the following forms:

$$\{\dot{\varepsilon}\} = \begin{Bmatrix} \dot{\varepsilon}_1 \\ \dot{\varepsilon}_2 \\ \dot{\varepsilon}_3 \\ \dot{\varepsilon}_4 \end{Bmatrix} = [B]\{u\} \tag{3.31}$$

B is the strain rate matrix, and

$$[B] = \begin{bmatrix} X_1 & 0 & X_2 & 0 & X_3 & 0 & X_4 & 0 \\ 0 & Y_1 & 0 & Y_2 & 0 & Y_3 & 0 & Y_4 \\ P_1 & 0 & P_2 & 0 & P_3 & 0 & P_4 & 0 \\ Y_1 & X_1 & Y_2 & X_2 & Y_3 & X_3 & Y_4 & X_4 \end{bmatrix} \tag{3.32}$$

In the equation, X_i, Y_i (i = 1, 2, 3, 4) represent the partial derivatives of the shape functions with respect to the global coordinates. For plane strain problems, when considering the axisymmetric case, $N_i = N_i/r$, the values of X_i and Y_i are as follows:

(1)(1)(1)(1)(1)(1)(1)(1) Plane strain problem:

$$\begin{Bmatrix} X_1 \\ X_2 \\ X_3 \\ X_4 \end{Bmatrix} = \frac{1}{8|J|} \begin{Bmatrix} y_{24} - y_{34}\xi - y_{23}\eta \\ -y_{13} + y_{34}\xi + y_{14}\eta \\ -y_{24} + y_{12}\xi - y_{14}\eta \\ y_{13} - y_{12}\xi + y_{23}\eta \end{Bmatrix} \tag{3.33}$$

$$\begin{Bmatrix} Y_1 \\ Y_2 \\ Y_3 \\ Y_4 \end{Bmatrix} = \frac{1}{8|J|} \begin{Bmatrix} -x_{24} + x_{34}\xi + x_{23}\eta \\ x_{13} - x_{34}\xi - x_{14}\eta \\ x_{24} - x_{12}\xi + x_{14}\eta \\ -x_{13} + x_{12}\xi - x_{23}\eta \end{Bmatrix} \tag{3.34}$$

where: $x_{ij} = x_i\text{-}x_j$; $y_{ij} = y_i\text{-}y_j$; $|J|$ i is the value of Jacobian determinant, and

$$|J| = \frac{1}{8}[(x_{13}y_{24} - x_{24}y_{13}) + (x_{34}y_{12} - x_{12}y_{34})\xi + (x_{23}y_{14} - x_{14}y_{23})\eta] \tag{3.35}$$

(B) Axisymmetric problem:

$$\begin{Bmatrix} X_1 \\ X_2 \\ X_3 \\ X_4 \end{Bmatrix} = \frac{1}{8|J|} \begin{Bmatrix} z_{24} - z_{34}\xi - z_{23}\eta \\ -z_{13} + z_{34}\xi + z_{14}\eta \\ -z_{24} + z_{12}\xi - z_{14}\eta \\ z_{13} - z_{12}\xi + z_{23}\eta \end{Bmatrix} \tag{3.36}$$

$$\begin{Bmatrix} Y_1 \\ Y_2 \\ Y_3 \\ Y_4 \end{Bmatrix} = \frac{1}{8|J|} \begin{Bmatrix} -r_{24} + r_{34}\xi + r_{23}\eta \\ r_{13} - r_{34}\xi - r_{14}\eta \\ r_{24} - r_{12}\xi + r_{14}\eta \\ -r_{13} + r_{12}\xi - r_{23}\eta \end{Bmatrix} \tag{3.37}$$

where: $r_{ij} = r_i\text{-}r_j$; $z_{ij} = z_i\text{-}z_j$; $|J|$ is is the value of Jacobian determinant, and

$$|J| = \frac{1}{8}[(r_{13}z_{24} - r_{24}z_{13}) + (r_{34}z_{12} - r_{12}z_{34})\xi + (r_{23}z_{14} - r_{14}z_{23})\eta] \tag{3.38}$$

The equivalent strain rate can be expressed as

$$\dot{\bar{\varepsilon}} = \sqrt{\frac{2}{3}\dot{\varepsilon}_{ij}\dot{\varepsilon}_{ij}} \tag{3.39}$$

then

$$(\dot{\bar{\varepsilon}})^2 = \frac{2}{3}\dot{\varepsilon}_{ij}\dot{\varepsilon}_{ij} = \{\dot{\varepsilon}\}^T[D]\{\dot{\varepsilon}\} \tag{3.40}$$

D is a constant diagonal matrix, and

$$[D] = \begin{bmatrix} \frac{2}{3} & 0 & 0 & 0 \\ 0 & \frac{2}{3} & 0 & 0 \\ 0 & 0 & \frac{2}{3} & 0 \\ 0 & 0 & 0 & \frac{2}{3} \end{bmatrix} \tag{3.41}$$

Substituting the Eq. (3.31) into Eq. (3.40) yields:

$$(\dot{\bar{\varepsilon}})^2 = \{u\}^T[B]^T[D][B]\{u\} = \{u\}^T[P]\{u\} \tag{3.42}$$

where **P** is the equivalent strain rate matrix, and

$$[P] = [B]^T[D][B] \tag{3.43}$$

The volumetric strain rate is

$$\dot{\varepsilon}_V = \varepsilon_{ij} = \dot{\varepsilon}_x + \dot{\varepsilon}_y + \dot{\varepsilon}_z \tag{3.44}$$

In the form of matrix:

$$\dot{\varepsilon}_V = \{\dot{\varepsilon}\}^T\{C\} \tag{3.45}$$

In Eq. (3.45), {C} is the volumetric strain rate matrix, and

$$\left.\begin{array}{c} \{C\} = \{1\ 1\ 0\ 0\}^T \text{(plane problem)} \\ \{C\} = \{1\ 1\ 1\ 0\}^T \text{(Axisymmetrical problem)} \\ \{C\} = \{1\ 1\ 1\ 0\ 0\ 0\}^T \text{(3D problem)} \end{array}\right\} \tag{3.46}$$

3. Rigid Plastic/Rigid-Viscoplastic Finite Element Stiffness Matrix

The element stiffness matrix is composed of second-order partial derivatives of the element functional with respect to nodal velocity components. Substituting Eqs. (3.27), (3.31), (3.42) and (3.45) into the variational Eq. (3.16) and taking the partial derivative of the nodal velocities yields the following matrix form:

$$\frac{\partial \pi^e}{\partial\{u\}} = \int \frac{\overline{\sigma}}{\dot{\overline{\varepsilon}}}[P]\{u\}dV + \int \alpha[B]^T\{C\}\{C\}^T[B]\{u\}dV - \int_{S_F}[N]^T\{F\}dS \quad (3.47)$$

Then the second-order partial derivative of the functional to the node velocity can be expressed as

$$\begin{aligned}\frac{\partial^2 \pi^e}{\partial^2\{u\}} &= \int \frac{\overline{\sigma}}{\dot{\overline{\varepsilon}}}[P]dV + \int \alpha[B]^T\{C\}\{C\}^T[B]dV \\ &- \int \left(\frac{1}{\dot{\overline{\varepsilon}}}\frac{\partial \overline{\sigma}}{\partial \dot{\overline{\varepsilon}}} - \frac{\overline{\sigma}}{\dot{\overline{\varepsilon}}^2}\right)\frac{1}{\dot{\overline{\varepsilon}}}[P]^T\{u\}\{u\}^T[P]dV \end{aligned} \quad (3.48)$$

Equation (3.48) is the calculation formula of the general element stiffness matrix of the rigid plastic/rigid-viscoplastic finite element method. For rigid plastic materials, the equivalent stress $\overline{\sigma}$ is independent of the equivalent strain rate $\dot{\overline{\varepsilon}}$, and the $\frac{\partial \overline{\sigma}}{\partial \dot{\overline{\varepsilon}}}$ value in the equation is zero; for rigid viscoplastic materials, $\frac{\partial \overline{\sigma}}{\partial \dot{\overline{\varepsilon}}}$ depends on the specific material constitutive relation model $\overline{\sigma} = \overline{\sigma}(\overline{\varepsilon}, \dot{\overline{\varepsilon}})$.

3.2.2 *Numerical Simulation of Superplastic Forming Process of Typical Structures*

Based on the basic theory of numerical simulation of superplastic forming, the finite element modeling method is introduced using typical single-layer and two-layer sheet structures as examples. The modeling methods for three-layer and four-layer sheet structures are similar. Starting with the SPF simulation of a single-layer sheet, we will gradually delve into the SPF/DB process simulation of multi-layer sheets, providing references for numerical simulation of SPF/DB processes for various typical structures.

1. Finite Element Modeling

(1) Establish Geometric Models

Geometric analysis models are established for the TC4 titanium alloy single-layer sheet structure and the typical two-layer sheet SPF/DB structure. The geometric model includes two components: the mould and the TC4 sheet. During the forming process, the sheet is in a superplastic deformation state. Compared to the sheet, the mould is much stiffer, so it can be treated as a rigid body when building the model. The outer surface of the typical part is used as the mould shape and saved in IGS format for modeling in ABAQUS. In the component entity definition, the mould is defined as an analytical rigid body, and R3D4 rigid body element is used. The sheet is defined as a deformable body and S4R shell element is used. For the typical

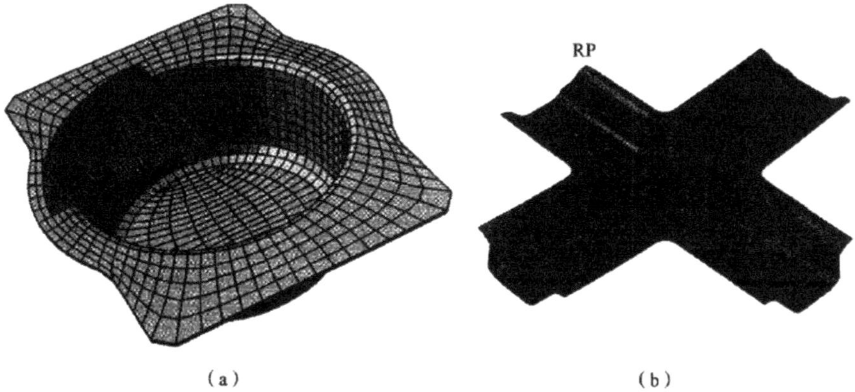

Fig. 3.7 Finite element geometric model. **a** Single-layer sheet structure; **b** Two-layer sheet structure

symmetric structure, a 1/2 or 1/4 model is selected for analysis, and the finite element analysis model is shown in Fig. 3.7.

(B) Material Constants

Superplastic forming is performed under isothermal conditions, so temperature changes during deformation are ignored. For simulation analysis, a strain hardening constitutive model is used, and the Backofen rheological equation [Eq. (3.1)] is used for the strain hardening model. The material constants for TC4 titanium alloy are obtained from velocity jump tests, with a value of K = 1315.2 MPa • s and m = 0.63. In ABAQUS/Standard creep analysis, a time-hardening model is used at a certain temperature, and the material constants are shown in Table 3.1. The constitutive equation is as follows:

$$\dot{\varepsilon}_c = A\sigma^n t^m$$

(C) Contact Definition

The sheet material is in contact with the mould surface during forming. Due to the difficulty in measuring and observing the friction state (including friction force and friction coefficient) between the sheet material and the mould or between the sheet materials during the superplastic forming process, only estimation based on experience or assumption is possible. Therefore, a penalty function model is used to calculate the contact stress to solve the convergence problem that may arise from the discontinuity between the two states of bonding and sliding. The Coulomb friction law is used for the contact friction model, and the friction coefficient is set to 0.1.

Table 3.1 Material constants in the constitutive equation

A/MPa^{-1}	n	m
1.12×10^{-5}	1.5873	0

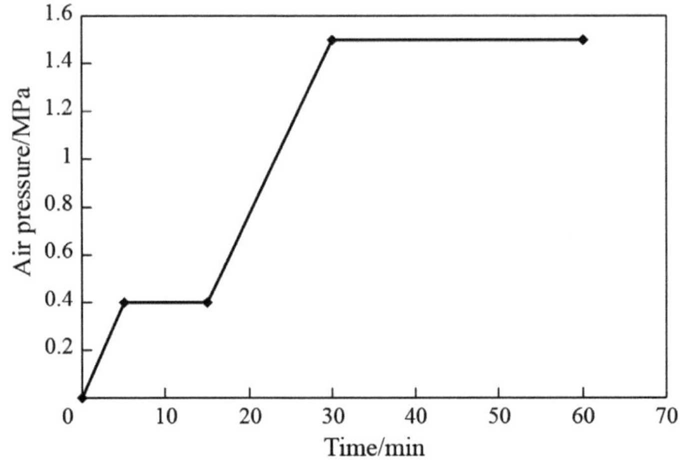

Fig. 3.8 Empirical pressurization curve

(D) Boundary Conditions and Loads

The mould is fully constrained, and in ABAQUS, a reference point needs to be defined for applying the constraints. For the TC4 sheet, the two ends are clamped by the mould, and the Tie constraint is used to bind the displacement of the sheet to the mould. At the joint, the displacement of the sheet in the X and F directions is constrained. The gas pressure load is a distributed surface load, and it needs to be applied in a direction consistent with the normal direction of the element in order to follow the element during the forming process. The pressure–time curve used in the part production is based on empirical studies, as shown in Fig. 3.8. The empirically determined pressurization path can meet the deformation requirements of the sheet during the superplastic forming stage.

2. Finite Element Analysis Results

The ABAQUS/Standard finite element solver is used to numerically simulate the superplastic forming process of typical single-layer and two-layer TC4 titanium alloy sheet structures, and predict the thickness distribution of the TC4 titanium alloy sheet after superplastic forming at 910 °C, as shown in Fig. 3.9. The initial thickness of the single-layer sheet structure was 1.5 mm, and the minimum thickness after forming was 0.52 mm. The initial thickness of the two-layer sheet structure was 1.2 mm, and the minimum thickness after forming was 0.75 mm.

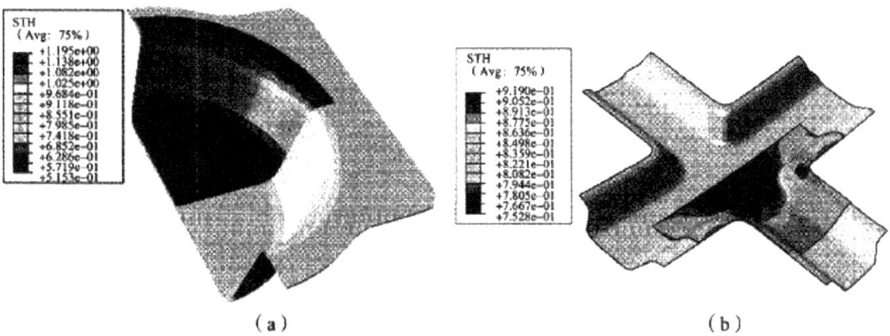

Fig. 3.9 Wall thickness distribution of parts after superplastic forming. **a** Single-layer sheet structure; **b** Two-layer sheet structure

3. Optimized Load Control Results

During the superplastic forming process, in order to avoid severe thinning of the sheet, the structure of the part is generally optimized based on the empirical pressure–time curve. To optimize the forming process and reduce the unevenness of wall thickness in the superplastic forming of the sheet, it is important to control the. The main goal of finite element analysis control is to use the predicted pressure loading history to ensure that the superplastic strain rate does not exceed the maximum optimized strain rate value at any stage of superplastic deformation for any element.

In this book, ABAQUS/Standard is used to realize the above load process under the condition of maximum strain rate through load control algorithm. The loading pressure is a physical quantity that varies throughout the entire simulation process to maintain the strain rate near a predetermined value. For TC4 titanium alloy, the optimized strain rate is about 0.001/s. Some functions in ABAQUS cannot be realized in the CAE module pre-processing module and must be modified and added through input files to achieve complex functions. The load optimization control is obtained by defining the magnitude of the load variable and controlling the creep strain rate:

*CREEP STRAIN RATE CONTROL, elset = Set-1,

AMPLITUDE = AUTO.

0.01

Based on the optimization load control method proposed above, the finite element analysis of the superplastic forming process of a single-layer sheet structure was carried out using ABAQUS/Standard, and the optimized load-time curve under certain conditions was obtained as shown in Fig. 3.10. From the optimized load curve, it can be seen that the required pressure in the initial stage of forming is relatively small and gentle, but as the forming process progresses to the corner, the hardening of the sheet causes the required forming pressure to rise rapidly. Figure 3.11 shows the thickness distribution after forming is completed. With the original thickness of the sheet material being 1.5 mm, the thinnest part of the formed structure appears

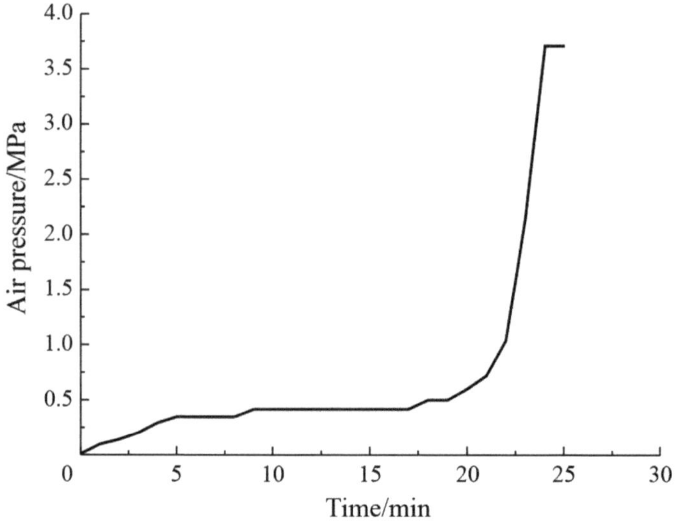

Fig. 3.10 Optimized load-time curve

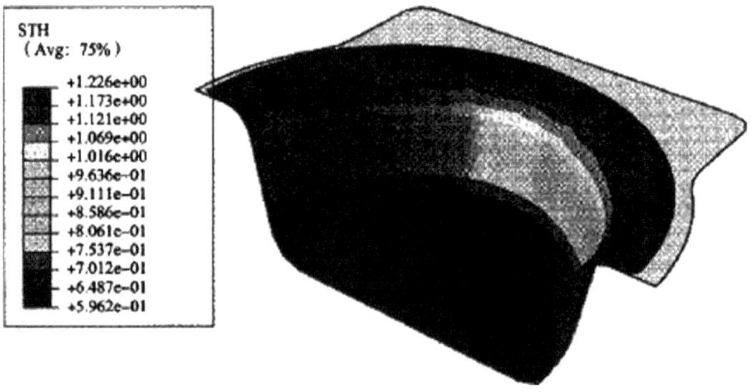

Fig. 3.11 Wall thickness distribution under optimized load

near the bottom corner with the thickness of about 0.59 mm. The calculation shows that the time is reduced compared to the conventional empirical time, taking 25 min.

3.3 Typical Structure Forming Process Design

The SPF/DB process provides more freedom for the structural design of parts. Designers can achieve various forms of structural design such as local reinforcement, truss support, rib support, and grid reinforcement according to requirements such as load-bearing, thermal insulation, and noise reduction. For a specific SPF/

DB structure, the design of geometric parameters of the structure directly affects the material flow and stress–strain distribution during the forming process, thereby affecting the forming quality of the structure. At the same time, the selection of forming process parameters and the design of forming moulds need to fully consider the entire forming process to avoid the occurrence of various forming defects.

3.3.1 Structural Parameter Design Based on Process Feasibility

When designing structural parameters, not only the requirements for structural strength and stiffness need to be considered, but also the requirements for process feasibility. Structural parameters such as the width-to-depth ratio of the part, the size of the fillet, the thickness ratio of skin/core, the width of the diffusion bonding, and the angle of the truss will all affect the forming quality of the part. Based on fundamental research and engineering experience, the design principles of parameter for typical SPF/DB structures have been summarized.

By designing the width-to-depth ratio of the part, the deformation of the superplastic-formed part can be controlled. Figure 3.12 shows the wall thickness reduction during the forming process of structures with the same initial sheet thickness but different width-to-depth ratios. As the width-to-depth ratio increases, the deformation of per unit sheet gradually decreases. When the width-to-depth ratio is 1, the wall thickness reduction in the fillet area at position 6 is 30%. When the width-to-depth ratio decreases to 0.5, the wall thickness reduction in the fillet area at position 6 is 70%.

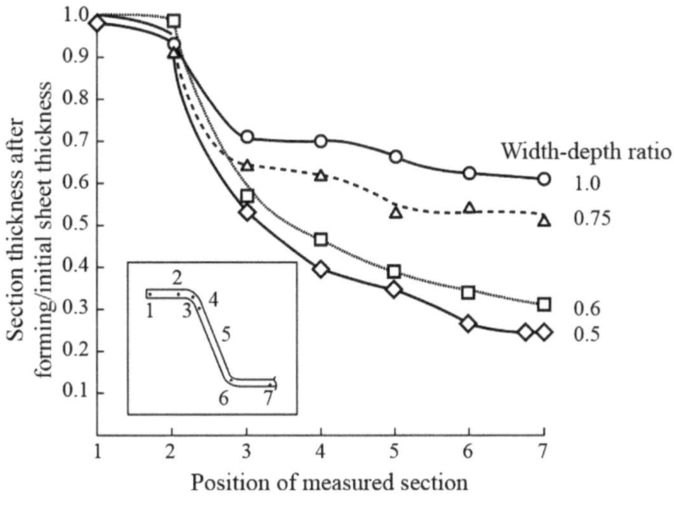

Fig. 3.12 Effect of width-depth ratio on wall thickness after part forming

The design of the fillet size directly affects the thinning of the sheet during superplastic forming. Titanium alloy has good fluidity in the superplastic state and can use a fillet radius can be used, but the thinning of the sheet at the small fillet radius is severe and can become a weak link in the part under stress. The choice of fillet radius size is related to the position of the fillet on the part. When designing fillets, the following principles can be followed:

(1) For fillet radii that are in a coordinated relationship, the fillet radius should be taken as (2 ~ 3)t where t is the thickness of the sheet.

(B) The rounded corners without coordinated relationships should have their radii enlarged as much as possible.

Structural parameters such as the thickness ratio of skin/core, the width of the diffusion bonding, and truss angle have an interactive effect on forming quality. Taking the three-layer sheet structure as an example, this chapter analyzes the influence of structural parameters on surface grooves and core thinning through numerical simulation, thus determining the design criteria of structural parameters. The profile shape of the three-layer sheet structure is generally a W shaped corrugated truss structure, as shown in Fig. 3.13. The structural design parameters include the upper and lower panel thickness d_1, truss thickness d_2, the width of the diffusion bonding d_3, and the angle θ between the truss and the upper and lower panels or between trusses. The three-layer sheet SPF/DB structure increases the corrugated angle and diffuser connection width compared to the two-layer sheet structure, making the influence of geometric parameters on formability more complex. By numerically simulating the forming process of the three-layer sheet SPF/DB structure with different geometric parameters, the influence of structural geometric parameters on the forming quality of the three-layer sheet SPF/DB structure is obtained, providing a reference for structural parameter optimization.

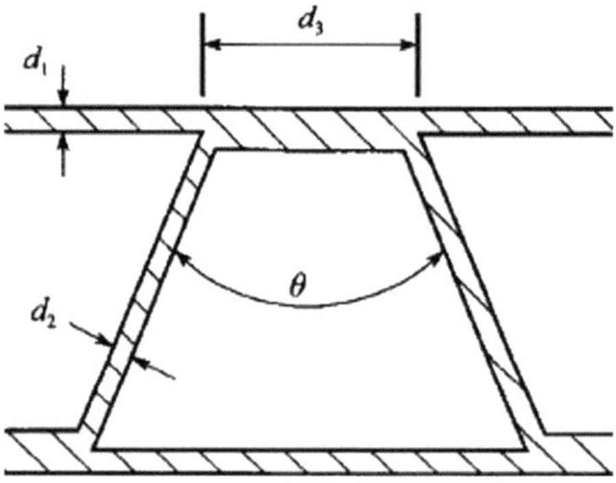

Fig. 3.13 Profile of three-layer sheet SPF/DB structure

3.3.1.1. The Influence of Structural Parameters on Surface Groove Defects of the Skin

After superplastic forming of the three-layer sheet SPF/DB structure with corrugated structure, surface grooves are prone to occur on the outer surface of the skin at the diffusion bonded zone, as shown in Fig. 3.14. The direct cause of this defect is that the deformation resistance of the intermediate core sheet during forming causes reverse tension on the skin at the diffusion bonded zone, making it difficult for the skin to conform to the mould and resulting in grooves.

1. **The effect of skin/core thickness ratio on groove defects**

Using the maximum groove depth as the evaluation index, the effect of skin/core thickness ratio on groove defect was measured. Figure 3.15 shows the relationship between maximum groove depth and skin/core thickness ratio for corrugated angles of 60° and 90°. As can be seen from the figure, when the skin/core thickness ratio is 1.0, obvious groove defects will occur after forming under both corrugated angles, especially for 60°. When the skin/core thickness ratio is greater than 3, the maximum groove depth of 60° is less than 0.1 mm, which can be considered that the groove defects are basically eliminated. No grooves appear when the skin/core thickness ratio is greater than 3 under 90 corrugated angles. Therefore, after forming, the three-layer SPF/DB structure basically does not have groove defects when the skin–core thickness ratio is greater than 3. From Fig. 3.15, it can also be seen that the data points with the same thickness ratio and different absolute thicknesses do not overlap, indicating that the influence of skin–core thickness on groove defects does not depend entirely on the thickness ratio of skin and core, but also on their absolute thicknesses. For example, when the skin/core thickness ratio is 2, skin and core with larger absolute thicknesses are more prone to generate groove defects after forming.

2. **The effect of diffusion bonding (DB) width on groove defect**

Figure 3.16 shows the relationship between maximum groove depth and DB width under 60° and 90° corrugated angles. It can be seen that under the same conditions, as the DB width increases, the depth of groove defects decreases significantly. This

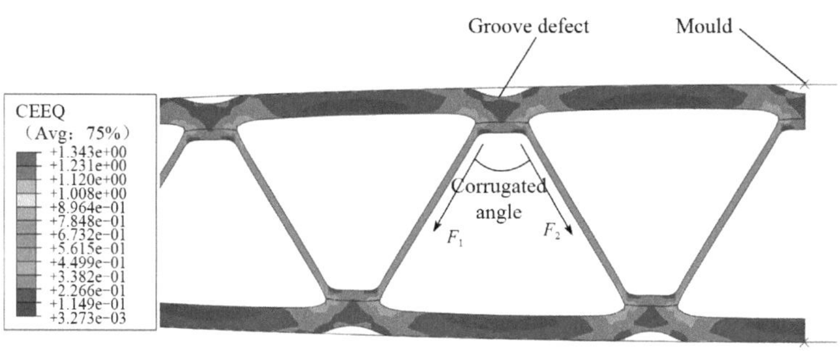

Fig. 3.14 Groove defect

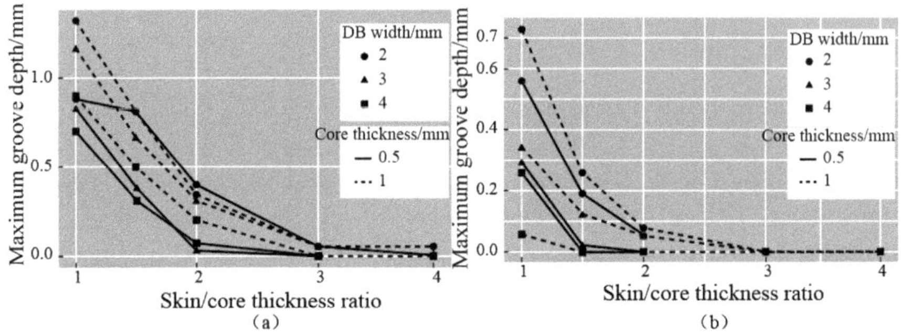

Fig. 3.15 Influence of skin/core thickness ratio on groove defects. **a** 60° corrugated angle; **b** 90° corrugated angle

indicates that the groove defects of the three-layer sheet SPF/DB structure can be reduced by properly increasing the diffusion bonding width.

3. **The Influence of Corrugation Angle on Groove Defects**

The corrugation angle has a significant influence on the groove defects of the SPF/DB three-layer sheet structure, especially when the skin/core ratio is small and groove defects are easily produced. Increasing the corrugation angle appropriately can significantly reduce the occurrence of groove defects. Figure 3.17 shows the groove defects under 60°, 90°, and 120° corrugation angles. It can be seen that when the corrugation angle is greater than 120°, groove defects will not occur. For situations that are prone to introduce groove defects, the groove depth decreases significantly with increasing corrugation angle.

4. **Optimization of Structural Parameter Combinations**

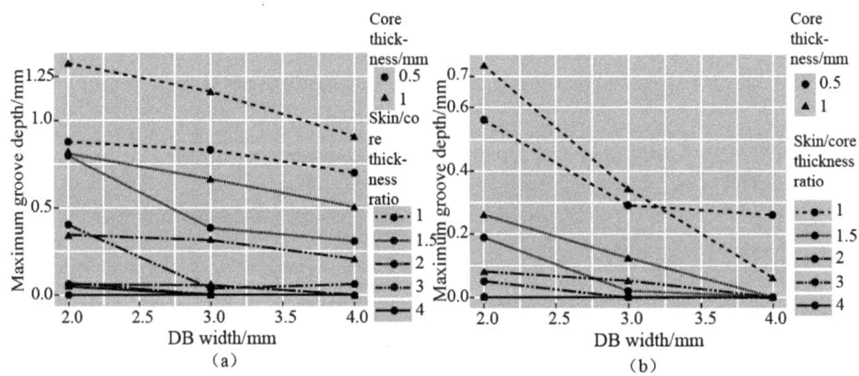

Fig. 3.16 Influence of DB width on groove defects (DB: diffusion bonding). **a** 60° corrugated angle; **b** 90° corrugated angle

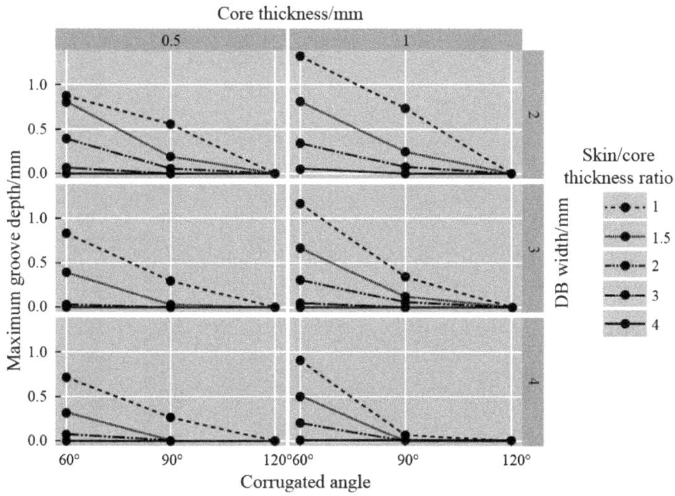

Fig. 3.17 Influence of corrugated angle on groove defects

The maximum groove depth corresponding to different structural parameter combinations is given through simulation, as shown in Table 3.2, which provides a reference for the design of the SPF/DB three-layer sheet structure. In this experiment, if the maximum groove depth is less than 0.01 mm, it is considered that no groove is generated. From Table 3.2, it can be seen that when the corrugation angle is high (>90°), it is suggested that skin/core ratio > 2, DB width > 3 mm. When the corrugation angle is low (<90°), it is suggested that skin/core ratio > 3, and DB width > 3 mm. Under these conditions, it is unlikely to introduce groove defects.

3.3.1.2. The Influence of Structural Parameters on Local Thinning of the Core Sheet

After the three-layer sheet SPF/DB structure is formed, the core sheet in the cavity triangle area exhibits thinning, especially near the diffusion bonded zone, which will deteriorate the formability and mechanical performance of the SPF/DB structure. Simulation of the thinning of the core sheet can provide important reference for the design and manufacturing of the SPF/DB structure of the three-layer sheet.

Before the formation of the SPF/DB structure of the three-layer sheet, the skin thickness was 1 mm, the core sheet thickness was 0.5 mm, and the rib width was 4 mm. Figure 3.18 shows the distribution law of the thickness of the core sheet after forming under the conditions of 60° and 75° corrugation angles. It can be seen from the figure that the core sheet at the diffusion bonded zone hardly exhibits thinning, and even thickening may occur when the corrugation angle is small. This is related to the order of the skin sticking to the mould at various locations during forming and its effect on the flow of the material. The core sheet in the cavity triangle area exhibits

Table 3.2 Groove depths obtained by simulation under different structural parameters

Core thickness/mm	Corrugated angle/(°)	Skin thickness/mm	Rib width/mm		
			2	3	4
0.5	60	0.5	0.88	0.83	0.7
		1	0.4	0.032	0.064
		1.5	0.06	0	0
		2	0.003	0	0
	90	0.5	0.56	0.29	0.26
		1	0.05	0	0
		1.5	0	0	0
		2	0	0	0
	120	0.5	0	0	0
		1	0	0	0
		1.5	0	0	0
		2	0	0	0
1	60	1	1.32	1.16	0.9
		2	0.34	0.31	0.2
		3	0.053	0.05	0.003
		4	0.052	0	0
	90	1	0.73	0.34	0.058
		2	0.079	0.05	0
		3	0	0	0
		4	0	0	0
	120		0	0	0
		2	0	0	0
		3	0	0	0
		4	0	0	0

Unit mm

overall thinning except for the locally thinned area, but the thickness distribution is relatively uniform, that is, the thinning of the core sheet is divided into uniform thinning and local thinning.

Figure 3.19 shows the thinning of the core sheet under the conditions of corrugation angles of 60°, 75°, 90°, 105°, and 120°. It can be observed that as the corrugation angle increases, the local thinning of the core sheet becomes significantly weaker. When the corrugation angle is 60°, the thinnest part of the core sheet after forming is about 0.24 mm, while when the corrugation angle is 120°, the thinnest part of the core sheet is about 0.44 mm.

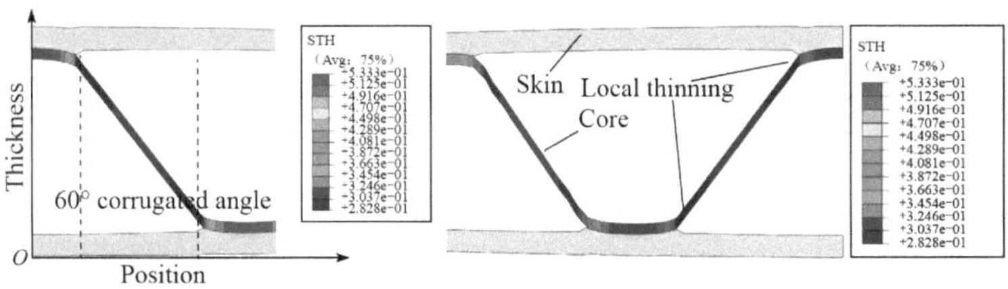

Fig. 3.18 Core thickness distribution after forming

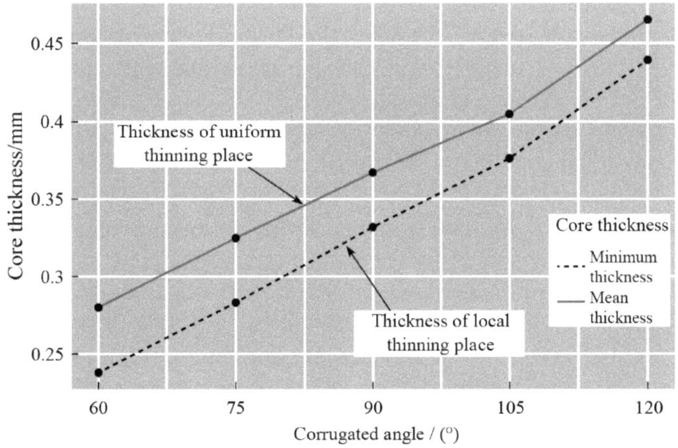

Fig. 3.19 Relation between core thickness and corrugated angle

3.3.2 Selection Criteria of Superplastic Forming/Diffusion Bonding Forming Process Parameters

The selection of superplastic forming/diffusion bonding parameters directly determine the forming performance, as well as the microstructure and mechanical properties of the parts. The following principles should be followed when selecting process parameters:

(1) Select a reasonable diffusion bonding temperature, pressure, and time to ensure a high bonding rate.
(2) Choose a reasonable superplastic temperature to reduce flow stress and suppress grain growth.
(3) Choose a reasonable forming rate to improve forming performance and shorten the high-temperature dwell time.
(4) Select a reasonable lubrication condition to improve part surface quality and increase wall thickness uniformity.

The main process parameters for diffusion bonding of titanium alloy include temperature, pressure, and time. The reasonable matching of these three parameters can avoid diffusion bonding defects such as macroscopic non-welding, point defects, and weak bonding. The temperature depends on the material's yield strength and atomic diffusion behavior and is generally taken near the recrystallization temperature. Gas pressure is the main influencing factor for plastic deformation at the diffusion bonding interface and the formation of the diffusion interface. Time is related to the selected temperature and pressure. Within a certain range, the bonding strength increases continuously with the diffusion time, but when the diffusion time exceeds a certain value, the grain will excessively grow, leading to a decrease in joint strength. The diffusion bonding temperature for TC4 titanium alloy is 900–930 °C, pressure is 1–3 MPa, and diffusion holding time is 1–4 h.

The main process parameters for superplastic forming include forming temperature and strain rate, which interact with each other and jointly affect the superplastic forming process. The forming temperature directly affects the material's flow stress, determines the gas pressure magnitude, and also affects the grain size. The strain rate determines the material's forming performance and depends on the variation of deformation pressure over time, which should be controlled within the allowable range of the superplastic forming material. When the forming temperature is low, the required forming pressure is high, and the high forming pressure leads to a high forming rate, uneven material deformation, and the easy occurrence of local thinning or even rupture.

Under the optimal superplastic forming temperature, the sheet can be deformed at the ideal strain rate by controlling the relationship between gas pressure and time. Taking the conical part as an example, an analytical algorithm is established to build the gas pressure loading curve. The superplastic forming process of the conical part is shown in Fig. 3.20, which is divided into two stages: free bulging and die pressing. Mechanical analysis is performed on these stages to obtain the optimized gas pressure loading curve.

The first stage is the superplastic free bulging stage. Assuming that the sheet is isotropic, the wall thickness of the free deformation part is uniform, and the free bulging profile is spherical. As shown in Fig. 3.21, it can be inferred from the balance of internal pressure and radial stress in the axial direction that:

$$\pi p(R\sin\varphi)^2 = \sigma_m \sin\varphi \cdot \delta 2\pi R \sin\varphi \tag{3.49}$$

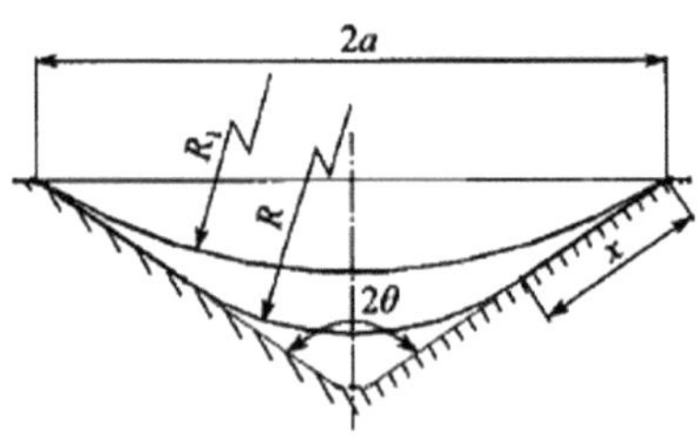

Fig. 3.20 Superplastic forming process of conical parts

$$\sigma_m = \frac{pR}{2\delta} \tag{3.50}$$

In the equation, δ represents the instantaneous wall thickness of the spherical shell. Since the formed material is a thin sheet, the through-thickness stress can be neglected. Due to the spherical shape of the free bulging profile, the relationship between the equivalent stress σ_e, circumferential stress σ_θ, and radial stress σ_m is:

$$\sigma_e = \sigma_\theta = \sigma_m \tag{3.51}$$

According to the geometric relationship in Fig. 3.21, the instantaneous curvature radius of the spherical shell is

$$R = \frac{a^2 + h^2}{2h} \tag{3.52}$$

where h is the instantaneous arch height of the spherical shell. According to the constancy of volume, the instantaneous wall thickness of spherical shell is

$$\delta = \frac{a^2 \delta_0}{a^2 + h^2} \tag{3.53}$$

where δ_0 is the initial wall thickness. Substitute Eq. (3.52) and Eq. (3.53) into Eq. (3.50) to get:

$$\sigma_e = \sigma_\theta = \sigma_m = \frac{p\left(a^2 + h^2\right)^2}{4a^2\delta_0 h} \tag{3.54}$$

As the circumferential strain rate $\dot{\varepsilon}_\theta$ is equal to the radial strain rate $\dot{\varepsilon}_m$, the equivalent strain rate $\dot{\varepsilon}_e$ under the volume-constancy condition is

$$\dot{\xi}_e = 2\dot{\xi}_m = 2\dot{\xi}_\theta = -\dot{\xi}_\delta = -\frac{d\delta}{sd\delta} \tag{3.55}$$

where t is the time. Substitute Eq. (3.53) into Eq. (3.55) to get

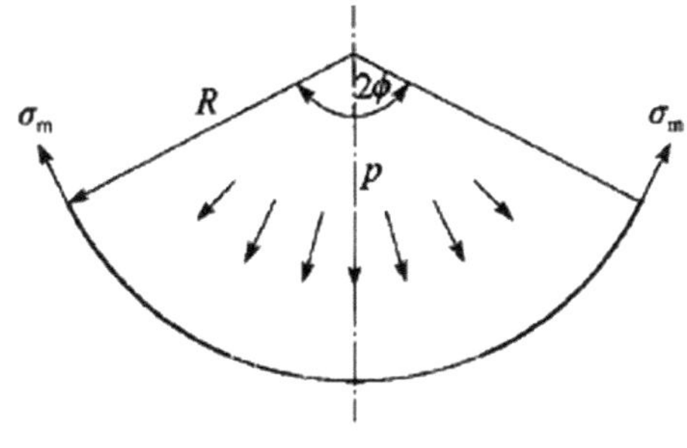

Fig. 3.21 Stress analysis of superplastic free bulging

$$dt = \frac{d\ln\left(a^2 + h^2\right)}{\dot{\xi}_e} \tag{3.56}$$

When the equivalent strain rate is the optimal value, it can be obtained by integrating Eq. (3.56) that

$$t = \frac{\ln\left(a^2 + h^2\right) - \ln a^2}{\dot{\xi}_e} \tag{3.57}$$

When the spherical shell bulging is tangent to the mould surface, the first stage is completed. At this time, according to the geometric relation, the curvature radius R_1 and the arch height h_1 are

$$R_1 = \frac{a}{\cos\theta} \tag{3.58}$$

$$h_1 = \frac{a(1 - \sin\theta)}{\cos\theta} \tag{3.59}$$

According to Eqs. (3.53) and (3.59), the part wall thickness δ_1 at this time is

$$\delta_1 = \frac{\delta_0(1 + \sin\theta)}{2} \tag{3.60}$$

According to Eqs. (3.57) and (3.59), the time t_1 used at the end of the first stage is

$$t_1 = \frac{\ln\frac{1+\sin\theta}{2}}{\dot{\xi}_e} \tag{3.61}$$

The superplastic constitutive equation is

$$\sigma_e = k\,\dot{\xi}_e^m \tag{3.62}$$

According to Eqs. (3.50), (3.57) and (3.62), the air pressure under the optimal equivalent strain rate is

$$p = \frac{4k\delta_0\dot{\xi}_e^m\left(e^{\dot{\xi}_e t} - 1\right)}{ae^{2\dot{\xi}_e t}} \tag{3.63}$$

The second stage is the mould adhering. The upper part of the spherical shell gradually adheres to the mould to form a part of the conical surface. The adhered length x gradually increases until it is completely adhered.

At time t, the instantaneous radius of curvature of the freely inflated spherical shell is R, the instantaneous wall thickness is δ, and the adhered length is x. From the geometric relationship, we can know that:

$$x = (R_1 - R)ctg\theta \tag{3.64}$$

At t + dt, the wall thickness is δ + dδ, the curvature radius is R + dR, it can be seen obtained according to consistency of volume, it is obtained that

$$\pi\left[R^2\cos^2\theta + R^2(1-\sin\theta)^2\right]\delta = 2\pi R\cos\theta\delta dx$$
$$+\pi\left[(R+dR)^2\cos^2\theta + (R+dR)^2(1-\sin\theta)^2\right](\delta + d\delta) \tag{3.65}$$

By simplifying simultaneous Eqs. (3.64) and (3.65), it can be obtained that

$$\frac{d\delta}{\delta} = \frac{(1-\sin\theta)}{\sin\theta}\cdot\frac{dR}{R} \tag{3.66}$$

Suppose $\frac{(1-\sin\theta)}{\sin\theta} = \tau$, the result of integral Eq. (3.66) is

$$\delta = \delta_1\left(\frac{R}{R_1}\right)^{\tau} \tag{3.67}$$

According to Eqs. (3.55) and (3.67), it can be obtained that

$$t = \frac{\tau}{\dot{\xi}_e}\ln\frac{R_1}{R} \tag{3.68}$$

According to simultaneous Eqs. (3.50), (3.51), (3.60), (3.62) and (3.67), it can be obtained that

$$p = k\dot{\xi}_e\delta_0(1+\sin\theta)\left(\frac{\cos\theta}{a}\right)^{\tau}R^{\tau-1} \tag{3.69}$$

According to Eqs. (3.68) and (3.69), the pressure loading curve under the optimal equivalent strain rate in the second bulging stage is

$$p = k\dot{\xi}_e\delta_0(1+\sin\theta)\left(\frac{\cos\theta}{a}\right)e^{\frac{(1-\tau)\dot{\xi}_e t}{\tau}} \tag{3.70}$$

By substituting the material parameters and related geometric quantities, the pressure loading curve of conical parts can be obtained at the optimum equivalent strain rate through computer data processing.

3.3.3 Superplastic Forming/Diffusion Bonding Mould Design

1. Material Selection

The diffusion bonding mould in superplastic forming needs to be subjected to a high-temperature environment of around 920 °C for a long time, facing the harsh environment of cyclic heating and cooling. Therefore, in addition to considering the smooth surface of the mould cavity, accurate dimensions, high-temperature air tightness, and sufficient rigidity in the structure, there are special requirements for the mould material.

(1) High-temperature mechanical properties and creep resistance

Superplastic forming technology uses air pressure forming, usually with an air pressure of 1–3 MPa. Air pressure forming requires mould sealing, which is generally achieved by convex-stem sealing. The convex-stem gradually embeds the sheet under the action of the press to form a gas sealing strip during the heating process. In this process, since the convex-stem acts on the sheet in a localized deformation manner, it requires that the mould material has good red hardness and creep resistance.

(B) Oxidation resistance

The mould should have good oxidation resistance. A dense oxide film is generated on the mould surface to protect the inner layer of the metal from further oxidize, maintaining the original size and shape of the mould cavity. Generally, the oxidation weight gain rate at high temperature is required to be less than 0.3 $g/(m^2 \cdot h)$.

(C) Thermal shock resistance

Superplastic forming is carried out at high temperature. When placing the blanks or extracting the parts, the surface temperature of the mould drops sharply, causing a temperature difference between the inside and outside of the mould, and further leads to thermal stress and mould deformation or even cracking. Therefore, it is required that the mould material has good resistance to thermal shock. The higher the thermal conductivity of the material, the smaller the temperature difference between the inside and outside of the mould, and the smaller the thermal stress.

(D) Thermal stability

The mould should have good dimensional stability at high temperature, and its coefficient of thermal expansion should be close to that of titanium alloy. Moreover, contamination from the mould material on the titanium alloy parts should be avoided.

(E) Casting and machinability and weldability

The moulds are generally made of cast steel. To ensure the quality of the mould, vacuum casting is preferred. The mould cavity of mould is complex and requires high dimensional accuracy, which needs to be achieved by machining. However, machinability and heat resistance are contradictory. The better the heat resistance, the worse the machinability. Therefore, both aspects must be considered when selecting materials. In addition, gas inlets is installed on the mould to facilitate vacuum pumping and nitrogen filling. Conventional thread sealing is difficult to work at high temperature, so welding is generally used to connect the gas pipes and the mould, which requires that the mould material has good weldability.

(F) Material supply and price

Considering production costs and production cycles, it is better to use materials for batch production with stable quality and low cost. There are several common mould materials in China, including acid-resistant stainless steel (1Cr18Ni9Ti), nickel–chromium superalloy (K11 and GH140), and high-temperature heat-resistant steel (R45 and Ni7N). The mechanical properties of these materials are compared and analyzed in Table 3.3. It can be seen that the high-temperature performance of 1Cr18Ni9Ti is poor and can only be used for experimental moulds for short-term use. The high-temperature performance of the other four materials can meet the requirements. Considering cost, K11 and GH140 are nickel–chromium high-temperature alloys, with nickel accounting for about 40% of the weight and the price being expensive. Their high-temperature strength far exceeds the requirements for titanium sheet superplastic forming. R45 and Ni7N are relatively low-priced. From the analysis of oxidation resistance, R45 has an oxidation weight gain rate of 0.08 $g/(m^2 \cdot h)$ at 900 °C and 2.4 $g/(m^2 \cdot h)$ at 1000 °C, and the average oxidation weight gain rate of Ni7N is less than or equal to 0.3 $g/(m^2 \cdot h)$ at 1100 °C (within 150 h). After comprehensive analysis, Ni7N, also known as ZG35Cr24Ni7SiN, is selected as the material for the SPF/DB mould.

2. Structural Design

SPF/DB moulds are divided into integral moulds and modular moulds. Integral moulds are used for small or simple-shaped components, while modular moulds are used for complex components. When designing moulds, the first step is to correctly select the parting surface, so that the mould structure is simple and the removal of the part after superplastic forming is convenient. The weight of the mould should be minimized, the speed of heating and cooling should be increased, and the quality of the mould should be uniform to reduce thermal deformation.

When designing moulds, it is necessary to fully consider the characteristics of the SPF/DB process and the various structural elements of the mould. Taking wall panel components as an example, the key elements of SPF/DB mould design are introduced below.

(1) Mould surface

During the high-temperature forming process, due to the different linear expansion coefficients of the sheet and the forming mould, if the mould surface is the same as the part's outer surface, there will be a deviation between the actual size of the part after forming and the theoretical value. To eliminate this size deviation, the mould

Table 3.3 Comparison of mechanical properties of mould materials

Material names	Normal temperature			High temperature			High temperature durability
	Tensile strength σ_b/MPa	Elongation δ/%	Percentage reduction of area Ψ/%	Tensile strength σ_b/MPa	Elongation δ/%	Reduction of area Ψ/%	Breakdown time/h
Ni7N	770 ~ 820	20 ~ 40	16 ~ 40	53 ~ 88 (1000 °C)	58 (1000 °C)	29	259 (temperature 1000 °C, loading stress σ = 30 MPa)
R45	740	13.6	14	163 (1000 °C)	–	–	30 (temperature 900 °C, loading stress σ = 50 MPa)
GH140	630	50	–	78	58 (1000 °C)	–	–
K11	450	5 ~ 12	6	300 (800 °C)	6 ~ 12 (800 °C)	29	200 (temperature 800 °C, loading stress σ = 120 MPa)
1Cr18Ni9Ti	550	40	60	180(800°C)	—	–	–

size should be corrected and compensated according to the difference in expansion coefficients. The mould design size can be determined by the following equation:

$$\begin{cases} L_{\mathrm{gj}} = L_{\mathrm{cj}}\,(1 + \alpha_j \cdot \Delta t) \\ L_{\mathrm{gm}} = L_{\mathrm{cm}}(1 + \alpha_{m\cdot} \cdot \Delta t) \\ L_{\mathrm{cm}} = L_{\mathrm{cj}}\,(1 + \alpha_j \cdot \Delta t)/(1 + \alpha_m \cdot \Delta t) \end{cases}$$

where L_{cm} is the nominal size (mm) of the mould at room temperature, L_{cj} is the nominal size (mm) of the part at room temperature, L_{gm} is the size of the mould at forming temperature, L_{gj} is the size (mm) of the part at forming temperature, α_m is the thermal expansion coefficient of the mould at forming temperature (°C^{-1}), α_j is the thermal expansion coefficient of the part at forming temperature (°C^{-1}), and Δt is the temperature difference between the forming temperature and room temperature.

In the above formula, $\alpha_j \cdot \Delta T$ and $\alpha_m \cdot \Delta T$ are the relative elongations of the part and the mould at the forming temperature, $D = \alpha_j \cdot \Delta t - \alpha_m \cdot \Delta t - \alpha_j \cdot t - \alpha_m \cdot t$, D is the scale factor. As a function of temperature, D is the first parameter determined in the hot mould design. For commonly used metal mould materials, the scale factor D used is between 0.004 and 0.008. In general, the expansion coefficient of titanium and its alloy is (8 ~ 10) × 10^{-6} °C, the expansion coefficient of stainless steel and other heat-resistant metal materials is (15 ~ 20) × 10^{-6} °C^{-1}, and the expansion coefficient of ceramic materials is (3.7) × 10^{-6} °C^{-1}. The expansion coefficients of TC4, Ni7N and 1Cr18Ni9Ti at different temperatures are shown in Fig. 3.22. Suppose that when the part is formed at a certain temperature, it meets the requirements that the part and the mould cavity are fully fitted, that is, $L_{gj} = L_{gm}$. After the above analysis and calculation, the mould shape is determined.

It is difficult to obtain an accurate compensation amount based solely on the scaling factor in mould design. Leaving a certain amount of process margin can also eliminate the dimensional deviation. In addition, for concave mould forming, the demoulding temperature must also be considered because the expansion coefficient of stainless steel mould material is larger than that of titanium alloy. If the demoulding

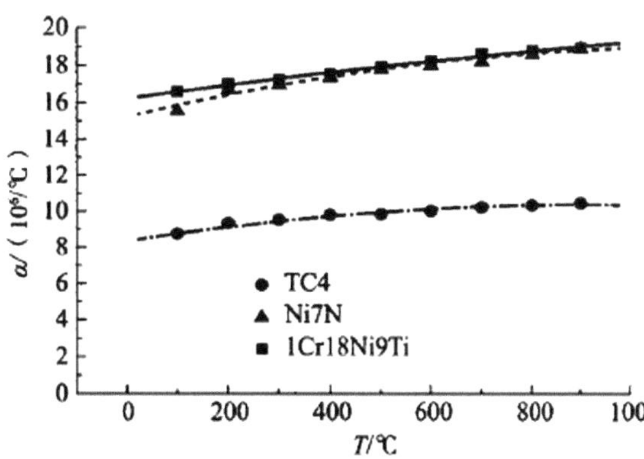

Fig. 3.22 Expansion coefficient α of several materials

temperature is too low, the large amount of shrinkage during the mould cooling process will cause the part to be trapped in the concave mould, leading to deformation of the part and distortion of internal features. For large-size concave moulds with deep cavities, mould materials with smaller expansion coefficients should be selected, while the opposite is true for convex mould forming.

(2) Thickness

During mould casting design, in addition to meeting the basic process needs, the requirements of mould rigidity and weight reduction shall also be considered. The heavy mould will not only bring a lot of inconvenience and difficulty to production, but also will increase the difficulty of casting and processing and increase the cost. Therefore, it is necessary to design lightening holes for mould casting. Considering that the mould surface is arc-shaped, large size castings are more likely to have various casting defects due to thickness changes, so the lightening holes are designed to maintain uniform thickness of the forming surface, as shown in Fig. 3.23, so as to make the forming surface heated and stressed evenly, and extend the service life of the mould. In addition, the lightening holes are reasonably arranged from the aspect of technological process, so that they can not only achieve the purpose of weight reduction, but also sufficient strength. Additionally, in combination with the stress condition of the equipment platform, the stiffening rib of the lightening hole shall not be pressed on the platform butt joint to prevent damage to the platform.

.

(3) Sealing

The role of the sealing stem is to realize the gas sealing of the mould cavity at high temperature, which shall be able to compensate the non-parallelism of the heating platform and the upper and lower surfaces of the mould. The convex bead is often used to seal the surface, which can reduce the alignment requirements during upper and lower mould positioning. The sealing stem shall be designed outside the effective

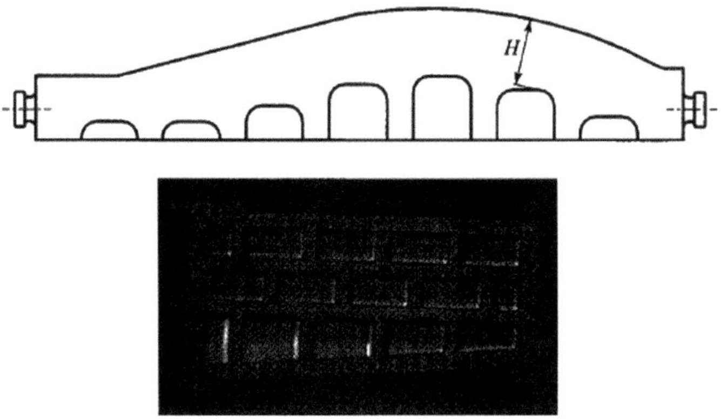

Fig. 3.23 Lightening hole of mould

area of the part, and its interface is usually rectangular or trapezoidal, with a height of $\leq$ 1.5 mm generally, at least 0.3 mm less than the total thickness of the sheet, and a width of $\leq$ 2 mm. In order to further improve the sealing effect, the design scheme of multiple convex beads or convex bead to groove can be adopted.

(4) Gas inlet and exhaust

For single-layer sheet structure, the inlet and exhaust can be designed by drilling holes at the bottom or side of the mould and welding air tubes on them. Among them, the exhaust hole should be designed in the last sticking mould area of the sheet, usually at the bottom corner of the mould. The diameter of the exhaust hole in contact with the mould surface should be as small as possible, generally 1.5–2 mm. When the exhaust hole is too large, the material is prone to rupture at the exhaust hole. When multiple exhaust holes need to be set, an additional bottom sheet can be used to allow the gas to be discharged from the edge gap of the bottom sheet to a larger exhaust hole. In the SPF/DB process of multi-layer sheet structure, the sheet blank is often sealed and welded into a pocket, and the air pipe is welded on the edge of the pocket.

(5) Hoisting

Different equipment requires different hoisting methods for moulds, and the requirements for mould shape are also different. Therefore, when designing a mould, equipment conditions should be fully considered. For example, as shown in Fig. 3.24, casting cylindrical lifting ears are used for hoisting, with 4 lifting ears for each mould, symmetrically distributed with 2 on each side, and the lifting ears smoothly transition to the mould body's large fillet. Due to the large weight of the mould, the lifting ears should have sufficient strength. The outer diameter of the lifting ears increases to ensure that the rope does not slip when using them for hoisting.

(6) Positioning

Fig. 3.24 Mould lifting setting

Determining the positioning relationship between the mould and the sheet metal is another key point in mould design, and the quality of positioning determines the dimensional and positional accuracy of the shape. Usually, positioning blocks or positioning pins can be used for positioning. As shown in Fig. 3.25, the mould uses positioning pins for positioning, and the four protrusions along the longitudinal edge of the mould are the positioning devices between the moulds. Two out of the four holes can be used for mould positioning.

The two protrusions along the transverse edge of the mould are the positioning devices between the mould and the sheet metal. The protrusions are located at the peak of the plane arc. The positioning points of the diffusion mould and the superplastic forming mould are consistent, so as to ensure the effective transmission of positioning. In addition, a pit for positioning is designed on the mould for positioning the benchmarks for scribing and cutting, etc.

(7) Thermocouple Holes

The thermocouple holes are used to place thermocouples to monitor the temperature of the mould in real-time. For large moulds, the number of thermocouple holes should be more than the number of heating zones, and the location should be close to the mould surface, near the sheet metal for superplastic forming. The depth of the thermocouple holes should be greater than 5 times the diameter of the hole to reduce the influence of hot air in the furnace.

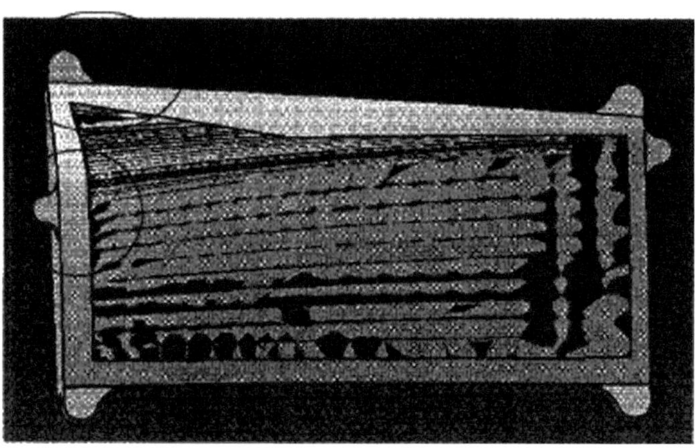

Fig. 3.25 Mould positioning device

3.4 Quality Control Technology of Superplastic Forming/Diffusion Bonding Process

There are many factors influencing the SPF/DB process of titanium alloy, which is easy to produce various quality defects. Table 3.4 lists the common forming defects and control methods of SPF/DB structural components.

In the following parts, several typical forming defects are selected, the existing forms and forming mechanism are analyzed in depth. In order to control the size, shape precision and mechanical properties of SPF/DB parts, SPF/DB process is improved, and the defects are eliminated.

3.4.1 Wall Thickness Uniformity

During the superplastic free bulging process, except for the biaxial tension state at the vertex of the spherical shell, the deformation zone is not in a biaxial tension stress state, resulting in uneven stress state and uneven thickness distribution. In addition, in the commonly used mould-restrained bulging, the material and the mould surface friction, and the deformation mainly concentrates in the uncontacted area, which aggravates the uneven thickness of the formed parts. The control of the wall thickness uniformity of the formed parts is a key technology of the superplastic forming process. The methods for controlling the uniformity of the wall thickness of superplastic formed parts usually include the movable male mould method, increasing static water pressure method, pre-forming method, and forward and reverse bulging. Here, we focus on the pre-forming method and the forward and reverse bulging method.

1. Pre-forming Method

The pre-forming method refers to the preliminary forming of the sheet material using methods such as spinning and hot pressing coupling to improve its wall thickness distribution to meet the requirements. Taking a semi-spherical part with a radius of 190 mm as an example (Fig. 3.26), the side wall of the semi-spherical part is pre-thinned by spinning, and the thickness of the original sheet material is retained at the bottom of the part. The pre-formed billet is used for superplastic forming.

Pre-forming method can effectively control the wall thickness uniformity of titanium alloy single-layer structure, it is simple and effective and convenient for production and promotion. Figure 3.27 shows the simulation of manufacturing a semi-spherical part using pre-forming method, because the thickness distribution of the part is uneven (1.2–5 mm), the thickness distribution of the pre-formed blank is also uneven, with the entrance corner area and the top area thicker, being 3–5 mm, and the sidewall area thinner, being 2.9–3.5 mm. During the forming process, the material in the sidewall area first contacts with the mould, resulting in the formation of closed

Table 3.4 Common defect and control methods of SPF/DB structure of titanium alloy

S/N	Defect forms	Specific descriptions	Control points
1	 Surface groove Surface groove	Grooves are formed at the junction of the structural component by diffusion bonding and superplastic forming, and the skin surface is incomplete	Measures such as adjusting the position of the anti-welding coating, the thickness ratio of the sheet metal, the corrugation angle, and the friction coefficient between the sheet metal and the mould can be taken
2	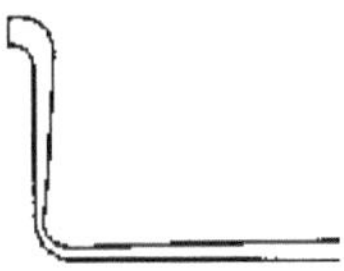 Wall thickness reduction Wall thickness reduction	There are differences in wall thickness reduction at various locations of the structural component, especially for complex shapes, where the wall thickness differences are more significant	Pre-deformation and optimization of the pressure–time curve be applied, so that the material can deform under conditions close to the optimal strain rate
3	 Triangular area Triangular area	After the formation of the four-layer sheet structure, a triangular area is formed at the corner	Under the optimal conditions of deformation (including temperature and strain rate), the pressure can be appropriately increased, and the holding time can be extended to reduce the formation of the triangular area
4	 Surface order difference Surface order difference	The two-layer sheet structure is prone to step differences on the external skin surface, especially for large-scale complex structural components	To address this issue, the processing accuracy of the sheet metal and the fit accuracy of the convex-concave mould should be improved. The pressure provided by the press should be greater than the gas pressure, to keep the upper and lower moulds in close contact during the forming process

(continued)

Table 3.4 (continued)

S/N	Defect forms	Specific descriptions	Control points
5	 Non-welding Non-welding	Insufficient welding or low welding rate may occurring at the interface of the diffusion bonding	The diffusion bonding temperature, pressure, time, and vacuum degree should be increased, and the surface condition of the sheet metal should be improved
6	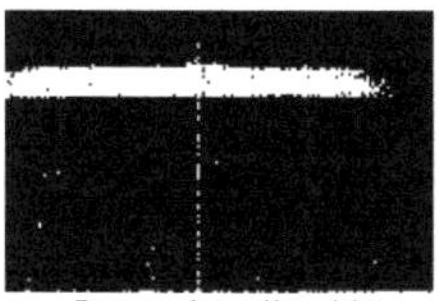 Incomplete rib grid Incomplete rib grid	The phenomenon of missing or broken reinforcement in the internal structure of multilayer sheets	When preparing the anti-welding coating, protect the surface of the diffusion region from contamination. Increase the diffusion bonding temperature, extend the diffusion bonding time, and increase the gas pressure
7	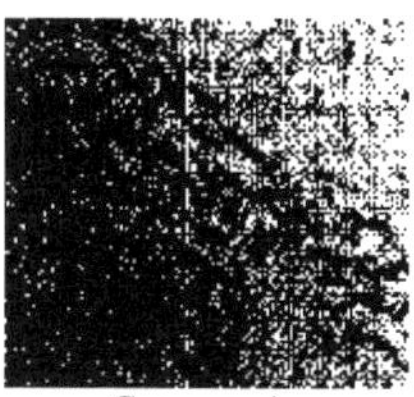 Coarse grain Coarse grain	The original equiaxed structure of the titanium alloy transforms into a coarse Widmanstätten structure	On the one hand, reasonable layout of the heating area and selection of appropriate heating parameters ensure a suitable heating rate and high temperature uniformity. On the other hand, the method of multi-point temperature measurement is adopted to achieve real-time monitoring of the forming temperature of the part and avoid grain growth
8	 Micro-crack Micro-crack	There are micro cracks on the surface, generally found in the corners of the parts or heavily oxidized areas	Before performing the SPF/DB process, remove the surface oxide layer. During the SPF/DB process, especially during the removal process from the furnace, the oxidation degree of the outer surface of the part be reduced by using oxygen protection
9	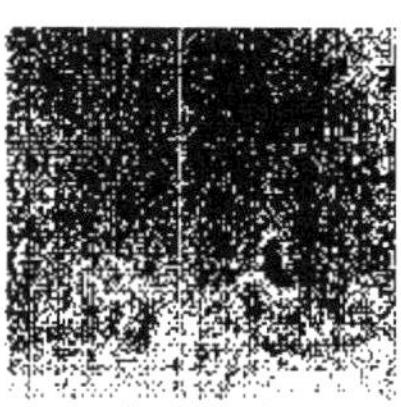 Surface pit Surface pit	There are pits on the surface of the part	Before installing the blank into the mould, use high-pressure air to blow the mould surface. Blow the upper mould first, and then blow the lower mould, taking care not to blow off the oxide or other impurities on the furnace wall onto the mould surface

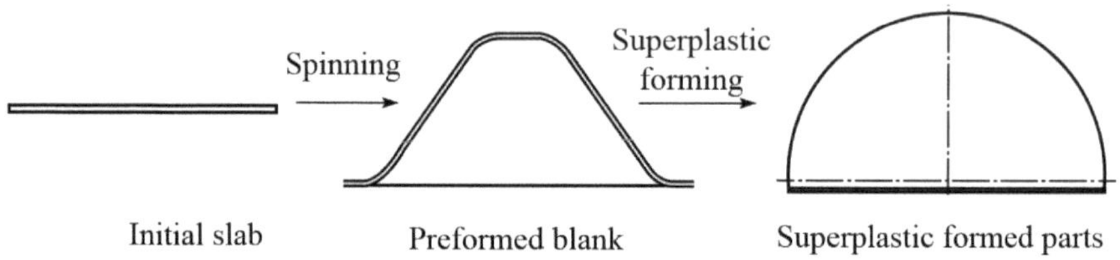

Fig. 3.26 Part forming process

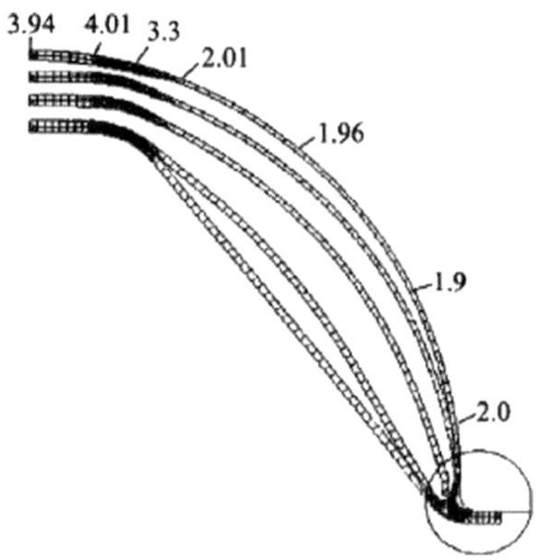

Fig. 3.27 Process simulation of manufacturing hemispherical parts by preforming method

cavities in the contact area and the entrance corner of the mould, which hinders the flow of material in the area and may cause wrinkles and other defects.

Figure 3.28 shows the superplastic forming process and wall thickness distribution of the finished parts for four pre-fabricated blanks of different thickness distributions. The preformed blanks are designed to have non-uniform thickness distribution, with the entrance corner area and the top area thicker, being 3–5 mm, and the sidewall area thinner, being 2.9–3.5 mm. During the forming process, the material in the sidewall area of blanks 2 and 4 also first contacts with the mould, resulting in the formation of closed cavities in the contact area and the entrance corner of the mould, which hinders the flow of material in the area and may cause wrinkles and other defects.

2. Forward and Reverse Bulging Method

The reverse mould preforming method is one of the commonly used methods to control the distribution of wall thickness. Taking the hemisphere as an example, the influence of various parameters of the reverse blow mould surface on the distribution of wall thickness of the part was analyzed by finite element simulation, and the optimal design of the reverse blow mould surface is realized through orthogonal experimental method.

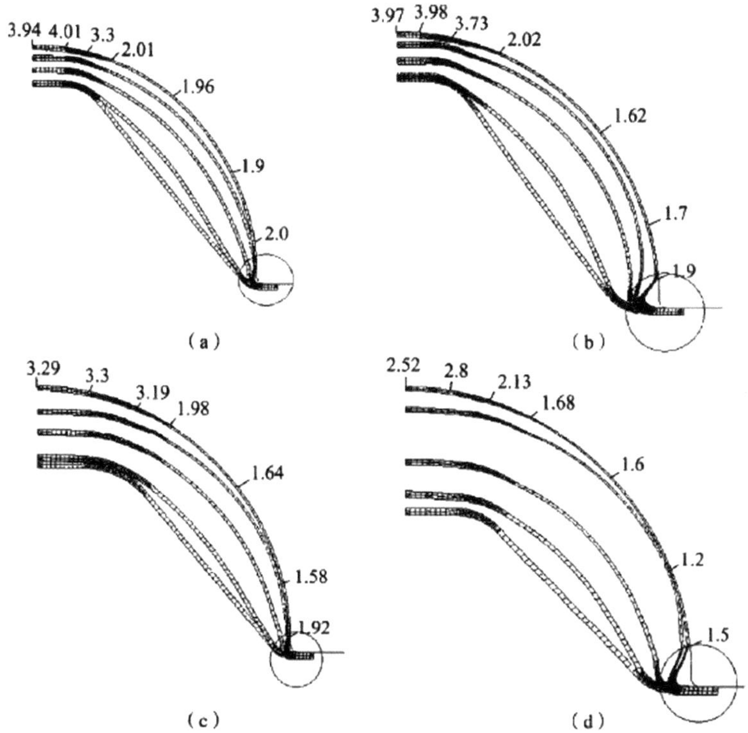

Fig. 3.28 Superplastic forming process of different sizes of preformed blanks. **a** Forming process of No. 1 part; **b** Forming process of No. 2 part; **c** Forming process of No. 3 part; **d** Forming process of No. 4 part

As shown in Fig. 3.29, the reverse blow mould surface contains constraints such as chamfer, height H, width L, and inclination angles α and β of the two sides of the triangle. β can be determined by the dimensions of H, L, and α, so they are the three main parameters, and the three chamfers are secondary parameters.

Orthogonal experiment is conducted to analyze the influence of each parameter on the wall thickness distribution, the inclination angle α, width L, and height H are given four numerical values each, as shown in Table 3.5.

Using the established finite element model and the numerical values given by the orthogonal experimental method, the influence of each parameter on the wall thickness distribution was analyzed. Figure 3.30a shows the deviation distribution of the angle α, width L, and height H, indicating that the height H has the greatest impact on the wall thickness distribution, followed by width L, while angle α has the smallest effect. Figure 3.30b–d show the wall thickness distributions of the parts at different angles, widths, and heights, respectively. In the process of blow forming with forward pressure, the wall thickness decreases monotonically. After reverse blow forming is used, the wall thickness distribution is significantly improved. Different angles lead to different positions experiencing reduction in wall thickness. The larger the angle,

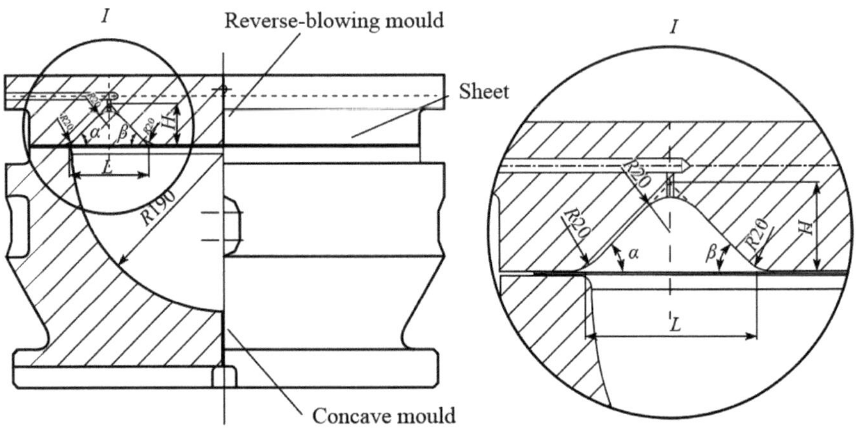

Fig. 3.29 Hemispheric superplastic forming mould

Table 3.5 Factor levels

Factors	Levels
α	30°, 45°, 60°, 90°
L	90 mm, 110 mm, 140 mm, 170 mm
H	40 mm, 50 mm, 80 mm, 100 mm

the closer the thinnest position is to the equator. When α = 45°, the range of wall thickness distribution is narrower, and the wall thickness is more uniform. Similarly, different widths also affect the area where the wall thickness is reduced. The larger the width, the closer the thinnest position is to the bottom of the sphere. Overall, when L = 140 mm, the wall thickness distribution is relatively uniform. Different heights have little effect on the position where the wall thickness is reduced, but have a greater impact on the amount of wall thickness reduction. The larger the height, the smaller the thinnest wall thickness. Overall, when H = 40 mm, the wall thickness distribution is relatively uniform. Based on the above analysis, the structural parameters for the optimized design of the pre-deformed surface are α = 45°, L = 140 mm, and H = 40 mm.

According to the simulation results above, a titanium alloy hemispherical shell mould was designed for superplastic forming experiments. Based on the structural characteristics of the part, the forming temperature range was selected as 890 ~ 895 °C, and the pressure–time curve for hemispherical shell forming was obtained through simulation calculations and experimental corrections, as shown in Fig. 3.31.

The superplastically formed part is shown in Fig. 3.32a, and its actual thickness distribution and theoretical thickness are shown in Fig. 3.32b, where A represents the design-required wall thickness, B represents the target wall thickness, and C represents the actual part thickness. The wall thickness at position 1 is slightly lower than the process requirement, where the amount of reverse pre-deformation is the

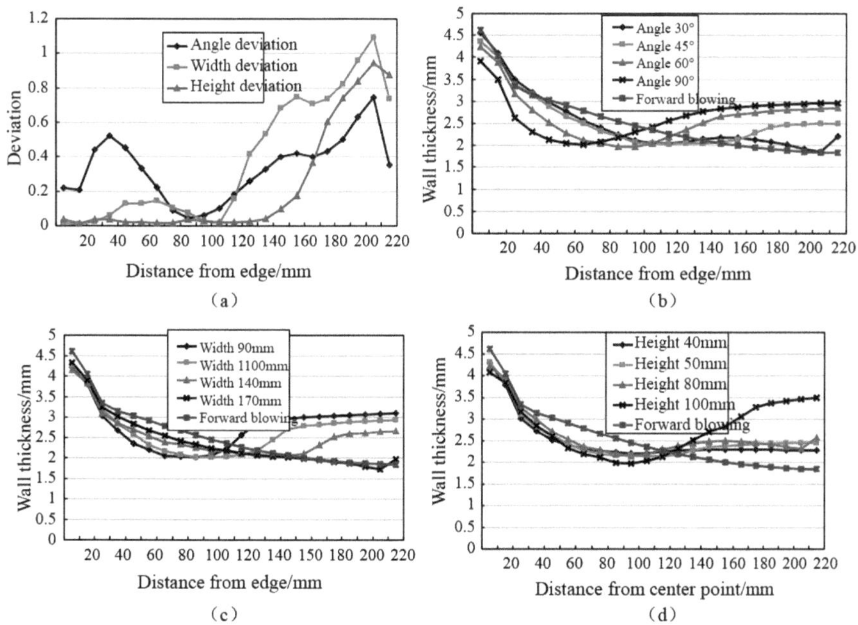

Fig. 3.30 Wall thickness distribution of formed parts. **a** α, L and H deviation distribution; **b** Wall thickness distribution under different angles; **c** Wall thickness distribution under different widths; **d** Wall thickness distribution under different heights

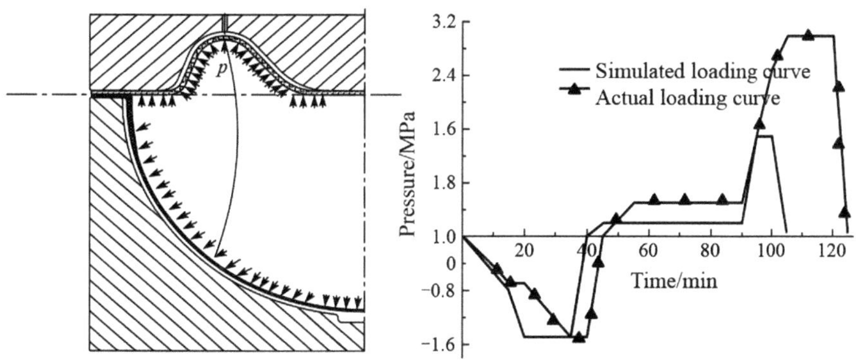

Fig. 3.31 Loading curve of superplastic forming

largest. Comparative analysis shows that the actual hemispherical part's spherical wall thickness distribution is basically consistent with the process requirements, with a wall thickness deviation of less than 0.3 mm, and the top thickness is about 1 mm thicker than the process requirement.

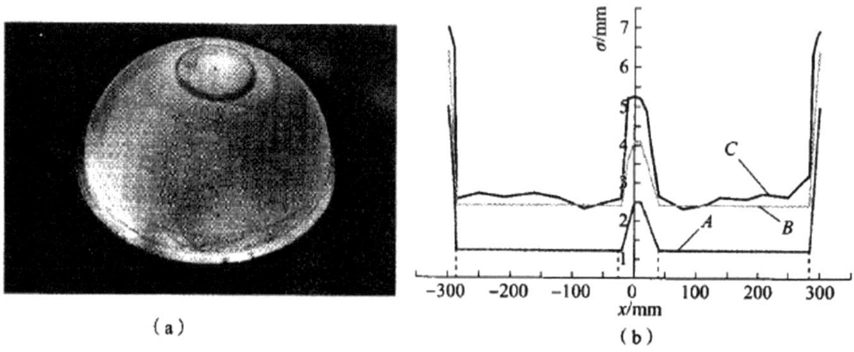

Fig. 3.32 Hemispherical shell forming parts and their thickness distribution. **a** Superplastic formed hemispherical shell; **b** Comparative analysis of thickness at location 1

3.4.2 Surface Wrinkles

Skin surface wrinkles are common defects in SPF/DB parts, which are influenced by many factors, and are closely related to structural parameters, forming process parameters, and material mechanical properties.

Taking the four-layer panel structure as an example, during the forming process, diffusion bonding is carried out between the two inner layers and between the inner and outer layers. Due to the special nature of the SPF/DB structure, a triangular area is formed at the diffusion bonding location between the two inner layers and between the inner and outer layers (as shown in Fig. 3.33a and b). During the superplastic forming of the inner layer, the triangular void gradually decreases as the deformation progresses, and when the pressure and time accumulated between the inner and outer layers reach a certain value, diffusion bonding occurs between the inner and outer layers. Theoretical analysis and experimental results show that the frictional force at the triangular void between the inner and outer layers is one of the main causes of surface wrinkles in four-layer SPF/DB structures.

The formation of surface wrinkles (as shown in Fig. 3.34) follows the criteria:

$$(f_2 - f_1) \geq \frac{\sigma_c + \sigma_s}{p}$$

Fig. 3.33 Position of triangular area of four-layer sheet structure

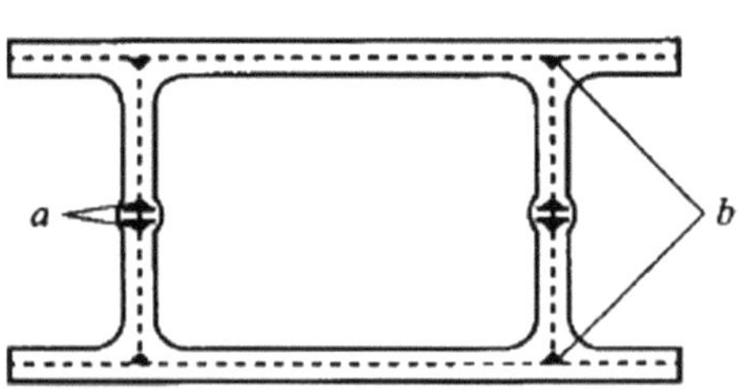

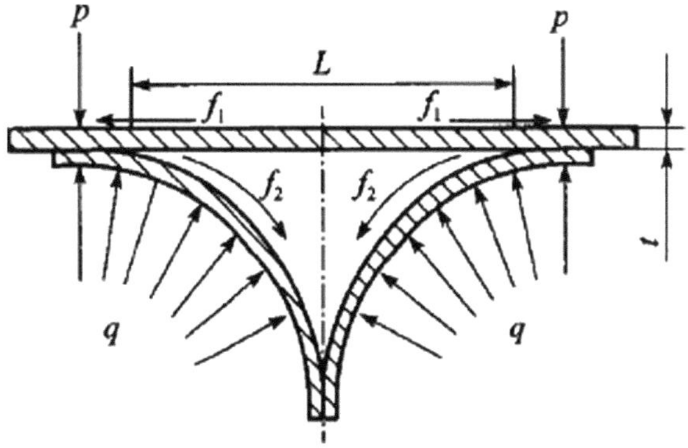

Fig. 3.34 Forming process stress

In the formula, σ_c is the critical stress for the instability of the outer layer sheet, $\sigma_c = \frac{\pi^2}{2} E_r \left(\frac{t}{L}\right)^2$, where t is the thickness of the sheet, L is the length of the sheet, E_r is the reduced modulus of the sheet, σ_s is the yield stress of the sheet at the superplastic temperature, p is the superplastic forming pressure, f_1 is the friction coefficient between the outer layer sheet and the mould, and f_2 is the friction coefficient between the inner layer and the outer layer.

Through analysis of the various parameters in the equation, it is believed that the outer layer can only avoid instability and surface wrinkling when f_1 is larger than. The following measures should be taken:

(1) Increase f_1: ensure smooth exhaust, vacuum the mould cavity, and roughen the mould surface.
(2) Reduce f_2: adding backing pressure, increase lubrication between the inner and outer layers.
(3) Reduce the superplastic forming pressure p.

Furthermore, σ_s and σ_c are related to material properties, grain size, instantaneous geometry and deformation temperature. σ_s can be increased by increasing the strain rate. It also helps prevent surface wrinkles.

In addition, whether it is a unidirectional or cross truss structure, the neutral layer in the inner layer shifts at the position of the wrinkle. The side where the wrinkle appears is the neutral layer that diffuses to the interface connecting the outer skin. The mould temperature unevenness, poor exhaust, and gravity creep of large components cause uneven forming rates on both sides of the neutral layer. The side that first comes into contact with the outer skin connects with the mould and the outer skin to form a whole, while the other side of the neutral layer continues to form. The tensile stress generated by the formation of the stretched area is transmitted through the diffusion joint to drive the movement of the inner layer and the outer skin in the diffusion bonding area. The positive pressure between the outer skin and the mould in the moulding area decreases, and the frictional force decreases, causing the skin to wrinkle. On the contrary, the inner layer gradually adheres to the mould on the forming side, and the volume of the compression triangular cavity between the inner layer and the outer skin decreases, and the local pressure increases. The frictional force between the outer skin and the mould increases, which helps to eliminate the surface wrinkle phenomenon, as shown in Fig. 3.35.

The process and reasons for wrinkle generation under neutral layer deviation in a four-layer sheet structure are analyzed using numerical simulation methods, as shown in Figs. 3.36 and 3.37. Under ideal conditions, the forming process of the four-layer sheet structure is shown in Fig. 3.36. The inner skin is located at the symmetric center of the outer skin under uniform air pressure loading. During the forming process, the inner skin deforms in a way close to mirror deformation, and then diffusion bonded to the outer skin without forming wrinkles on the part surface.

During the forming process, if the inner layer sheet deviates from the theoretical position within a certain range and does not undergo diffusion bonding with the outer skin, the diffusion bonded joint can still move to the geometric center of the outer blank under the forming pressure. If the initial position of the inner layer sheet severely deviates from the central plane and an abnormal diffusion connection occurs between the inner and outer skins, the outer blank that first contacts one side of the inner layer sheet will be wrinkled due to the traction force in the normal direction, as shown in Fig. 3.37.

As shown in Fig. 3.38, when the initial deviation of the middle blank is increased and diffusion bonded with the outer blank, the inner layer sheet pulls one side of the outer blank outward in the normal direction under internal load. In the subsequent forming process, the outer blank and the inner side further come into contact, causing deformation of the skin. In the later stage of deformation, the inner skin produces

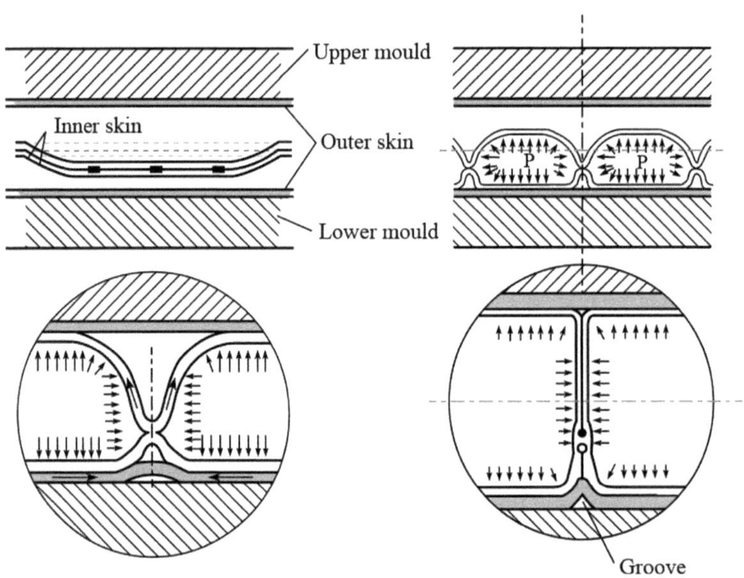

Fig. 3.35 Wrinkle formation process caused by neutral layer shift

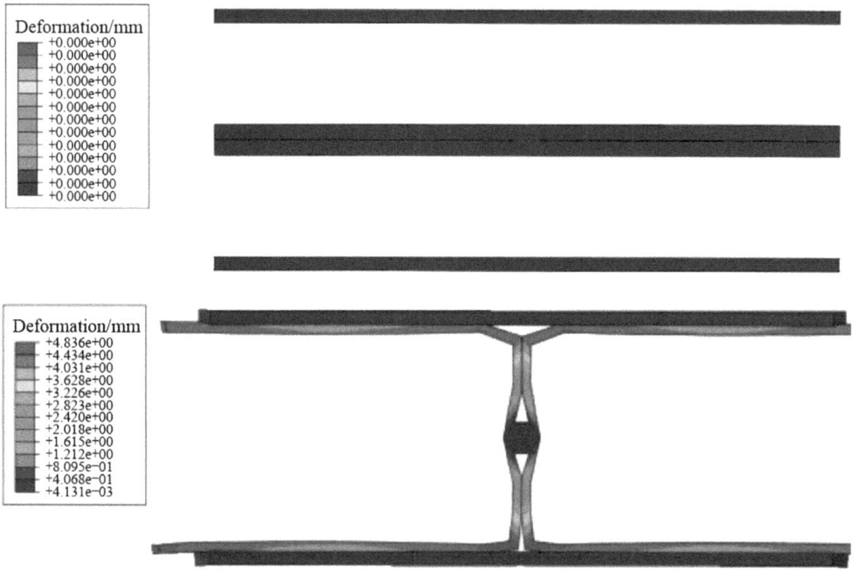

Fig. 3.36 No wrinkles formed when middle blank is located on the symmetrical plane

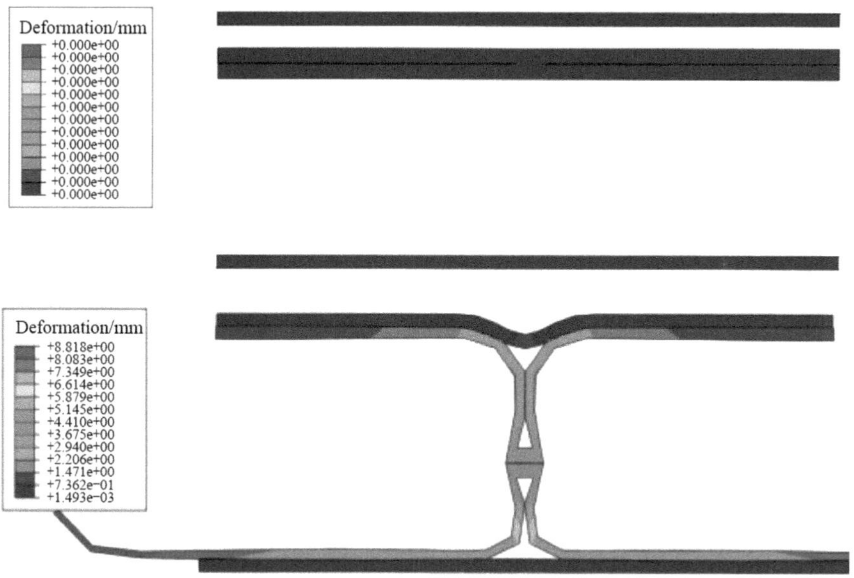

Fig. 3.37 The middle blank deviates from the symmetrical plane without contacting the outer skin

Fig. 3.38 Abnormal diffusion bonding between middle-layer blank and skin, resulting in severe wrinkles

a squeezing effect on the outer blank, resulting in severe wrinkling on the surface of the part. Due to the large deformation resistance of the material when the inner layer sheet contacts the skin, the final displacement of the wrinkled side of the truss is obvious.

Therefore, controlling the forming rate of the inner layer panels on both sides of the neutral layer is key to suppressing wrinkle formation. For large components, the main measures to control the forming rate of the inner layer panels are:

(1) Shortening the high-temperature residence time of the material to reduce the influence of gravity creep;
(2) Apply back pressure between the inner and outer layers to increase the positive pressure and friction between the skin and the mould, and prevent the outer skin from becoming unstable and deforming;
(3) Ensure unobstructed air passage, optimize the pressure–time curve, and increase the deformation rate as much as possible within the formable range.

For small components with high grid density and small size, the main measures are:

(1) Structural design should ensure even distribution of grid sizes;
(2) Ensuring unobstructed air passage, small grid sizes can be uniformly and stably formed under lower pressure, and the forming rate of the inner and outer skins should be increased as much as possible within the allowable range;
(3) Adopte back pressure forming.

Figure 3.39 shows the pressure–time curve for the four-layer panel structure formed by back pressure. Using back pressure forming can effectively solve the problem of outer skin instability during deformation, and make the forming rate of the inner layer panels more even and stable, significantly suppressing surface wrinkles.

3.4.3 Local Fracture

Compared with SPF/DB unidirectional truss structure, the formation of the four-layered cross truss structure is more difficult. In the unidirectional truss structure, gas enters through the inlet and reaches the middle layer of each cavity through channels. The smoothness of the inlet and outlet airflow can be easily ensured. However, in the

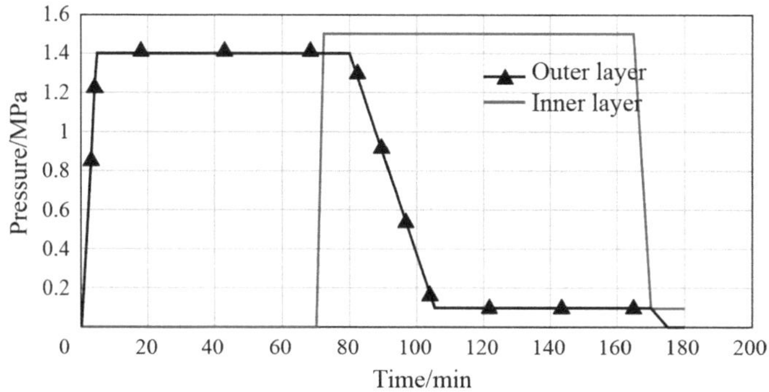

Fig. 3.39 Pressure–time curve of forming four-layer sheet structure using back pressure method

cross truss structure, each grid is relatively independent, and the grids are connected by vent holes. Only when the surrounding grids expand and form an inlet path, can the airflow enter the adjacent grid for forming. The airflow during the forming of adjacent grids is random, which makes it difficult to control. Figure 3.40 shows the phenomenon of inner skin rupture during the forming process of several typical four-layered structures.

The reasons are as follows:

(1) Uneven grid size, local deformation, and severe thinning of the wall thickness at the corners.

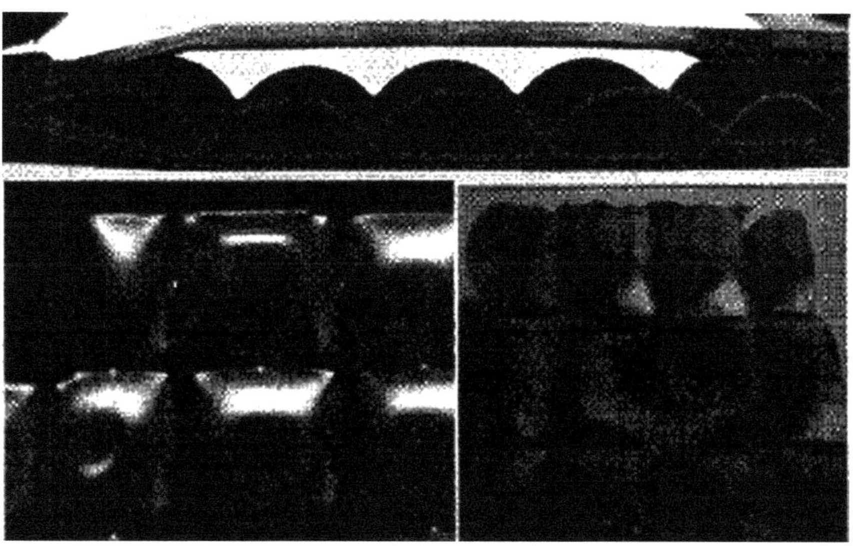

Fig. 3.40 Local rupture of the cross truss structure

(2) Inadequate airflow during the forming process results in rapid formation of small-area grids under high pressure, causing strain hardening and rupture. Titanium alloy sheets are very sensitive to strain rate. If there is insufficient airflow, the local strain rate increases, and the grid forming exits the superplastic forming region and directly enters the high strain rate forming region. As the strain rate increases, the material's elongation significantly decreases, resulting in the occurrence of rupture in areas with large local deformation, especially at the corners.

(3) The inner layer sheet in the area of the embedded instance frame is pulled and torn during the external turning process.

Therefore, in the structural design and forming process of cross truss structures, the following notes shall be taken:

(1) The size distribution of the truss should be as uniform as possible;
(2) Ensure the smooth design of the inlet airway, avoid problems such as airway blockage caused by unreasonable design;
(3) Adopt a staged pressure method, control the strain rate in the low-pressure stage to ensure the uniform formation of each grid, and ensure the airway smoothness. In the high-pressure stage, the strain rate should be increased as much as possible to shorten the forming time;
(4) Apply a static water pressure to use the reverse pressure difference method to form and further improve the uniformity and stability of wall thickness during the deformation process.

3.4.4 Cooling Deformation

For TC4 titanium alloy, the temperature for superplastic forming is generally above 900°C. If the high-temperature furnace discharge method is used, in addition to preventing oxidation of the parts, it is also necessary to prevent part deformation caused by rapid cooling. Especially for thin-walled hollow structures with complex shapes, control of the cooling deformation law is the key to obtaining parts with high-precision. Taking the three-layer sheet hollow structure as an example, the deformation law and control method during the cooling process are analyzed.

Based on the classical heat transfer theory, a heat transfer model for the cooling process of the hollow structure is established as shown in Fig. 3.41. The cooling mechanism is divided into convection, radiation, and heat conduction. Heat transfer between the hollow structure and air is achieved through convection and radiation, while heat transfer inside the hollow structure is mainly conducted by heat conduction, transferring heat from the high-temperature region to the low-temperature region.

Figure 3.42 shows the temperature variation with time during the 30-min air cooling process at room temperature (20 °C) of a three-layer sheet hollow structure formed at 920 °C. It can be seen from the figure that the thin area cools rapidly

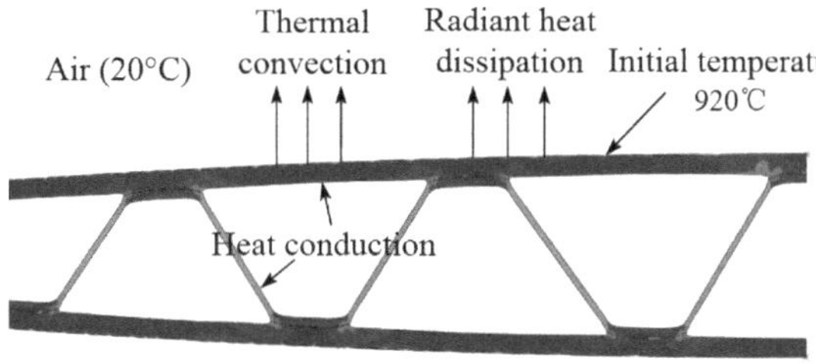

Fig. 3.41 Cooling heat transfer model of three-layer sheet structure

after the start of air cooling, and there is a significant temperature difference among different regions of the hollow structure, with the highest temperature at the thick section near the cavity and the lowest temperature at the edge of the cavity. After 1 min of room temperature cooling, the highest temperature of the thick section near the cavity is 656 °C and the lowest temperature at the edge of the cavity is 367 °C.

Figure 3.43 shows the distribution of residual stress and maximum strain of the hollow structure after cooling to 20 °C (room temperature). It can be seen from the figure that the residual stress after cooling is small, and most of the residual stresses are less than 20 MPa. The maximum strain after cooling is negative, and the cooling strain is between 0.88 and 1.1% with uneven strain distribution at different locations.

Based on the above analysis of the deformation reasons and experimental measurements mentioned, the method of storing the hollow structures in a thermal insulation

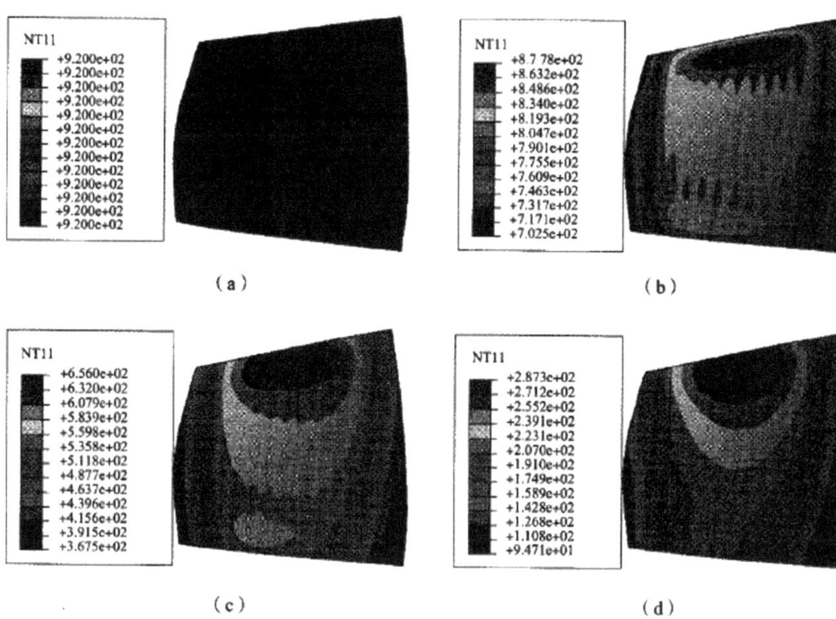

Fig. 3.42 Temperature change of hollow structure during cooling. **a** t = 0 s; **b** t = 10 s; **c** t = 1 min; **d** t = 5 min

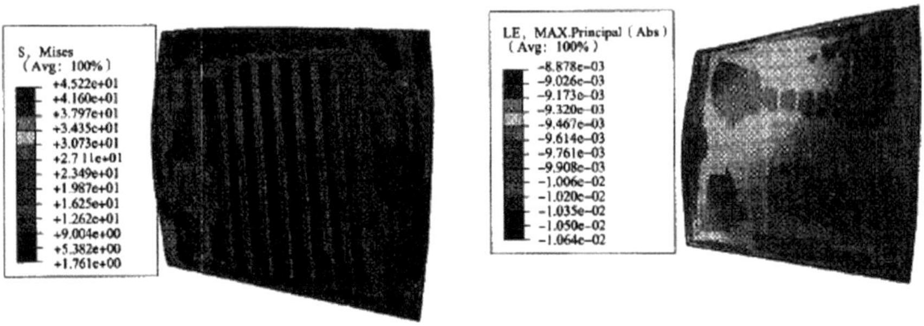

Fig. 3.43 Distribution of residual stress and maximum strain of the hollow structure after cooling

box after taken out of the furnace is proposed, in order to reduce the temperature gradient. Experiments are carried out and the deformation of hollow structures under slow cooling is measured. As shown in Fig. 3.44, the maximum deformation of the hollow structure decreased from 0.4 to 0.8 mm in the middle convex part to within 0.2 mm, which verified the effectiveness of this approach. The three-dimensional shape of the formed hollow structure was measured, and the contour accuracy of the profile is within the required specifications, ranging from −0.169 to 0.078 mm.

3.4.5 *Surface Steps*

The outer skin surface of large two-layer SPF/DB structures is prone to surface steps (as shown in Fig. 3.45), which directly affects the surface quality. For parts with large dimensions and large changes in curvature, surface steps are more likely to occur.

The surface step is caused by the actual gap between the upper and lower moulds in the high-temperature forming process of the sheet metal being greater than the design value, resulting in an increase in the deformation height of the sheet in the deformation zone, causing steps between the deformed and undeformed regions of the sheet. The cause of this problem is shown in Fig. 3.46:

(1) There is a fitting error in the mould. Especially for the mould surface of complex double-curved shape, it is difficult to ensure the machining accuracy of the mould, resulting in incomplete mating of the mould and a deviation between the gap between the upper and lower moulds and the design value.
(2) The processing accuracy of the sheet is low. Especially for large-sized parts, the sheet is prone to exhibit low processing accuracy, and the local high points on the sheet lead to the actual gap between the upper and lower moulds being larger than the design value.
(3) The pressure loading curve is unreasonable. During the forming process, when the pressure acting on the sheet and the upper mould is greater than the sum of the machine tool pressure and the weight of the upper mould, the upper and

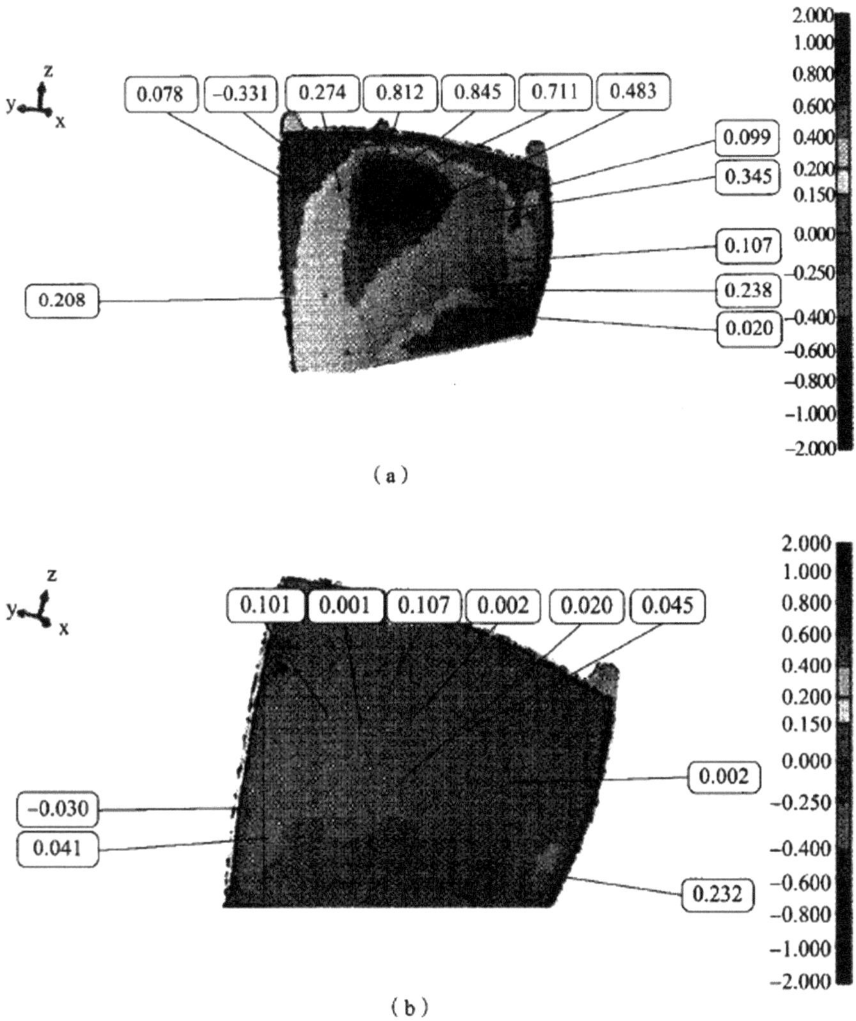

Fig. 3.44 Cooling deformation characteristics of the hollow structure under different cooling conditions. **a** Air cooling; **b** slow cooling conditions

lower moulds will separate, causing an increase in the gap between the upper and lower moulds.

To avoid the problem of surface step, on one hand, it is necessary to improve the processing accuracy of the sheet metal, and on the other hand, a reasonable design of air pressure—machine tool pressure—time curve, and strictly control of equipment parameters during the forming process are suggested. Based on this, the surface step can be effectively eliminated through the matching of the upper and lower moulds. For larger-sized moulds, there is a larger area that needs to be matched and

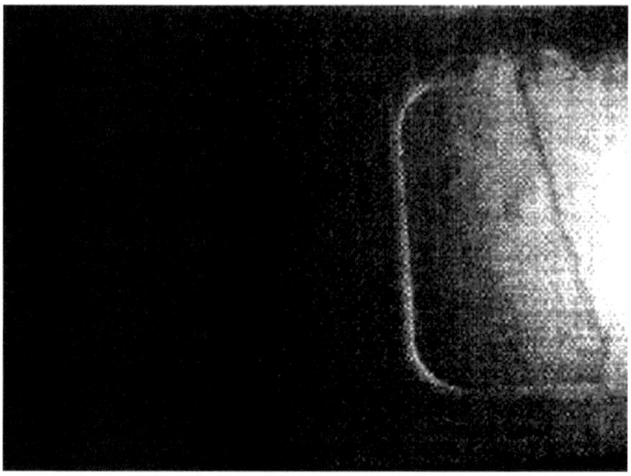

Fig. 3.45 Order difference of skin surface

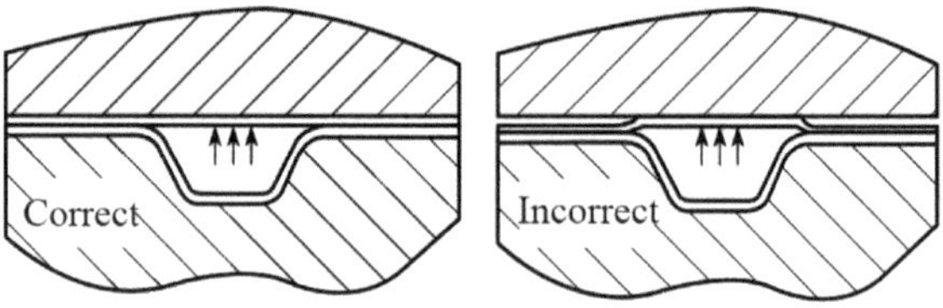

Fig. 3.46 Diagram of the generation of order difference

repaired. It is difficult to match all fitting surfaces by hand, and a combination of precision machining and hand grinding methods is suggested. The hand grinding area is the black solid strip area in Fig. 3.47. As inspection of the formed part shows, the maximum value of the step difference decreased from 0.8 mm to 0.3 mm after fixing, showing a significant improvement.

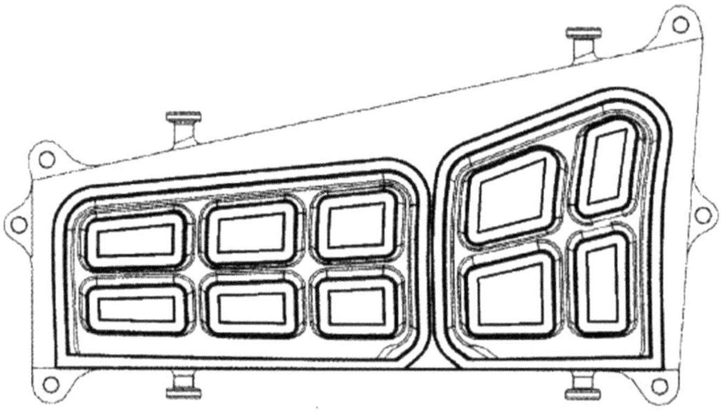

Fig. 3.47 Diagram of hand lapping area of the mould

$$f_1 - f_2 > k\sigma_f \varepsilon \tag{3.71}$$

$$f_2 = 2\int_0^{\frac{\pi}{2}} P \times \omega \times r \times \sin\theta d\theta \tag{3.72}$$

$$\begin{aligned} f_2 &= 2\int_0^{\frac{\pi}{2}} P \times \omega \times \frac{L_2 - 2\delta_2}{2} \times \sin\theta d\theta \\ &= -P\omega(L_2 - 2\delta_2)\cos\theta|_0^{\pi/2} \\ &= P\omega(L_2 - 2\delta_2) \end{aligned}$$

$$f_1 = 2\int_{\frac{3\pi}{2}}^{2\pi} P \times \omega \times r \times \sin\theta d\theta$$

$$\begin{aligned} f_1 &= 2\int_{\frac{3\pi}{2}}^{2\pi} P \times \omega \times \frac{L_1 - 2\delta_2}{2} \times \sin\theta d\theta \\ &= -P\omega(L_1 - 2\delta_2)\cos\theta|_{1.5\pi}^{2\pi} \\ &= -P\omega(L_1 - 2\delta_2) \end{aligned} \tag{3.73}$$

References

1. Cui, Yuanjie. 2011. *Numerical Simulation and Experimental Study of Superplastic Forming/Diffusion Bonding Process for TC4 Multi-Layer Structures*. Nanjing: Nanjing University of Aeronautics and Astronautics.
2. Guan, Zhiping. 2008. *Quantitative Mechanical Analysis of Superplastic Tensile Deformation*. Changchun: Jilin University.
3. Ning, Huiyan. 2008. *Finite Element Simulation and Analysis of Superplastic Bulging Forming of Fine Crystal Magnesium Alloy Sheet*. Harbin: Harbin University of Science and Technology.
4. Chen, Changping. 2009. *Numerical Simulation and Experimental Study on Superplastic Bulging Forming of AZ31 Magnesium Alloy Heart-Shaped Parts*. Harbin: Harbin University of Science and Technology.
5. Shidun, Wu. 1997. *Theory of Superplastic Deformation of Metals*. Beijing: National Defense Industry Press
6. Jian, Zhao. 2014. *Numerical Simulation of Superplastic Free Bulging Under Typical Loading Paths*. Changchun: Jilin University.
7. Jianwei, Chen. 2007. *Superplastic Forming and Precision Control of TC4 Titanium Alloy Cylindrical Workpieces*. Harbin: Harbin Institute of Technology.
8. Zhao, Yi. 2007. *Study on SPF-DB Numerical Simulation of Titanium Alloy Multi-Layer Plate Based on Optimized Load Control*. Xi'an: Northwestern Polytechnical University.
9. Zhiqiang, Li, and Guo Heping. 2004. Application and development of superplastic forming/diffusion bonding technology. *Aeronautical Manufacturing Technology* (11): 50–52.
10. Bing, Zhao, Li Zhiqiang, Hou Hongliang, et al. 2015. Research progress on preparation technology of metal three-dimensional lattice structures. *Rare Metal Materials and Engineering* (6): 83–96.

11. Bing, Zhao, Li Zhiqiang, Hou Hongliang, et al. 2017. Study on preparation process and compression properties of titanium alloy three-dimensional lattice structures. *Rare Metals 41* (3): 258–266.
12. Li, Z.Q., B. Zhao, J. Shao, et al. 2019. Deformation behavior and mechanical properties of periodic topological Ti structures fabricated by superplastic forming/diffusion bonding. *International Journal of Lightweight Materials and Manufacture* 2: 1–30.
13. Han, Xiaonong, Jie Shao, Xuepiao Bai, et al. 2013. Study on design and bearing capacity analysis of SPF/DB three-sheet cylindrical slewing structure. *Aeronautical Manufacturing Technology* 16: 69–71.
14. Cao, Yunhong, et al. 2002. Study on application of titanium alloy forming process to aircraft missile. *Aerodynamic Missile* (7a): 50–60.
15. Yang, Qinxin, Guoquan Tong, and Zezhou He. 2017. Experimental study on superplastic forming/diffusion bonding of titanium alloy two-layer plate structure. *Rare Metals 41* (12): 1305–1310.
16. Zhang, Xuexue. 2012. *Experimental Study on Superplastic Forming/Diffusion Bonding of TC4 Multilayer Plate*. Nanjing University of Aeronautics and Astronautics.
17. Alabort, E., D. Putman, and R.C. Reed. 2015. Superplasticity in Ti-6Al-4V: Characterisation, modelling and applications. *Acta Materialia* 95: 428–442.
18. Liu, Shengjing, Hailong Lei, and Fulong Chen. 2014. Analysis of bulging mechanism of superplastic forming/diffusion bonding four-layer sandwich structure. *Forging & Stamping Technology* 39 (8): 30–36.
19. Du, Lihua, Xingzhen Zhang, Xiaoning Han, et al. 2018. Study on the influence of geometric parameters on the surface quality of SPF/DB three-layer structure. *Aeronautical Manufacturing Technology 61* (10): 100–103.
20. Liu, Shengjing, Jun Zhang, Hailong Lei, et al. 2014. Study on laminated-layer superplastic forming process of TC4 titanium alloy conical parts. *Hot Working Technology43* (17): 155–159, 164.
21. Xin, Qibin, and Linlin Wang. 2013. *Computer Simulation of Material Forming*. Beijing: Metallurgical Industry Press.
22. Liu, Xianghua. 1994. *Rigid-Plastic Finite Element Method and Its Application in Rolling*. Beijing: Metallurgical Industry Press.

Chapter 4
Microstructure and Properties of Superplastic Forming/Diffusion Bonding Process

During superplastic forming, the microstructure of the material undergoes changes such as alternations in grain shape and size, grain sliding and rotation, dislocation, and void formation. These changes have a significant impact on the mechanical properties of the material after superplastic forming/diffusion bonding. Diffusion bonding involves metallurgical bonding at the material surface under the effect of pressure at certain temperatures. Non-welding and weak bonding at the interface are common defects in the diffusion bonding process, these defects reduce the effective bearing area and create stress concentration near the interface, leading to varying degrees of impact on the mechanical properties of the structure.

This chapter provides a summary of the microstructure changes, pore evolution, and interface defects that occur during the superplastic forming/diffusion bonding process. The static and fatigue properties of these changes are characterized, and the effect of microstructure and defects on mechanical properties is analyzed.

4.1 Microstructure in Superplastic Forming/Diffusion Bonding Process

During superplastic deformation, materials undergo microstructure changes that are different from those observed during traditional plastic deformation. In the process of superplastic deformation, grains are usually coarsened due to the influence of various factors such as strain rate, deformation temperature and total deformation. Additionally, the grains become equiaxed and elongated due to the effects of grain sliding, rotation, and dynamic recrystallization. These changes may also induce voids and dislocations, which arise from grain boundary sliding or insufficient coordinated deformation. By studying the microstructure changes during superplastic deformation, we can better understand the micro-mechanisms of superplastic deformation and predict the changes in mechanical properties of the material after superplastic deformation.

Z. Li, *Superplastic Forming/Diffusion Bonding Technology of Titanium Alloys*,
https://doi.org/10.1007/978-981-99-3909-1_4

4.1.1 Material Microstructure

The changes in microstructure during superplastic deformation primarily involve the alteration of grain shape and size. Typically, the grains in the material tend to grow larger after superplastic deformation due to the thermal effect, resulting in grain coarsening. Several factors affect the rate of grain growth during superplastic deformation, including the strain rate, deformation temperature, and total deformation. Additionally, the rate of growth is influenced by the alloy composition, particularly in two-phase alloys, where it is related to the phase proportion and the precipitation of second-phase. For instance, when a Zn-Al eutectoid alloy undergoes superplastic deformation, the grain size of the specimen at the gauge length is found to be 1.7 times larger or more than that at the clamping position. This observation suggests that the growth of grain size is not solely due to thermal effect, but also due to superplastic deformation.

In addition to grain coarsening, superplastic deformation also leads to refinement and globularization of grains under certain conditions, which are typically related to the occurrence of dynamic recrystallization. Dynamic recrystallization is a common phenomenon observed during superplastic deformation of materials. Generally, alloys with an original fiber microstructure or elongated grains are prone to dynamic recrystallization during superplastic deformation, resulting in equiaxed fine-grained microstructures. Research conducted by T.H. Alden revealed that Sn-5% Bi alloy with a distinct fiber or lamellar initial structure transformed into a uniform fine-grained equiaxed microstructure after superplastic deformation.

Moreover, the sliding, rotation, and transposition of grains are commonly observed during superplastic deformation of materials, and these phenomena endow the materials with larger elongation. Although the morphology and size of the grain change during sliding and rotation, the grain will still maintain a good equiaxed shape. In a study, researchers marked lines on the specimensurface and observed that some of the marking lines changed from the original straight line to a broken line at the grain boundary after undergoing superplastic tensile, indicating relative displacement between grains. Additionally, some marked line segments were at an angle with the original straight line, indicating grain rotation. Furthermore, Naziri et al. observed the superplastic deformation process of Zn-Al eutectoid alloy films in situ and found grain boundary sliding, grain movement, and transposition, as well as the coupling process of grain boundary sliding, dislocation movement, and diffusion creep during superplastic deformation.

1. **TC4 alloy**

The TC4 alloy is a commonly used titanium alloy in the aerospace industry due to its excellent overall properties. It is a typical $\alpha + \beta$ two-phase titanium alloy that not only exhibits a good match of strength and toughness and capable of working at 400 °C for a long duration, but also has good processing properties and superplasticity, which make it suitable for various The microstructure of TC4 alloy with different grain sizes is obtained by low-temperature and long-time treatment, whereby no

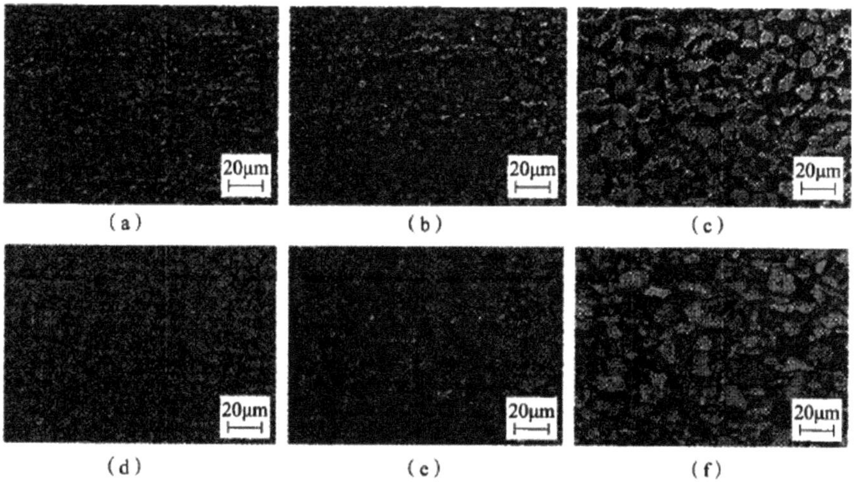

Fig. 4.1 Microstructure of TC4 alloy before and after superplastic deformation, before deformation with the grain size of **a** 5μm; **b** 8μm; **c** 12μm; after deformation with the grain size of **d** 5μm; **e** 8μm; **f** 12μm

phase transformation occurs during the process and grain size is controlled by heat treatment time. The specific process parameters are 920 °C/0.5 h, 920 °C/2 h, and 920 °C/24 h, respectively. Under these conditions, the proportion of the primary α phase after treatment remains nearly unchanged at around 92%, whilethe size of the primary α phase grows continuously as the heat treatment time is extended, as seen in Fig. 4.1 a–c. The average grain size of the primary α phase is about 5 μm after 0.5 h heat treatment at 920 °C. When the heat treatment time is extended to 2 h, the grain size increases to about 8 μm, and when extended to 24 h, it grows to about 12 μm.

Figure 4.1d–f show the microstructures of the TC4 alloy with original microstructures of different grain sizes after superplastic deformation. It is clear that the microstructure of the original sheets changes significantly after superplastic deformation. Recrystallization occurs and the primary α phase is coarsened obviously. The average grain size of the primary α phase after deformation is about 9 μm, 11 μm, and 18 μm, respectively. Results show that superplastic tensile elongation decreases while the peak stress increases with an increase in primary α phase size. After superplastic deformation of the TC4 alloy with different primary α phase sizes, the primary α phase sizes grow, and the original lamellar microstructure of the primary α phase changes to a needle-like microstructure during deformation. This illustrates that grain growth is one of the remarkable characteristics of the microstructure evolution during superplastic tensile deformation of the TC4 alloy.

Figure 4.2 shows the transmission electron microscope (TEM) microstructure of TC4 alloy after superplastic deformation at 920 °C and strain rate of $1 \times 10^{-3}s^{-1}$. The graph reveals that the α and β phase interfaces are arc-shaped and severely twisted, indicating that phase boundary sliding is the primary mode of deformation in the superplastic process. Dislocations were not found in the α-grain interior or near the α-phase grain boundary, and the grain boundary between α-phase and α-phase is narrow and straight. However, dislocations were observed originating from the boundary between the α and β phases, indicating that this boundary is the dislocation source. The sliding of phase/grain boundary is coordinated by the slip or climb of dislocations near the boundary/grain interface, induced by interface sliding that is more likely to occur in the boundary between the α and β phases rather than the boundary between α phases. Moreover, it can be observed that during deformation, β phases squeeze into the grain boundary between α phases and rearrange along the grain boundary of α phases. Numerous small "protrusions" on the α phase grain boundary result in the redistribution of crystal groups of the same phase, increasing the phase interface. This suggests that the β phase is prone to deformation during superplastic deformation of fine-grained TC4 alloy, and the sliding mainly occurs at the boundary between the α phase and β phase.

Figure 4.3 displays the transmission electron microscope (TEM) microstructure of TC4 alloy with relatively coarse grains after superplastic deformation at 920 °C and strain rate of $1 \times 10^{-3} s^{-1}$. In specimens with coarser grains, a large number of dislocations are found not only near the boundary between α phase and β phase, as

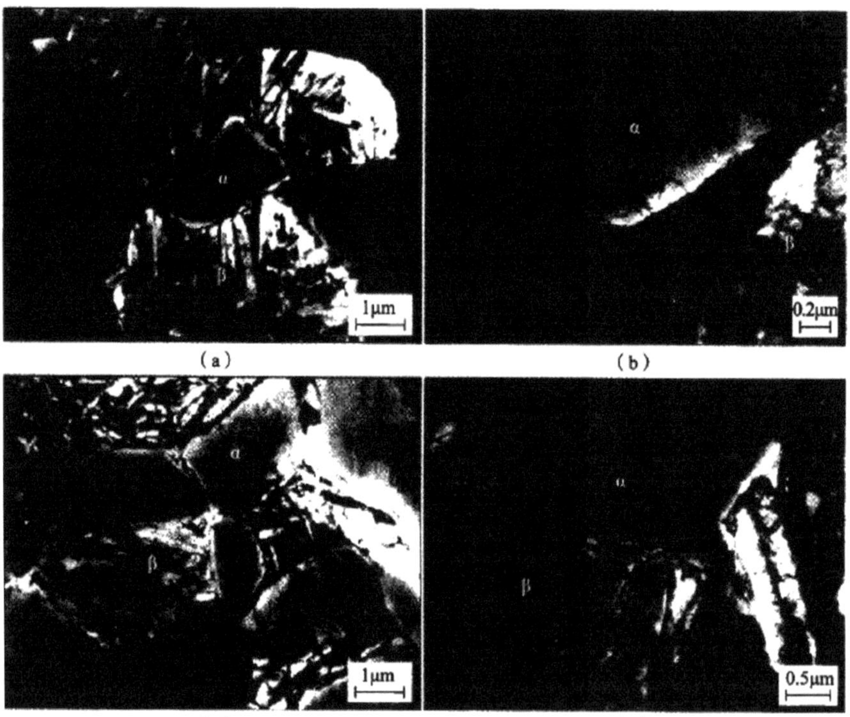

Fig. 4.2 TEM microstructure of fine-grained TC4 alloy after superplastic tensile deformation

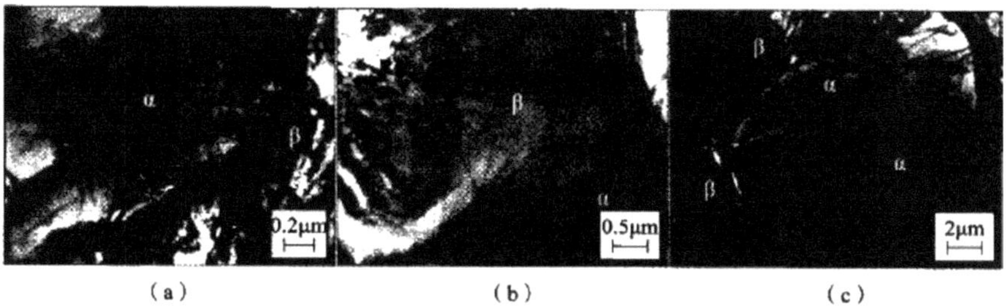

Fig. 4.3 TEM microstructure of coarse-grained TC4 alloy after superplastic tensile deformation

shown in Fig. 4.3a, but also near the grain boundary between α phase and α phase and inside α grains, as shown in Fig. 4.3b. Moreover, there are sub-grain boundaries in α grains with dislocations near them, as shown in Fig. 4.3c. These findings suggest that dislocation movement during superplastic deformation leads to the formation of a sub-grain boundary consisting of dislocation walls. These sub-grain boundaries may further lead to dynamic recrystallization as deformation continues.

Based on the results presented above, it is clear that there are significant differences in the microstructure of TC4 alloy after superplastic deformation with varying initial grain size. In the case of fine-grained TC4 alloy, the boundary between the α and β phases is severely distorted and arched, leading to dislocation origin from the boundary. The main mechanism of deformation involves interface sliding with coordinated dislocation movement. There is almost no dislocation found in the α-phase grains except for the dislocation near the interface. However, for the coarse-grained TC4 alloy, a large number of dislocations are found not only near the boundary between the α and β phases but also inside the α-phase grains. This indicates that intragranular dislocation movement also plays a significant role, in addition to grain boundary sliding. Therefore, at this point, the superplastic deformation mechanism can be considered as the combined action of both grain boundary sliding and intragranular dislocation movement.

Furthermore, the study also investigates the impact of nanocrystalline materials on the superplastic deformation of TC4 alloy. To obtain nanocrystalline TC4 alloy material, the cold rolled α + β phase alloy (Fig. 4.4a) is solution treated to achieve a fully lamellar microstructure (Fig. 4.4b). Severe plastic deformation under high pressure torsion is then used to produce the desired material. The TEM structure, depicted in Fig. 4.4c, shows diffraction spots polymerizing into an annular shape, indicating the formation of ultrafine-grained structures with large angular grain boundaries. The average size of uniformly distributed nanocrystals is approximately 70nm.

The prepared nanocrystalline TC4 alloy is subjected to superplastic tensile at a low temperature of 650 °C, and it exhibits high elongation. Specifically, Fig. 4.4d shows that at the strain rate $5 \times 10^{-4}s^{-1}$, an elongation of 820% is obtained. This suggests that the nanocrystalline structure contributes to the excellent superplasticity of the material at low temperature and high strain rate.

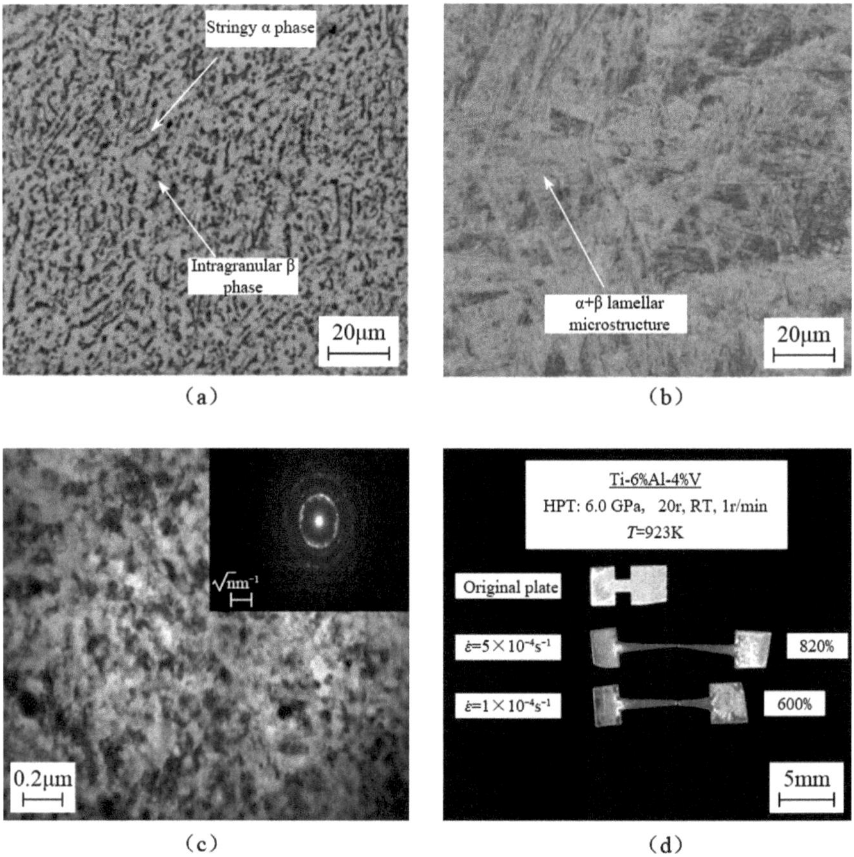

Fig. 4.4 TC4 alloy microstructure morphology **a** Cold-rolled state; **b** Lamellar microstructure; **c** TEM microstructure after severe plastic deformation under high pressure torsion; **d** Morphology of superplastic tensile specimen

2. **TA32 alloy**

TA32 alloy is a newly developed high-temperature resistant near α-type titanium alloy. This alloy is based on TA12 alloy with changes made to the content of hot-strength elements, such as Nb, and by removing the rare earth element Nd. TA32 alloy has a low content of β-phase stable elements, which results in good thermal strength and stability at 550 °C. However, due to its high deformation resistance and low thermal conductivity, it is difficult to deform TA32 alloy into parts with complex geometries at room temperature. Therefore, studying the superplastic deformation of TA32 alloy under different deformation conditions is of great significance as it provides a feasible technical approach for forming complex parts.

The experimental material used is a 2mm sheet of TA32 alloy with a β-transformation temperature of about 1005 °C. The original microstructure of TA32 alloy is shown in Fig. 4.5a using optical microscope (OM). The black β phase are either equiaxed or elongated along the boundary of the gray α phase. Both of the two phases are elongated along the sheet rolling direction (RD), and the proportion of primary α phase is about 85%. Additionally, a large number of small circular dots of $(TiZr)_6Si_3$ silicides are dispersed at the boundaries. The EBSD images and grain size distribution of the original microstructure of TA32 alloy are shown in Fig. 4.5b, c. It can be seen that most grains are equiaxial, some grains are blocky, and small grains are distributed among large grains, showing uneven distribution. The Energy Dispersive Spectrometer (EDS) analysis results of the original microstructure of TA32 alloy are shown in Fig. 4.5d. The spectral peak of element C is caused by the carbon spraying treatment of the specimen, and the spectral peak of element La is the noise peak. The elements at the overlapping peak are Sn and Mo.

In Fig. 4.6, we can see the microstructure of TA32 alloy at a temperature range of 880–940 °C and an initial strain rate of $1 \times 10^{-3}s^{-1}$. At 880 °C, fine β-phase grains appear at α grain boundaries or α/β phase boundaries, indicating transformation from β to α phase. The microstructure morphology of the specimen is similar to the original microstructure, as shown in Fig. 4.6a. As the temperature increases, the

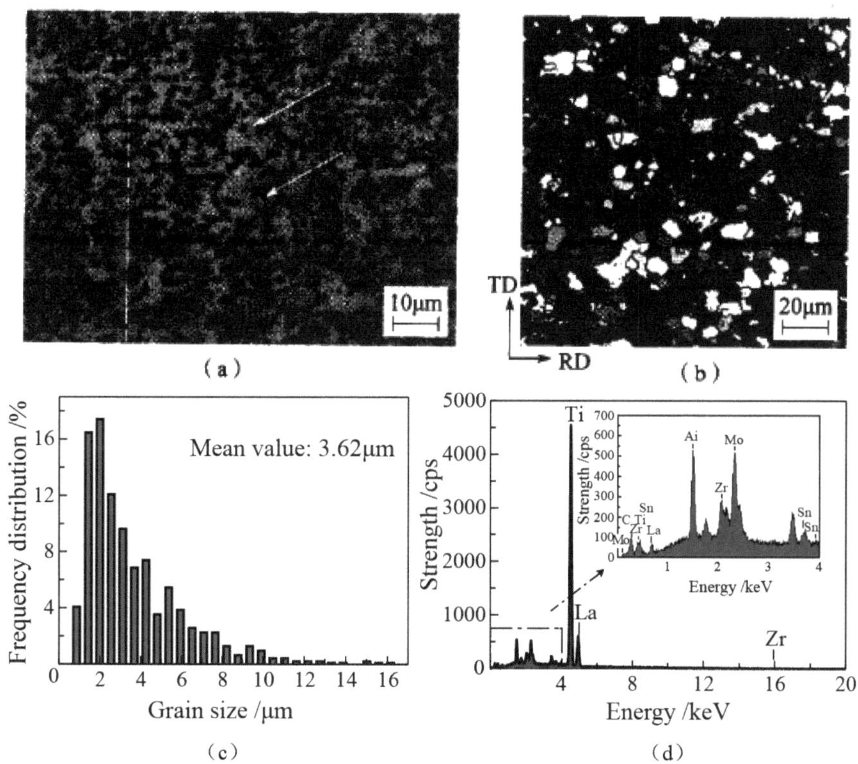

Fig. 4.5 Original microstructure of TA32 alloy (cps: counts per second) **a** Metallographic picture; **b** EBSD picture; **c** Grain size distribution; **d** EDS result

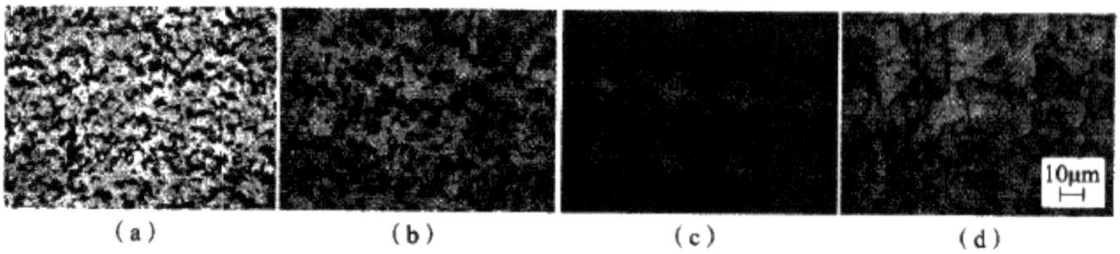

Fig. 4.6 Microstructure of TA32 alloy with the initial strain rate of 1 × 10−3 s−1 at **a** 880°C; **b** 900°C; **c** 920°C; **d** 940°C

α-phase grains become larger while maintaining an equiaxed morphology. The size of β-phase grains also increases significantly, with the morphology changing from strip-shaped to equiaxed due to the increased amount of α transforming to β phase. As β-phase grains combine continuously combine with each other, they become larger. Meanwhile, non-equiaxed grains break and re-nucleate to become equiaxed grains due to dynamic recrystallization. The recrystallized β phase coarsens until connecting with each other. With a relatively long heat preservation time at the initial strain rate of $1 \times 10^{-3}s^{-1}$, atomic diffusion migration and grain boundary movement become stronger, which promote the growth of αand β phase grains. Furthermore, the grain boundary diffusion rate of β phase is higher than α phase at high temperatures, leading to relatively rapid growth of β-phase grains. During superplastic deformation, both α phase and β phase deform simultaneously, and competitive dynamic recrystallization occurs between them.

At a temperature range of 920–940 °C, the elongation after fracture is relatively large, especially at 920 °C. The appropriate two-phase proportion restrains the grain growth of the two phases, which is conducive to superplastic deformation. At this point, the maximum elongation after fracture is 774%.

Figure 4.7 displays the microstructure of TA32 alloy at 900 °C and an initial strain rate ranging from 5×10^{-4} to $1 \times 10^{-2}s^{-1}$. The figure shows that the grain size of both phases increases as the strain rate decreases. This is due to dynamic recrystallization, which fully occurs at low initial strain rates (5×10^{-4} to $1 \times 10^{-3}s^{-1}$), leading to the nucleation of more undistorted fine equiaxed grains near the grain boundary. However, long-term high-temperature deformation accelerates atomic movement and grain boundary movement, causing dynamic recrystallized grains to continuously aggregate and grow. At higher initial strain rates (5×10^{-3} to $1 \times 10^{-2}s^{-1}$), the accumulation of stored energy increases the rate of dynamic recrystallization, but the rapid deformation rate prevents dynamic recrystallized grains from fully merging and growing.

The proportion of β phase slightly increases as the strain rate decreases, indicating that the change of strain rate is not the primary inducement for the α → β phase transformation. Additionally, comparing Figs. 4.6 and 4.7 shows that the strain rate has little effect on the morphology and size of β-phase grains, whereas the impact of temperature is more significant.

3. **TNW700 alloy**

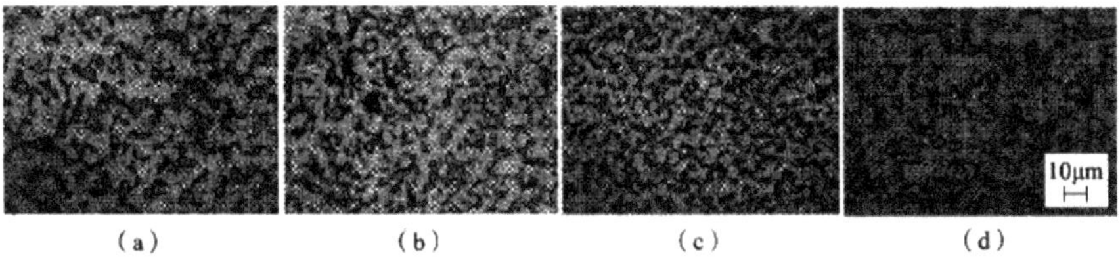

Fig. 4.7 Microstructure of TA32 alloy at 900 °C with different initial strain rates **a** 1×10^{-2} s^{-1}; **b** 5×10^{-3} s^{-1}; **c** 1×10^{-3} s^{-1}; **d** 5×10^{-4} s^{-1}

The TNW700 alloy is a high-temperature titanium alloy that belongs to the Ti–Al-Zr-Sn-Nb-W series and is strengthened by multiple elements. It can be used as a bearing material for a short period of time under 600–750 °C conditions. The experimental material used in this study is a TNW700 alloy plate in the M-state. The β-transformation temperature measured by differential thermal analysis is about 1000 °C. Figure 4.8 shows the original microstructure of the TNW700 alloy, which is composed of closed-packed hexagonal (hcp) α phase (black area in Fig. 4.8a) andbody-centered cubic (bcc) β phase (white area in Fig. 4.8a) with a volume fraction of α phase of about 94%. The EBSD test results show that the raw material of TNW700 alloy has a basal texture characteristic, with the c axis of α phase nearly vertical to the rolling plane.

The superplastic tensile test is conducted at a temperature of 925 °C and a strain rate of $0.001s^{-1}$ with different strain values ranging from 0.1 to 1.75. Tensile deformation with constant strain rate is adopted during the test, and after reaching the target strain, the specimen is immediately quenched in water to maintain the high-temperature deformation microstructure. Figure 4.9 shows the true stress—true strain curve of the superplastic deformation test and the corresponding specimen. The critical strain of strain hardening and flow softening is about 1.13. During the strain hardening stage, the gauge section of the specimen becomes longer and narrower uniformly without necking. When the deformation strain increases to 1.5, the gauge section obviously becomes narrower, but there is still no significant necking. This indicates that the flow softening of the TNW700 titanium alloy has little relationship with the local necking of the specimen but is caused by the change of microstructure. When the deformation strain reaches 1.7, the bearing capacity of the specimen decreases due to the large reduction of the cross-sectional area, leading to local necking and fracture.

The microstructure of the TNW700 alloy is examined in the central area of the specimen's gauge section. SEM images under various true strains (0.1, 0.25, 0.5, 0.75, 1.0, 1.3, 1.5, and 1.7) and at the point of fracture are presented in Fig. 4.10a–i. After superplastic deformation, the β phase is evenly distributed in the primary α phase matrix. As the strain increases, the volume fraction and grain size of the β phase increase. The volume fraction of β phase at each strain is 20.1%, 20.6%, 21.4%, 23.2%, 24.8%, 28.7%, 26.6%, 32.1%, and 33.2%, respectively. These results

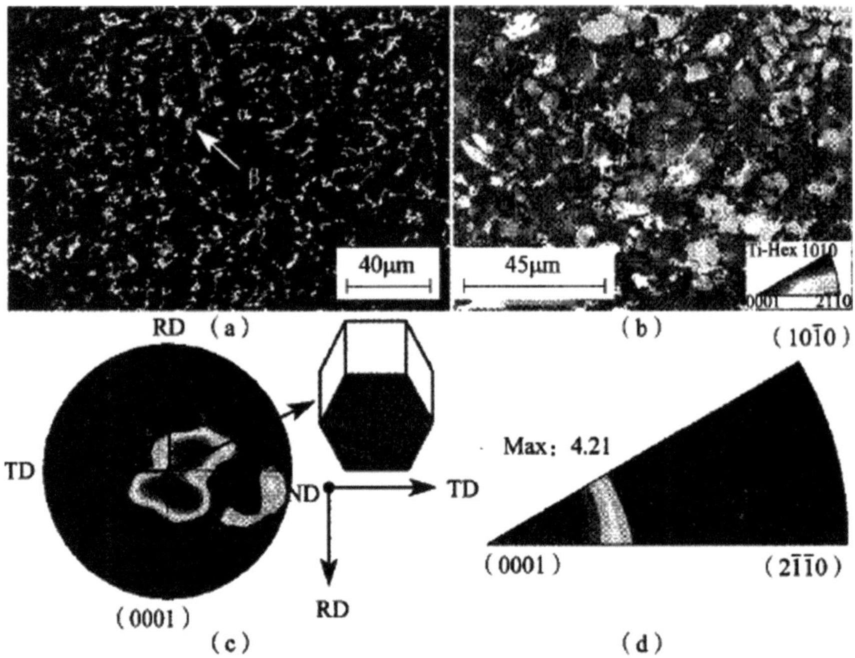

Fig. 4.8 Original microstructure of TNW700 alloy plate **a** SEM picture; **b** Grain orientation picture; **c** (0001) Pole picture; **d** Reverse pole picture

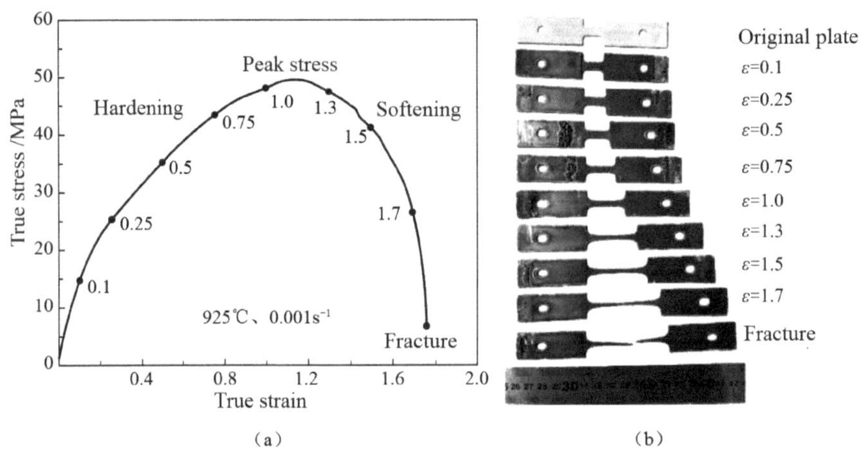

Fig. 4.9 True stress—true strain curve and corresponding specimen of TNW700 alloy superplastic deformation test at 925 °C and $0.001\,s^{-1}$ **a** True stress—true strain curve; **b** Specimen morphology

indicate that α → β dynamic phase transformation occurs during superplastic deformation, where the stress or strain can induce α phase to transform to β phase. This allotropic transformation of titanium and titanium alloys often occurs below β transformation temperature, and the driving force is related to the net softening between α phase (hardening phase) and β phase (softening phase). During the phase transformation, α phase and β phase follow the Burgers orientation relationship, i.e., (110)β//(0002) α, [111]β//[1120] α. Dynamic phase transformation is an essential flow softening mechanism that coordinates deformation, releasing stress concentration and restraining the formation of pores, thereby enhancing the superplasticity of the material. This mechanism is in contrast to the continuous strain hardening behavior of TNW700 titanium alloy when the deformation strain is less than 1.13. Therefore, it is hypothesized that β grain dynamic growth may be the reason for the strain hardening of TNW700 titanium alloy.

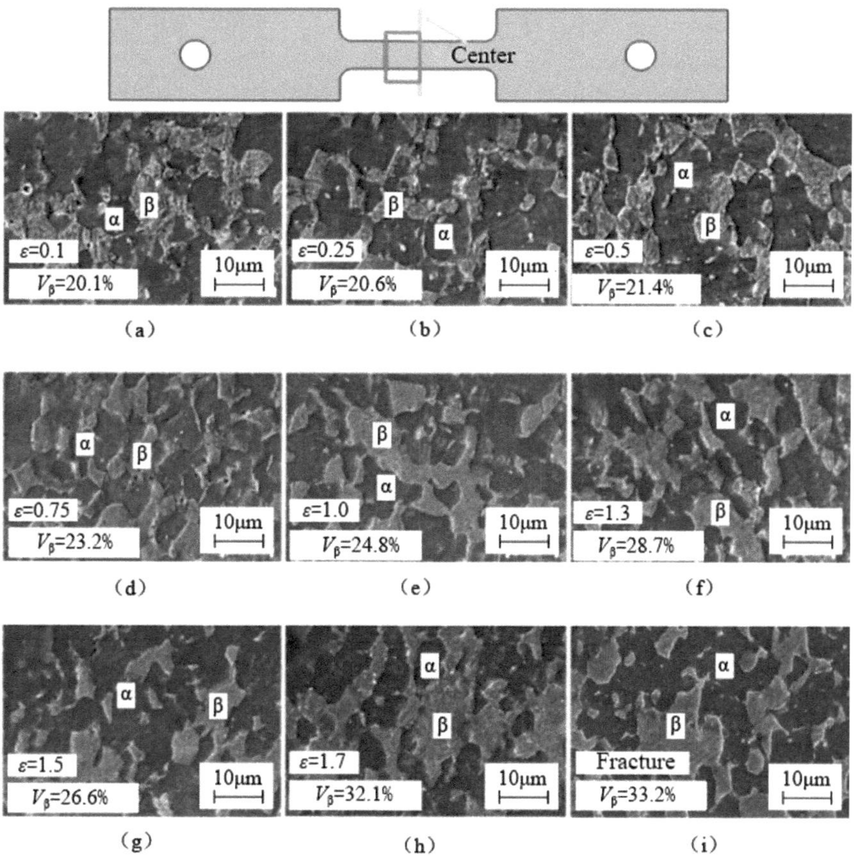

Fig. 4.10 SEM pictures of TNW700 alloy under different strains after superplastic deformation at 925 °C and $0.001 s^{-1}$

Figure 4.11 shows the microstructure of TNW700 alloy after superplastic deformation studied using TEM. The primary α phase in the specimen remains equiaxed after water quenching, while the prior β phase has transformed completely into martensitic α' lath, which is also called secondary α_s phase. This transformed β phase can be observed. The α' martensite includes {334} hexagonal martensite and {344} hexagonal martensite, which have the same HCP lattice as the primary α phase with similar lattice constants. The {334} hexagonal martensite generally has a lamellar morphology without twinning, and it is commonly found in pure titanium and titanium alloys with low element content. The {344} hexagonal martensite has a serrated morphology with high density of dislocations, stacking faults, and twinnings in the martensite plate, as frequently observed in titanium alloys with high β phase stable elements. Martensitic transformation is a non-diffusion phase transformation that occurs during cooling of titanium alloy from a high temperature, which is a solid phase transformation controlled by interface migration. During the phase transformation, there is only lattice reconstruction but no atomic diffusion. Additionally, α' phase and β phase still follow Burgers orientation relationship, and they are both considered as high-temperature β phase.

Figure 4.12 shows the EBSD graph (image quality) and grain boundary distribution characteristics of TNW700 alloy at 925 °C, $0.001\,s^{-1}$, and strains of 0.25, 1.0, 1.5, and 1.75 (fracture). The image uses circles, ellipses, and square boxes to represent small-angle (1°-5°) grain boundaries, medium-angle (5°-15°) grain boundaries, and large-angle (>15°) grain boundaries, respectively. Small-angle and medium-angle grain boundaries refers to the case when the angle between atomic arrangement of two adjacent grains are very small, and the grain boundary between two grains is composed of the fully matched part and the mismatched part. On the other hand, a large-angle grain boundary refers to the case that the atomic arrangement on the grain boundary is close to disorder. The grey and black areas in the graphs correspond to α phase transformed β phase. It is important to note that the secondary α_s phase and

Fig. 4.11 TEM picture of β phase after superplastic deformation at 925 °C and $0.001\,s^{-1}$

primary α phase in the transformed β phase have the same crystal structure. However, EBSD technology faces difficulty in accurately measuring the volume fraction and grain size of α phase and β phase in a water-cooled quenched specimen. Therefore, this book uses the secondary α_s phase to quantify the volume fraction and grain size of high-temperature β phase.

According to the IQ picture in Fig. 4.12, the average grain size of β phase increases with the increase in strain. Under four different strains, the average grain size of β phase is about 4.0 μm, 4.91 μm, 5.87 μm, and 5.9 μm. The reason for the large grain size of β phase at high strain is related to the increase in heat exposure time. The dynamic growth of β-phase grains in TNW700 alloy is the main cause of strain hardening of the material. However, excessive grain growth or coarsening will lead to decrease in superplasticity. The grain axial ratio (GAR) refers to the ratio of longitudinal grain size to lateral grain size, which is usually used to characterize the grain shape. Grains with a GAR less than 2 are defined as equiaxed grains. When strain is below 1.5, the average GAR of β-phase grains remains basically unchanged, with a value of about 2.2, which means that β-phase grains basically remain quasi-equiaxed. As the strain rises from 1.5 to 1.76, the average GAR of β-phase grains increases rapidly from 2.23 to 2.96. From the IQ picture, it can be seen that the increase in β-phase grain axial ratio is related to its aggregation and elongation along the tensile direction, as shown in the ellipse in Fig. 4.12d.

Compared to the primary α phase, the β phase is relatively soft and has more active slip systems, making it serve as a "lubricant" during superplastic deformation. Moreover, the presence of the β phase can restrict the dynamic growth of grains and

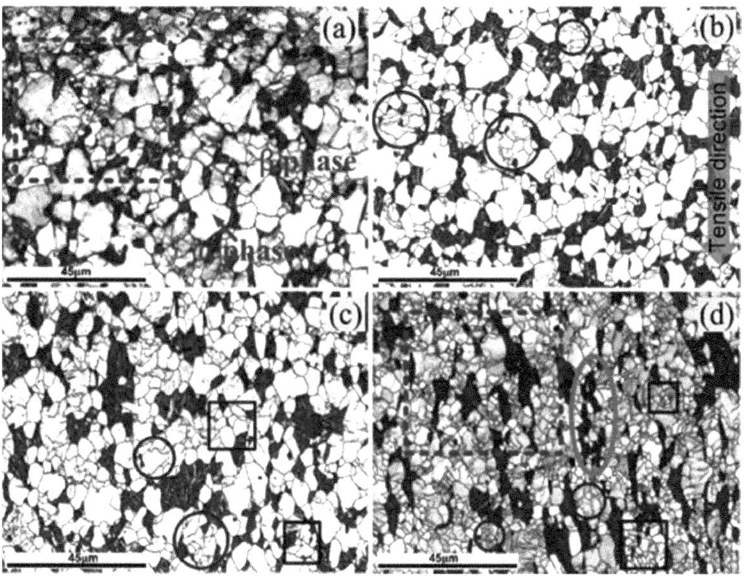

Fig. 4.12 EBSD forming quality (IQ) under different strains **a** 0.25; **b** 1.0; **c** 1.5; **d** 1.75

the aggregation and combination of holes. It can also assist the regulation of deformation inside the grains during superplastic deformation, by means of dislocation creep and diffusion creep. Therefore, the volume fraction, grain size, and distribution of the β phase are critical to the superplastic deformation mechanism and maximum elongation of titanium alloys. In Fig. 4.13, the nucleation and growth of the β phase in TNW700 alloy are illustrated. Figure 4.13a, e are schematic diagrams of nucleation and growth, while Fig. 4.13b–d and f–h are the corresponding TEM graphs. The β-phase grain nucleation is found in a small amount of α-phase grain, i.e., intragranular β phase, as shown in the ellipses in Fig. 4.13a, c. Furthermore, β-phase grains are found at the grain boundary (ellipse) or triple line boundary (circle) of α phase, which are called grain boundary β phase, as shown in Fig. 4.13a, b, d. The preferential nucleation of β-phase grains along the grain boundary is due to the small driving force required for nucleation.

Studies on the phase transformation mechanism of pure titanium using in-situ EBSD technology also found that β-phase grains tend to nucleate at the grain boundary or inside the α-phase grains. However, intragranular β-phase grains are unstable and will be gradually merged by grain boundary β phase during growth. Additionally, the nucleation of intragranular β phase often occurs near the {334} habit plane and maintains a Burger orientation relationship with the α-phase parent grain, while grain boundary β-phase has no specific habit plane and maintains a Burger orientation relationship with only one of the adjacent α-phase grains. The growth mode of grain boundary β-phase grain is shown in Fig. 4.13e–h): ① grow along the α phase grain boundary to the triple line boundary (Fig. 4.13f); ② grow by directly wedging into α-phase grain (Fig. 4.13g). Both modes depend on the disordered diffusion of atoms. Intragranular β-phase grain grows by dissolving and consuming the α-phase grains around it. Additionally, competitive growth exists between intragranular β phase and grain boundary β phase, and the high mobility

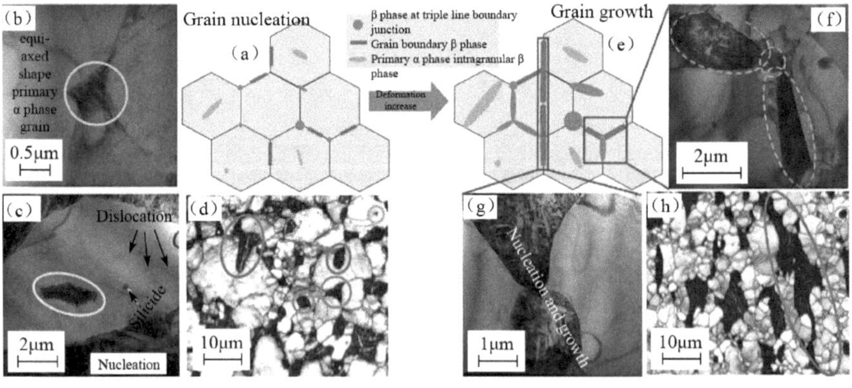

Fig. 4.13 Nucleation and growth of β-phase grain **a** Schematic of grain nucleation. β-phase grains nucleated at **b** triple line boundary junction, **c** inside α-phase grain, **d** at grain boundary of α phase. **e** Schematic of grain growth. Grain growth **f** at triple line boundary junction, **g** by wedging into α-phase grain and **h** by aggregation along the tensile direction

of grain boundary makes the latter more dominant. Diffusion and redistribution of β phase stable elements during high-temperature deformation are other possible causes of β-phase grain nucleation and growth. Furthermore, the grain boundary β-phase agglomerate along the tensile direction, as observed in Fig. 4.13h. This indicates that stress/strain can induce the nucleation, growth, and redistribution of the β phase during superplastic deformation.

To clarify the changes in texture of TNW700 alloy during superplastic deformation, Figs. 4.14 and 4.15 show the orientation map of primary α phase grains and the corresponding inverse pole figure (IPF) and (0001) pole figure (PF) under different total strain. Texture changes generally involve the gradual rotation and displacement of the slip plane or slip direction relative to the main deformation direction. IPF and PF are crystal texture characteristics obtained by projecting the crystal coordinate system into the specimen coordinate system. Figure 4.14 shows that the primary α phase grains of TNW700 alloy after superplastic deformation have a certain preferred orientation, with most of the grains being red, indicating that the base plane (0001) is perpendicular to the ND direction of the specimen and has the same texture as the original plate.

Figure 4.15 shows that under four strains, the maximum texture intensity of primary α phase grains is 7.22, 6.33, 6.34, and 4.53, respectively, which is lower than

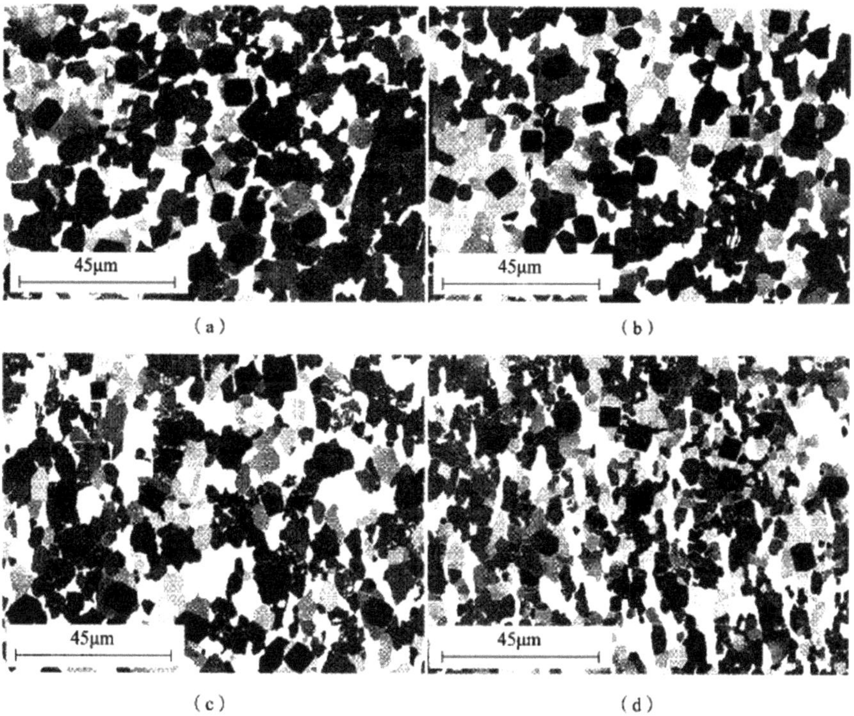

Fig. 4.14 Grain orientation of TNW700 alloy under different strais **a** 0.25; **b** 1.0; **c** 1.5; **d** 1.75

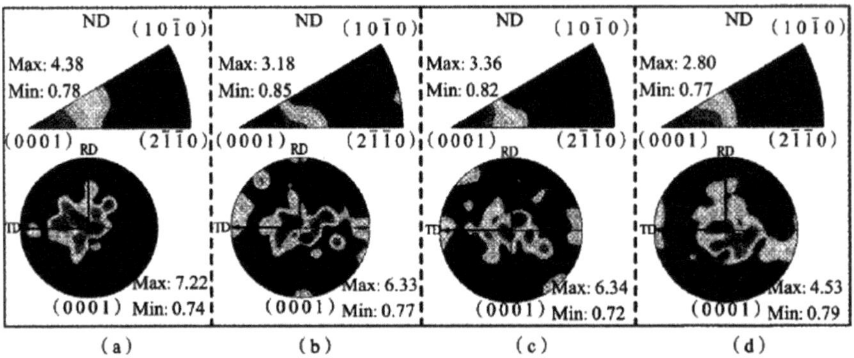

Fig. 4.15 Inverse pole figure (IPF) and (0001) pole figure of specimen under different strains **a** 0.25; **b** 1.0; **c** 1.5; **d** 1.75

that of the original material (8.47). Grain boundary sliding generally does not change the crystal orientation of the material. The decreasing maximum texture intensity of the material with increasing strain indicates that grain orientation tends to gradually randomize, which suggests that grain boundary sliding may be accompanied by grain rearrangement or rotation. This phenomenon has also been observed during superplastic deformation of Ti55 and TA32 alloys. Moreover, during the superplastic deformation of TC4 alloy, grain boundary/phase boundary sliding can increase the texture composition of α phase and further weaken the texture intensity.

The TEM analysis of TNW700 alloy after superplastic deformation provides further insight into the dislocation evolution. Figure 4.16 displays the bright field graph of the fracture zone after deformation at 925 °C and $0.001s^{-1}$. Square dislocation grids consisting of small angle grain boundaries are observed in Fig. 4.16a, which is a typical characteristic of early sub-grain formation. Additionally, dislocations emanating from grain boundaries or corners and terminating on the opposite side of grains are observed. Under the effect of thermal activation, dislocations intertwine to form a cellular structure, as shown in Fig. 4.16b, and a significant number of movable free dislocations can be seen in the cellular structure, which is another common feature of early sub-grain formation. As strain increases, the cell wall created by dislocation intertwining will evolve into a regular dislocation grid with lower energy or small angle grain boundaries. Free dislocations in the cell then enter the surrounding small-angle grain boundary via dislocation annihilation or rearrangement. The primary α-phase grain has some equidistant and parallel dislocation arrays, as indicated by the arrow in Fig. 4.16c, indicating active dislocation movement during superplastic deformation. The plane slip of dislocation is the primary deformation mechanism of the primary α-phase grain. Intragranular dislocation movement can provide strain capacity for deformation, which is beneficial to superplastic deformation. Moreover, parallel or intertwined dislocations can be rearranged to form LAGB grain boundaries. Figure 4.16c also shows dislocation walls and substructures formed by dislocation intertwining, indicating dynamic polygonization and recovery of primary α-phase grains. The formation of dislocation walls is generally linked to the slip or climb of dislocations around the dislocation band. Figure 4.16d reveals some equiaxed α-phase grains without dislocations, confirming the continuous dynamic recrystallization of α-phase grains. The evolution of dislocation in β-phase during superplastic deformation is challenging to analyze and determine, since β-phase undergoes a polymorphic transformation when cooled from high temperature.

4. **TC21 alloy**

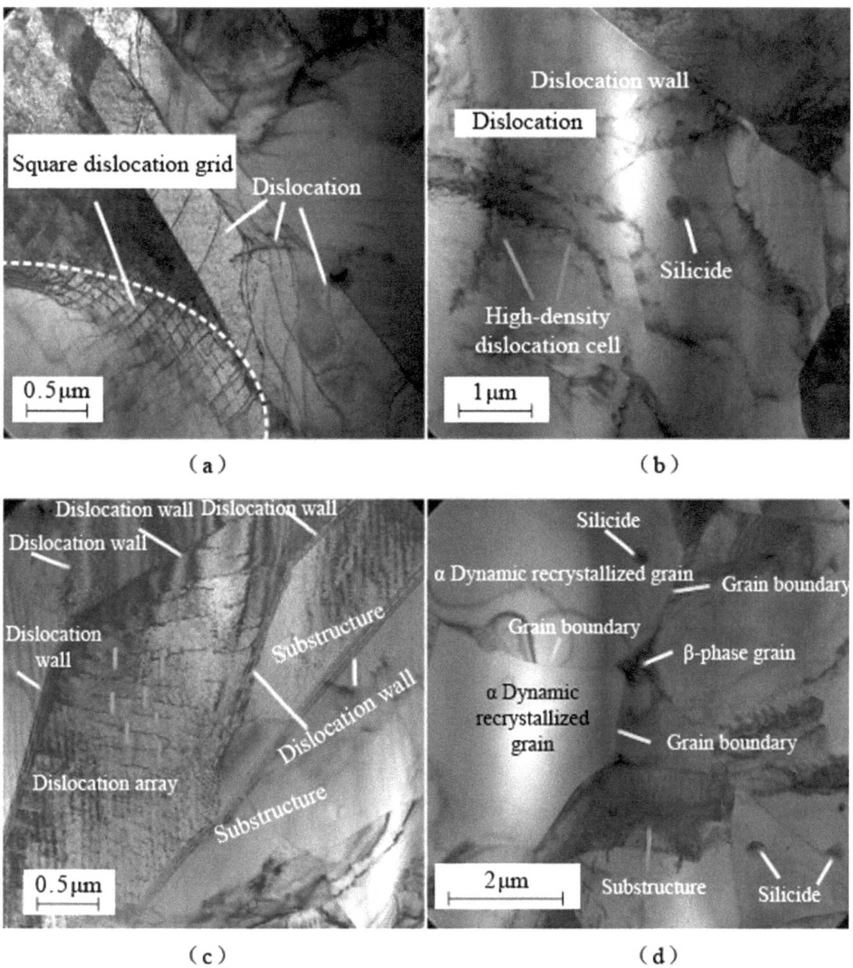

Fig. 4.16 High-resolution TEM graph of TNW700 alloy specimen (bright field image)

TC21 alloy is a titanium alloy that presents high strength, high fracture toughness, low crack growth rate, excellent fatigue and welding properties, making it suitable for manufacturing large integral components. In this research, a fine-grained TC21 with a grain size of about 4μm is used. Superplastic deformation tests are conducted at temperatures ranging from 860 to 950 °C and strain rates of $5 \times 10^{-4} \sim 1 \times 10^{-3} s^{-1}$. The results indicate that the TC21 alloy fine-grained material specimens did not fracture and experienced uniform deformation without necking, achieving an elongation of 1240% at 890 °C and a strain rate of $5 \times 10^{-4} s^{-1}$.

TEM observations of the microstructure of TC21 alloy at 860 °C and $1 \times 10^{-3} s^{-1}$ are shown in Fig. 4.17. During superplastic deformation, the grain boundary of α-phase grains widens and becomes arced, the grains are equiaxed but uneven in size (as indicated by the arrow in Fig. 4.17a), which are typical characteristics of dynamic recrystallization. Dislocations in grain 1 in Fig. 4.17a tend to move towards and aggregate at the grain boundary. The dislocation density at the grain boundary is higher than that inside the grain. When a dislocation encounters an obstacle during movement, climb generated by thermal activation helps it bypass the obstacle, and dislocation arrays form from slip dislocations under the action of climb (Fig. 4.17b). These arrays then polygonize and continuously absorb the distortion energy generated by high temperature deformation, evolving into small-angle grain boundaries or even large-angle grain boundaries. No obvious dislocation characteristics were observed in the deformed specimen under a low strain rate of $5 \times 10^{-4} s^{-1}$.

Fig. 4.17 TEM microstructure of TC21 alloy showing **a** Recrystallization; **b** Dislocation configuration

4.1.2 Diffusion Bonding Interface Microstructure

Diffusion bonding is a process that creates a strong and reliable bond between metal materials by fitting their surfaces closely and applying a specific temperature and pressure. The bonding occurs through the diffusion of atoms between the material surfaces, forming a bonded interface. The process involves complex physical and chemical changes that occur in several stages. Firstly, the surfaces to be diffused have an uneven profile. As the surfaces come into contact, the wave crests make contact first, and the non-contact parts form holes at the interface. Under pressure, elastic–plastic deformation and creep deformation occur on the contact points, causing the interface holes to shrink and the actual contact area to increase continuously. When the bonding interfaces are close enough, atomic interaction occurs, and the grains come into contact to form a relatively stable physical contact. The next stage involves the activation of surface atoms under the diffusion bonding pressure and temperature, causing the atoms to move from their original position to a new equilibrium position. This movement results in the formation of chemical bonding, such as metal bonds, between the interfaces. The mutual diffusion is then initiated under the ongoing effect of temperature and pressure, grains are continuously formed at the diffusion bonding interface, and grain boundaries are subjected to changes and migration. The contact interfaces gradually become unrecognizable and a strong bonding is achieved.

During diffusion bonding, the oxide film on the surface of titanium alloy can be dissolved in the base metal at high temperature, which does not hinder the bonding process. When titanium alloys with the same composition are diffusion bonded under appropriate process conditions, no original interface will be observed. The range of diffusion bonding parameters for titanium alloys is wide, with the pressure ranging from 1-10MPa, time ranging from several minutes to tens of minutes, and temperature generally between 800 and 1000 °C. If the diffusion bonding time is too short, many holes may remain in the interface. If the temperature is too high or the time is too long, the grains in the interface and base metal may grow, affecting the performance. Pressure is also an important factor, affecting the first and second stages of solid-phase diffusion bonding. If the pressure is too low, the plastic deformation is insufficient, the physical contact on the surface will be incomplete, and the residual pores on the interface will be too large and too many. A higher pressure produces greater surface plastic deformation, and lower the surface recrystallization temperature to accelerate grain boundary migration. Higher pressure is helpful to the shrinkage and elimination of micropores in the second stage of solid phase diffusion bonding, and can also reduce or prevent diffusion pores in the diffusion bonding of dissimilar metals. When other parameters are fixed, higher pressure can produce better diffusion bonding joint, and the upper pressure limit depends on the limit of total deformation of diffusion bonding part and equipment tonnage. On the premise of ensuring better interface joint microstructure and properties, factors such as temperature, pressure, time and surface state should be considered comprehensively to reduce costs and improve efficiency.

The original grain size of titanium alloy also affects diffusion bonding. A finer original grain size requires less time and lower pressure to achieve a good diffusion bonding interface. Therefore, the original microstructure of superplastic forming/diffusion bonding titanium alloys should be fine-grained. In the following section, the microstructures of diffusion bonding interfaces of some typical titanium alloys will be introduced.

1. **TC4 alloy**

The material used for diffusion bonding is a 1.5mm thick TC4 alloy plate, and the process parameters and welding rate results are shown in Table 4.1. The table reveals that as the diffusion bonding pressure or time increases, the welding rate also increases, provided that the temperature remains constant. At a higher temperature of 910 °C, lower pressure and less time are needed to achieve similar bonding rates. For instance, at 880 °C, 2.0MPa pressure, and 90 min heat preservation time, the welding rate exceeds 95%. By raising the temperature to 910 °C and maintaining the pressure for 90 min, a welding rate of over 95% can be achieved at a lower pressure of 1.5MPa. The interface microstructure shows no signs of the original interface, indicating complete welding has been achieved. Figure 4.18 shows the microstructure at the interface under different diffusion bonding parameters. The grain size of the plate is consistent, with obvious short black lines in the unwelded area of the interface and complete recrystallization in the welded area, forming the same grains as the base material. The grains have an equiaxed microstructure. The analysis indicates that to improve the diffusion bonding effect, the temperature should be maintained between 880 and 910 °C, the pressure should not exceed 2MPa, and the heat preservation time should not exceed 90 min. If the temperature is low, the pressure and heat preservation time should be increased, whereas if the temperature is high, the pressure or heat preservation time can be reduced accordingly.

2. **TA32 Alloy**

In the diffusion bonding test of TA32 alloy, the temperature was varied from high to low while keeping the diffusion bonding time and pressure constant at 2 h and 2 MPa, respectively. The temperature range for diffusion bonding was set between 880 and 940 °C. Figure 4.19 shows the microstructure at the diffusion bonding interface. The

Table 4.1 Relationship between diffusion bonding process parameters and welding rate

Temperature/°C	Pressure/MPa	Time/min	Welding rate/%
880	1.0	60	70
	1.5	60	85
	1.5	90	95
	2.0	90	> 95
910	1.0	30	85
	1.0	60	95
	1.5	90	> 95

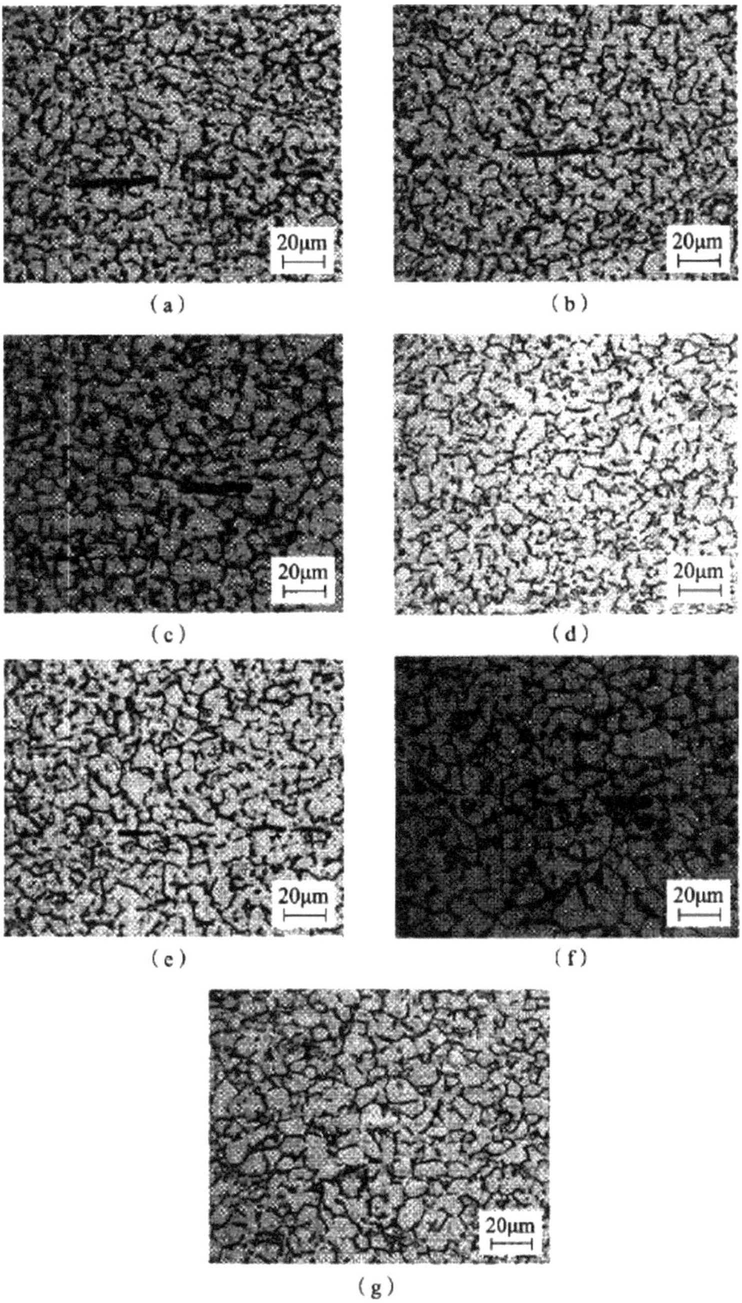

Fig. 4.18 Microstructure morphology at interface under different diffusion bonding conditions **a** 880 °C, 1.0MPa, 60min; **b** 880 °C, 1.5MPa, 60min; **c 880** °C, 1.5MPa, 90min; **d 880** °C, 2.0MPa, 90min; **e 910** °C, 1.0MPa, 30min; **f 910** °C, 1.0MPa, 60min; **g 910** °C, 1.5MPa, 90min;

Figure shows that at 880 °C, 2MPa, and 2 h, the TA32 alloy exhibited good diffusion bonding results, but some dispersed unwelded spot defects were present. At 900 °C, 2MPa, and 2 h, complete welding was mostly achieved, with only a few unwelded spot defects. At 940 °C, 2MPa, and 2 h, complete welding of the TA32 alloy was achieved. The grains at the interface were the same as those of the base metal on both sides of the interface, with no trace of the original interface observed in the interface microstructure. Observation of the microstructure under the above diffusion bonding process conditions revealed that the grain morphology at the welding interface was consistent with that of the base metal on both sides of the interface, still appearing equiaxed. However, as the diffusion bonding temperature increased, the grain size at both sides of the interface and at the diffusion bonding interface also increased, resulting in better diffusion bonding effects.

3. **Diffusion bonding of TC4/TB8 dissimilar titanium alloy**

Diffusion bonding of dissimilar titanium alloys involves the solid phase joining of two different titanium alloys by diffusion bonding. In this research, we examine the diffusion bonding between TC4 alloy and TB8 alloy. TB8 alloy has a nominal composition of Ti-15Mo-3Al-2.7Nb-0.25Si and is a metastable β titanium alloy characterized by good high temperature strength and creep resistance, high hardenability, excellent oxidation resistance and corrosion resistance, as well as good cold workability. After age strengthening, its tensile strength can be significantly improved. TB8 alloy is suitable for manufacturing aircraft or engine structural parts, honeycombs, fasteners, and hydraulic pipes working at elevated temperatures, and can also serve as the matrix of metal-matrix composites.

The diffusion bonding test is carried out under 870 °C, 2MPa, and 1.5h, and the microstructure of the diffusion bonding interface of the two different titanium alloys with a thickness of 1.0mm is shown in Fig. 4.20. The Figure shows that full diffusion bonding of the two titanium alloys is achieved under these parameters, and a reaction layer with a width of about 10μm appears on the TB8 alloy side. The reaction layer has a needle-like β transformed microstructure, which is different from the typical solid solution treated microstructure of TB8 alloy, represented by coarse β grains. The secondary α phase in the reaction layer, indicates change in chemical composition due to the diffusion of alloy elements. The α phase stable elements diffuse into the TB8 alloy side, forming $\alpha + \beta$ alloy with rich β phase. Generally, secondary α phase formed during furnace cooling is coarse in shape, as the slow cooling rate provides sufficient time for precipitation and growth. In the present study, however, secondary α phase precipitated in the reaction layer is relatively fine. Furthermore, a very thin α phase layer is observed on the TC4 alloy side, and there is almost no β phase. This is also attributed to the local composition change caused by alloy element diffusion.

Electron probe microanalysis (EPMA) is conducted to characterize the diffusion bonding interface as well as the base metals on both sides, with a focus on the distribution and content of Mo, Nb, Si, V, and other elements. Figure 4.21 shows the results, A1 and Ti are found to have a similar distribution, with higher content in TC4 alloy than in TB8 alloy. As a result, an element transition zone is formed at the interface. Meanwhile, a 10μm transition zone also appears at the TB8 alloy

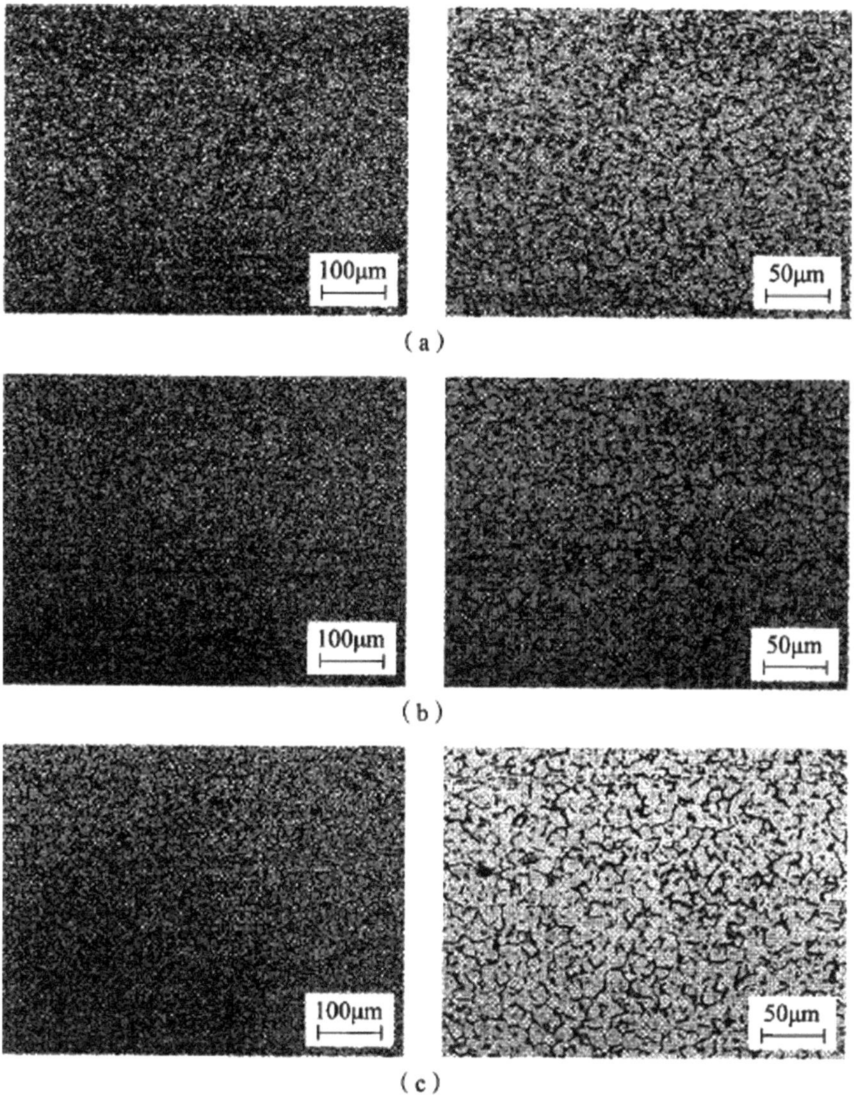

Fig. 4.19 Microstructure at diffusion bonding interface of TA32 alloy under different process conditions **a** 880 °C, 2MPa, 2h; **b** 900 °C, 2MPa, 2h; **c** 940 °C, 2MPa, 2h

side. V diffuses to the transition zone on the TB8 alloy side, and there is a narrow layer with poor V element at the interface, which is about 2μm wide. Si, Nb, and Mo only exist in TB8 alloy, Si diffuses from TB8 alloy to TC4 alloy with an uniform gradient, the distribution of which is not affected by the reaction layer. Si has an atomic size smaller than that of Ti can dissolve in α and β phases. The analysis results indicate that the solid solubility of Si is similar in both phases. The addition

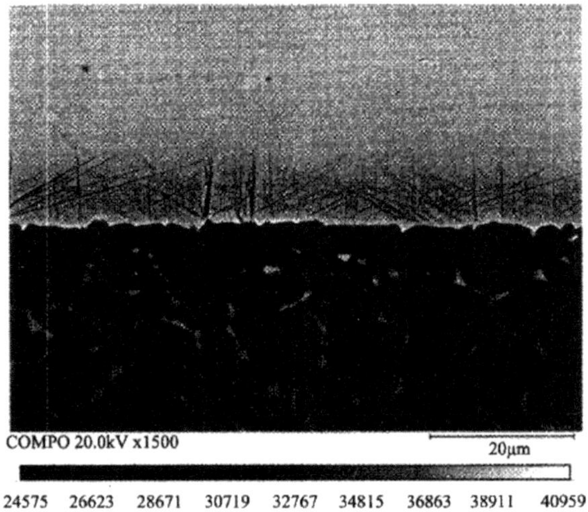

Fig. 4.20 Microstructure of diffusion bonding interface between TC4 alloy and TB8 alloy

of Si can improve oxidation resistance and form silicide to improve high-temperature strength. Compared with Si, there are significant differences in distribution in the reaction layer for Nb and Mo. A clear boundary is observed in the reaction layer, indicating that the ability of Nb and Mo to diffuse across the interface is relatively weak.

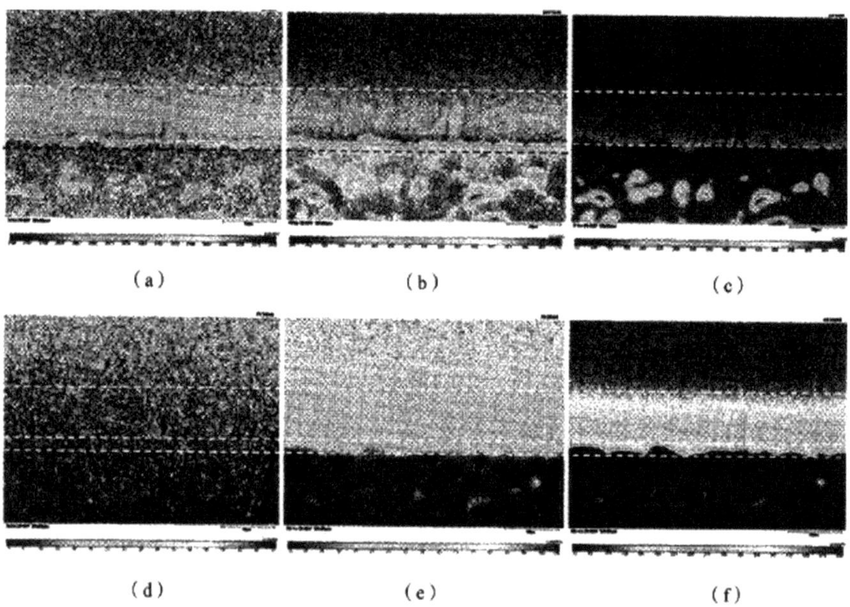

Fig. 4.21 EPMA results of diffusion bonding interface between TC4 alloy and TB8 alloy **a** Ti element; **b** Al element; **c** V element; **d** Si element; **e** Nb element; **f** Mo element

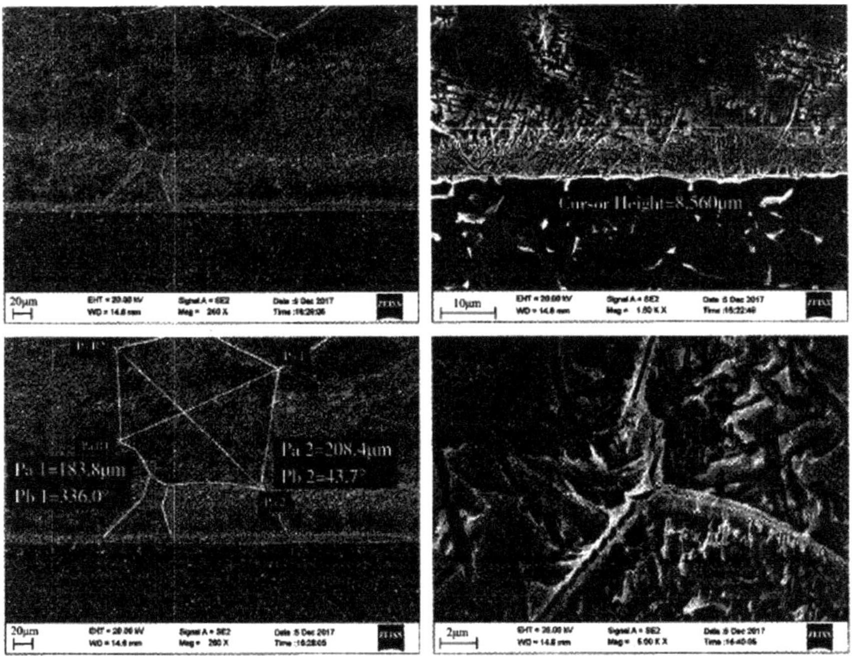

Fig. 4.22 SEM image of diffusion bonding interface of TB8/TC4 alloy

The microstructure of the diffusion bonding interface at a lower temperature of 850 °C is depicted in Fig. 4.22. It can be observed that TB8 and TC4 alloys are also completely bonded at this temperature, and the thickness of the interface reaction layer is around 8.5μm. Notably, the temperature is higher than the phase transformation temperature of TB8 alloy (830 °C), which results in the growth of TB8 alloy to a grain size of over 200μm, while the morphology of TC4 grains remains unchanged.

4.2 Defects in Superplastic Forming/Diffusion Bonding Process

Pores are microstructural defects that commonly occur during the superplastic deformation process. The behavior of pores generally involves four stages: nucleation, growth, bonding, and fracture. In diffusion bonding, interface defects such as lack of welding and weak bonding can occur, which can negatively impact the structure's bearing capacity. Therefore, it is crucial to study and control both pores and other diffusion bonding defects in order to optimize the material's performance.

4.2.1 Pores and Fractures

Pores can be categorized into two types based on their shapes: wedge-shaped pores, also known as V-shaped pores, which form at the junction of three grains due to stress concentration; and circular pores, also known as O-shaped pores, which appear along the grain boundary and phase boundary, often in a circular or elliptical shape. The O-shaped pores usually appear on the grain boundary or phase boundary perpendicular to the tensile stress and on the grain boundary with a ridge, formed by the converging and aggregation of supersaturated vacancies towards the grain boundary or phase boundary. V-shaped pores tend to appear under high stress, while O-shaped pores under low stress. The surface area of V-shaped pores is larger than that of O-shaped pores in the same volume, which means a larger stress is required for their formation, as the energy for their formation is proportional to the surface area. Once the V-shaped pore is formed, it transforms into the O-shaped pore by diffusion at high temperature, releasing some energy, since the energy of the V-shaped pore is higher than that of the O-shaped pore (in the same volume). Pores can be generated not only under tensile stress but also under pressure stress. This indicates that shear stress plays a significant role in the generation of pores. However, it is not easy to generate pores under pressure stress, especially in case of superplastic deformation under higher spherical tensor pressure stressThe nucleation of pores is a relatively complex process as pores may exist in the raw materials or be generated in the superplastic deformation process. Pores usually form on the grain boundary, especially at the triple line boundary junction or the second-phase particles. Local stress concentration caused by grain boundary sliding can be coordinated by diffusion or dislocation movement. However, if the generated stress exceeds the speed of stress relieving, pore nucleation will occur. This pore nucleation process at the triple line boundary junction caused by grain boundary sliding is depicted in Fig. 4.23. An increase in the external stress may exceed the critical stress of nucleation in many positions, thereby accelerating pore nucleation. Research on A1-6.3Cu-0.29Mn shows that when the superplastic tensile strain is about 20%, pores are found at the triple line boundary junction, large particles on the grain interface, and other grain boundary corners. These pores are considered as products of uncoordinated grain boundary sliding, as the hardness of the precipitated CUA12, T phase, and AlFeSi of the alloy is much higher than that of the matrix phase. Stress concentration is generated on the matrix interface around the large particles that are not easy to deform, resulting in uncoordinated deformation and pore formation when large particles are encountered in grain boundary sliding. Intragranular slip helps relax the stress concentration caused by grain boundary sliding and thus helps to restrain the generation and propagation of pores to a certain extent. The generation of pores can also relax the stress concentration at the grain boundary and coordinate the grain boundary sliding, as shown in Fig. 4.24.

To understand the pore nucleation conditions at the triple line boundary junction caused by grain boundary sliding, Stroh proposed the following conditions using Zener's hypothesis, which states that cracks are generated at the triple line boundary junction when grain boundary sliding occurs under shear stress:

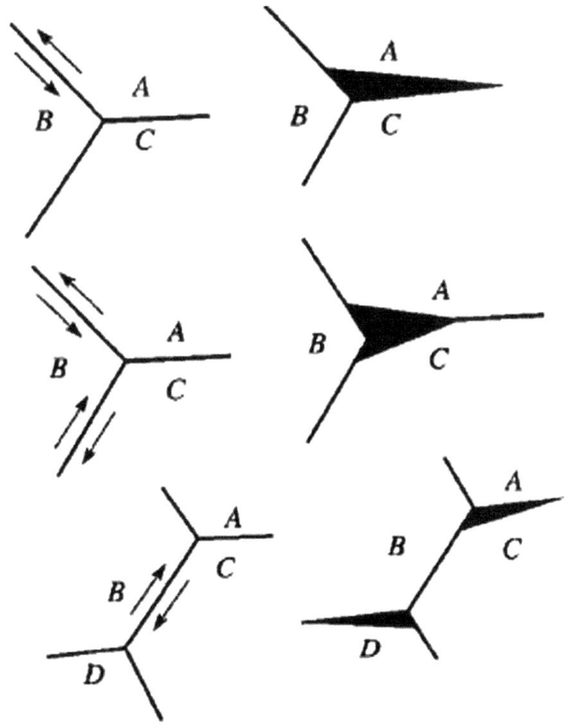

Fig. 4.23 Pore nucleation at triple line boundary junction caused by grain boundary sliding

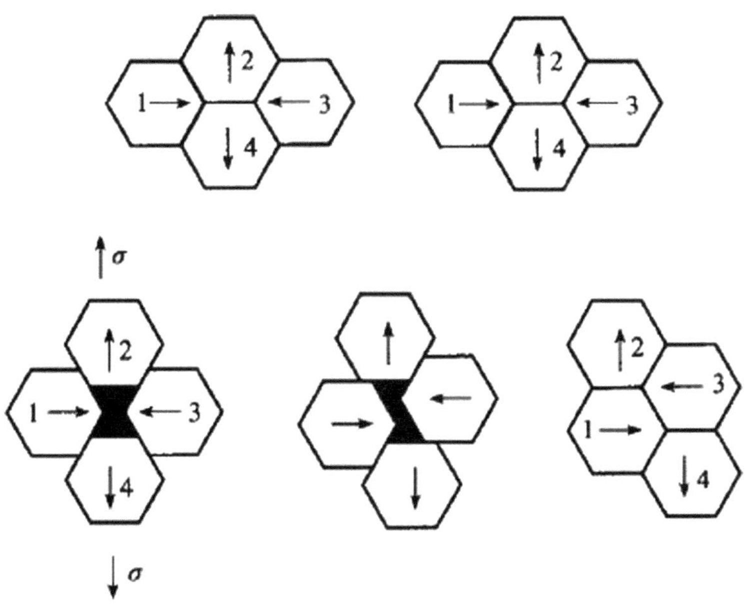

Fig. 4.24 Pore coordinating grain boundary sliding

$$\tau^2 > \frac{12\nu G}{\pi L}$$

where: τ is the shear stress; ν is the surface energy of the pore; L is the length of slip crystal face (i.e., the distance between the grain boundary protrusions or the distance between the grain boundary particles); G is the shear modulus.

Baj, Ashby, Yoo and Trinkaus proposed a simple high-temperature deformation pore nucleation rate function as follows:

$$\frac{\mathrm{d}N}{\mathrm{d}t} = c(N_{\max} - N)$$

where: c is the nucleation rate coefficient; $N_{\max}$ is the maximum number of possible nucleation; N is the number of newly nucleated pores; t is the time.

Raj and Ashby did a in-depth research on the pore nucleation using classical theory, and proposed a pore nucleation model:

Where: r_c is the critical pore nucleation radius; ν is the surface energy of the pore; σ is equivalent stress. Under superplastic deformation, the critical diameter of pore nucleation is about 0.2–1.0 μm.

During superplastic deformation of materials, it is widely accepted that the primary mechanisms of pore growth involve stress-induced growth along grain boundaries and plastic deformation of the material surrounding the pores. At the early stages of nucleation or growth, pores are primarily controlled by vacancy diffusion, while after the strain increases, pore growth may be mainly controlled by grain boundary sliding. For the growth of diffusion pores, Beere and Speight proposed a growth rate controlled by diffusion:

$$\frac{\mathrm{d}r}{\mathrm{d}\varepsilon} = \frac{2\Omega\delta_{gb}D_{gb}}{kT} \cdot \frac{1}{r^2} \cdot \left(\frac{\sigma - 2\nu/r}{\dot{\varepsilon}}\right) \cdot \alpha$$

where, Ω is the atomic volume; δ_{gb} is the grain boundary width; r is the pore radius; ν is surface energy; ε is true strain; σ is the flow stress; T is the thermodynamic temperature; $\dot{\varepsilon}$ is the strain rate; k is the Boltzmann constant; D_{gb} is the grain boundary diffusion coefficient; α is the coefficient of pore size and spacing, and the value is

$$\alpha = \frac{1}{4\ln\left(\frac{\lambda}{2r}\right) - \left[1 - \left(\frac{2r}{\lambda}\right)^2\right]\left[3 - \left(\frac{2r}{\lambda}\right)^2\right]}$$

where, λ is the distance between pores.

Chokshi conducted an extensive study on the role and impact of pores during the superplastic deformation process. He proposed a superplastic diffusion growth model of pores for situations where the pore size is larger than the grain size, based on the fact that grains of the material for superplastic deformation are generally fine:

$$\frac{\mathrm{d}r}{\mathrm{d}\varepsilon} = \frac{45\Omega\delta_{\mathrm{gb}}D_{\mathrm{gb}}}{d^2kT} \cdot \frac{\sigma}{\dot{\varepsilon}}$$

where, d is the grain size.

The equation above indicates that the change in pore size with strain is not dependent on the initial pore radius but is inversely proportional to the square of the grain size. However, this model has certain limitations, including the requirement for low strain rates, moderate test temperatures (as pores formed by vacancy diffusion are more likely to appear at the grain boundary rather than in the lattice), and grain sizes smaller than 5 μm. These constraints somewhat limit its applicability.

During superplastic deformation, pores can grow either by aggregating vacancies under diffusion control or by pore surface strain caused by the average stress without involving vacancy flowing to the pores. Mc-Clintock proposed a model for the growth of cylindrical pores in power-law hardening materials, which Hancock subsequently developed into an exponential growth equation and a pore growth model controlled by plastic deformation. The fundamental assumption of this model is that pore growth is determined by the plastic deformation of the matrix surrounding the pores. The formula for this model is as follows:

$$\frac{\mathrm{d}r}{\mathrm{d}\varepsilon} = r - \frac{3\nu}{2\sigma}$$

The former term of the equation reflects the growth of pores caused by plastic deformation around the pores, and the latter term reflects the effect of surface growth and shrinkage of pores under low stress level. This model is widely used in superplastic forming to accurately predict the large expanded pores formed in creep and superplastic deformation.

The diffusion pore growth model proposed by Beere et al. and the plastic deformation-controlled pore growth model proposed by Hancock are illustrated in Fig. 4.25. The critical pore radius, r_c, which increases with decreasing strain rate, is also shown in the figure. When the pore radius grows to $\mathrm{r_c}$, the pore growth mechanism changes from diffusion-controlled to plastic deformation-controlled. At the superplastic deformation strain rate that enables maximum elongation, it is generally accepted that plastic deformation-controlled pore growth becomes the dominant mechanism when the pore diameter is larger than 1 μm.

Based on the pore growth model proposed by Hancoclc, Stowell put forward a equation for volume increase of pores controlled by plastic deformation:

$$\frac{\mathrm{d}\upsilon}{\mathrm{d}t} = \eta\upsilon\dot{\varepsilon}$$

where: $\frac{\mathrm{d}\upsilon}{\mathrm{d}t}$ is the change rate of pore volume υ with time t; η is the pore growth rate parameter; $\dot{\varepsilon}$ is the true strain rate.

Based on the theoretical analysis of pore growth, Cocks and Ashby gave the following relation of η:

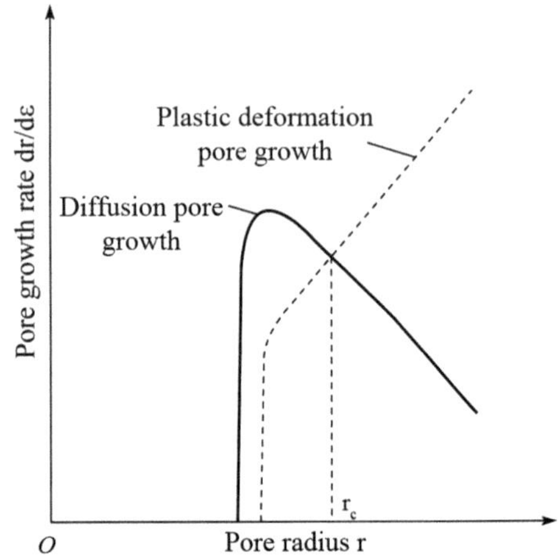

Fig. 4.25 Schematic diagram of pore growth mechanism

$$\eta = \frac{3}{2} \times \frac{m+1}{m} \times \sinh\left(\frac{2}{3} \times \frac{2-m}{2+m}\right)$$

where, m is the strain rate sensitivity index.

According to the research of Stowell, Cocks and Ashby, Pilling gave the following equation:

$$\eta = \frac{3}{2} \times \frac{m+1}{m} \times \sinh\left[\frac{2 \times (2-m)}{2+m} \times \left(\frac{K}{3} - \frac{\rho}{\sigma}\right)\right]$$

where: ρ is the additional stress; σ is the equivalent uniaxial flow stress; K is the constant varying with the stress state, which is 1–2 for unidirectional tensile and 2–2.5 for bidirectional tensile. The value of K depends on the degree of grain boundary sliding during deformation, with a lower value indicating no grain boundary sliding and a higher value indicating free grain boundary sliding. In superplastic forming processes, $K = 1.5$ and $K = 2.25$ generally represent uniaxial and biaxial deformation, respectively. During superplastic deformation, grain boundary sliding is the primary mechanism. However, if no other material flow process (such as diffusion creep or dislocation creep) can bridge the pores caused by grain boundary sliding, or if the bridging speed is slower than the pore generation speed, pores will form between the grains. Experiments show that the pore density increases with strain rate and grain size, indicating that higher strain rates and larger grain sizes make it more difficult to coordinate and adapt different deformation mechanisms during superplastic deformation. When a certain degree of small and dispersed pores are present in the material, it can be beneficial to the grain boundary slip. When the grain boundary slips to the triangular grain boundary, it becomes difficult to continue slipping, the pores can help with stress relaxation and improve plasticity. However, if there are numerous pores inside the material or if the pore size is relatively large, then there will be an increase in the number or intensity of stress concentration areas. If these stress concentrations cannot be relaxed in time, it will lead to a decrease in the m value, which characterizes the stress relaxation ability of the material. Consequently, the material's capacity to deform will be reduced.

Furthermore, a large number of holes, especially large V-shaped holes, in the material after superplastic forming can significantly reduce the strength performance and fracture toughness of the material. This poses a threat to the reliability of the formed part and can limit its application. Although the size and number of pores typically increase with the strain capacity, materials with superplasticity have high strain rate sensitivity, which allows them to resist necking. Therefore, even if there is a thin bonding between pores, they can still resist fracture and show high pore tolerance at the macro level. The final fracture of materials is usually caused by the aggregation or bonding of pores. Before large-area pore bonding occurs, pores can coordinate grain boundary sliding and improve elongation. However, when large-area pore bonding occurs and the specimen section is reduced beyond its capacity to withstand external stress, fracture will occur.

Taplin and Smith proposed a general model of superplastic fracture, as shown in Fig. 4.26, where pore nucleation is caused by stress concentration. As the pores grow and cracks form, further deformation may lead to fracture. The increase of the strain rate sensitivity index m can prevent necking and delay fracture. Lian Jianshe et al. theoretically studied the fracture process of materials in superplastic unidirectional tensile by using the nonlinear long-wave analysis method and proposed the superplastic fracture mechanism diagram, as shown in Fig. 4.27. They believed that the superplastic deformation and the resulting material fracture are controlled by the strain rate sensitivity index m and the pore growth rate. The proposed fracture modes include: for pore-sensitive materials, fracture is caused by the growth of internal pores without obvious external geometric necking or thinning; for pore-insensitive materials, fracture is caused by external geometric necking or thinning; and mixed fracture caused by the coexistence of the two.

It is generally accepted that there are two main fracture types of superplastic tensile specimens: one is that the specimen is gradually drawn into fine wires and fractures in the form of "dot," often exhibiting expanded necking and high elongation; the other is that after the size and number of pores in the specimen exceed critical values, fracture occurs without obvious necking due to the bonding and aggregation of pores. The fracture of metal polycrystals can be divided into transgranular fracture and intergranular fracture The fracture occurs during superplastic deformation is generally considered as intergranular fracture. Further research on the bonding and aggregation of pores and the final fracture of materials is needed to better understand the superplastic deformation law and the forming limit of materials.

The author's research results on the pores in some typical alloys are described as below:

1. **5083 Aluminum alloy**

The research used a cold-rolled sheet of 5083 aluminum alloy with a coarse dendritic elongated grain structure as seen in Fig. 4.28a. The material was subjected to heat

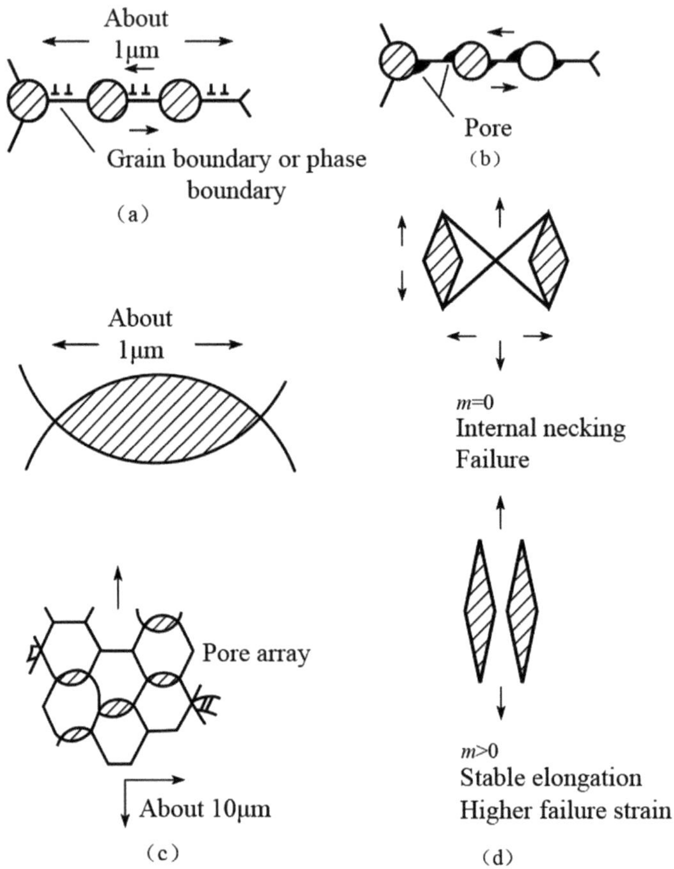

Fig. 4.26 Model for fracture process by pores

treatment at 555 °C for 40 min, resulting in an equiaxed microstructure with a grain size of approximately 17μm, as shown in Fig. 4.28b.

Unidirectional tensile tests were then performed at a temperature of 555 °C and a strain rate of $5 \times 10^{-4}s^{-1}$. Bidirectional equal-stress specimens are prepared from the bottom of the conical part and at the three-dimensional corner of the stepped box. Finite element modelling was used to calculate the pressure–time curve and ensure a constant strain rate during the test.

Following deformation, the unidirectional tensile specimen was cut parallel to the rolling direction, and the position of the bidirectional equal-stress specimen was selected based on change in grid and results of FEM calculation.

The results show that hard particles (Al_6Mn) in the material after thermomechanical treatment for grain refinement, were distributed along the rolling direction, with a radius of approximately 10μm. Upon closer inspection, it was discovered that the large hard particles are broken, forming pores between the broken parts (labeled A in

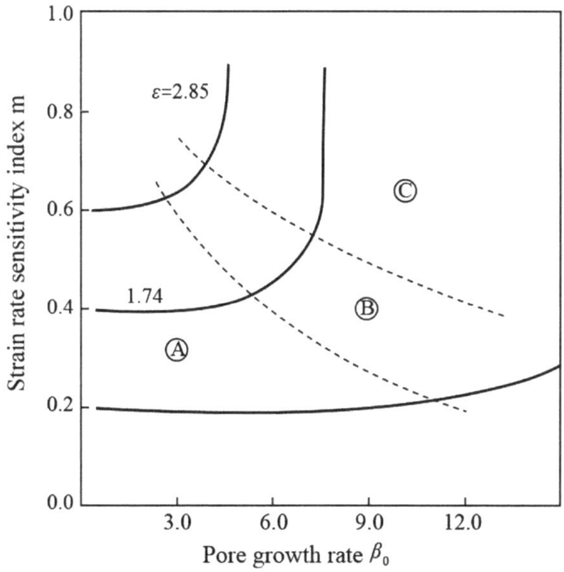

Fig. 4.27 Superplastic fracture mechanism under unidirectional tensile

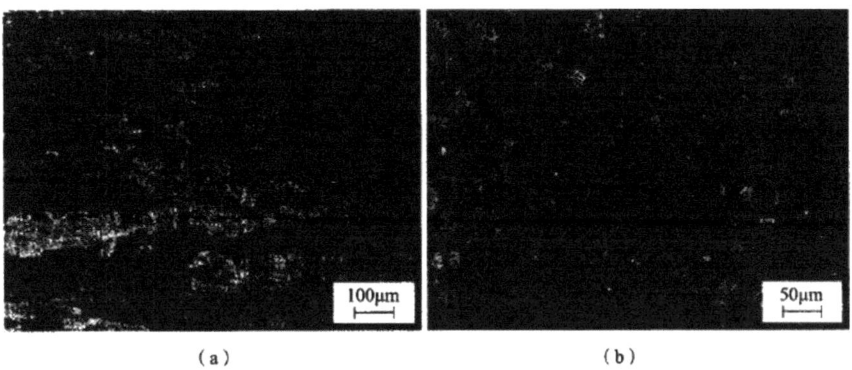

Fig. 4.28 Microstructure of 5083 aluminum alloy raw materials **a** Cold-rolled state (dendritic); **b** After recrystallization (equiaxed)

Fig. 4.29) and around the large hard particles (labeled B in Fig. 4.29). The pores were found to be distributed along the rolling direction. After recrystallization, pores are found to be around the large hard particles but not the small dispersed ones. Further observation indicated that the number of preexisting pores was significantly reduced due to the small number of large hard spots.

Figure 4.30 displays the metallograph at $\varepsilon = 0.4$ under bidirectional equal-stress deformation with a horizontal rolling direction. Noticeably, a prominent band of pores is formed, the pores are primarily spherical, possessing a radius of 8-10μm. At higher magnification, no apparent pore bonding can be observed. In Fig. 4.31, the pore growth in the unidirectional tensile specimen is demonstrated. Specifically,

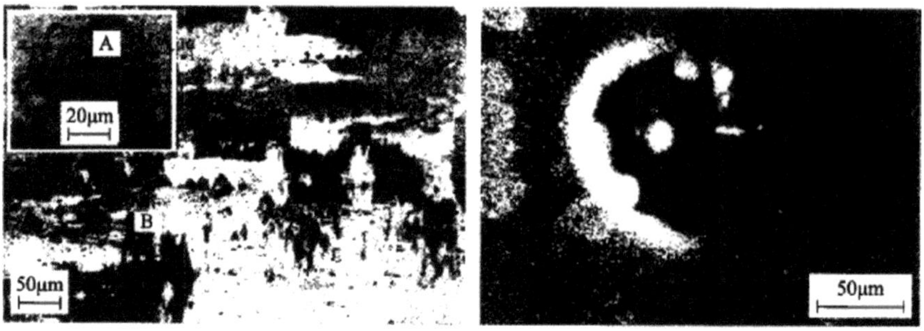

Fig. 4.29 Characteristics of pores in 5083 aluminum alloy raw materials and after recrystallization

in Fig. 4.31a, ε is 0.8 and the tensile axis is parallel to the rolling direction, while in Fig. 4.31b, ε is 0.6 and the tensile axis is perpendicular to the rolling direction, and the tensile axis is horizontal in the both graphs. The band of pores is always parallel to the rolling direction. At lower strain levels ($\varepsilon < 0.6$) in Figs. 4.30 and 4.31b, the pores are spherical, and a small amount of bonding can be identified along the rolling direction. However, at $\varepsilon = 0.8$, pore bonding occurs in both directions. Upon comparing the three images, it becomes clear that with an increase in stress level, the pore band becomes less distinct. At a higher strain level with $\varepsilon = 0.85$, the pore band is unrecognizable due to numerous bonding of pores both perpendicular and parallel to the rolling direction.

As scanning electron microscope graphs at higher magnification (Fig. 4.32a) show, the pores grow along the boundary of hard particles. It should be pointed out that the growth of pores in 5083 aluminum alloy (Fig. 4.32b) does not occur at the early stage of deformation, but only at the later stage. By comparing the two stress states, it is found that the intergranular pores of spider web shape appear at lower strain level during unidirectional tensile. Comparing the morphology of pores under different strain levels in the two stress states, it can be seen that more pores appear in the

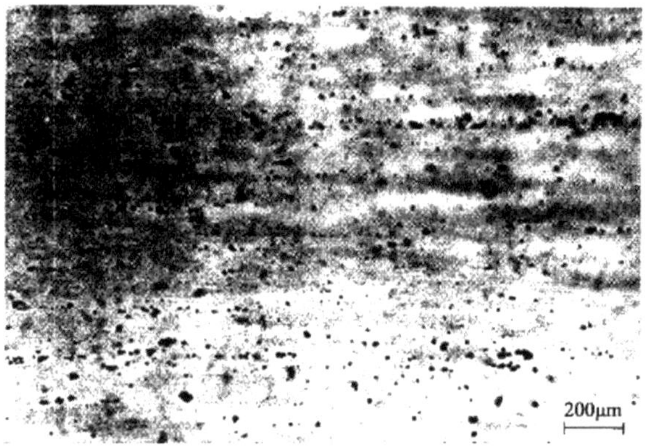

Fig. 4.30 Metallograph when the bidirectional equal-stress deformation reaches $\varepsilon = 0.4$

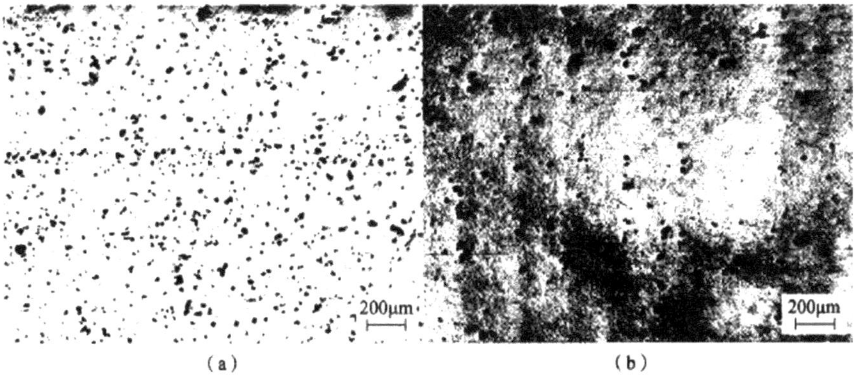

Fig. 4.31 Pore characteristics of unidirectional tensile specimen **a** Tensile axis is parallel to the rolling direction; **b** Tensile axis is perpendicular to the rolling direction

bidirectional tensile state, and the pore bonding occurs earlier in the unidirectional tensile state.

The results of the unidirectional tensile test show that changing the angle between the tensile axis and rolling direction results in a corresponding change in the direction of pore bands. In the case of bidirectional deformation, however, the pore band remains parallel to the rolling direction. This suggests that the dynamic nucleation of pores is still related to hard particles, as proposed by Chokshi et al. According to their theory, elastic stress concentration forms around hard particles during grain boundary sliding, hindering slip. However, the rapid intergranular diffusion creep prevent the formation of serious stress concentration around small hard particles, and therefore avoid pore nucleation. At superplastic temperatures, Coble diffusion creep around large hard particles involves interphase diffusion index instead of intergranular diffusion index, owing to the much smaller value of the former. This means

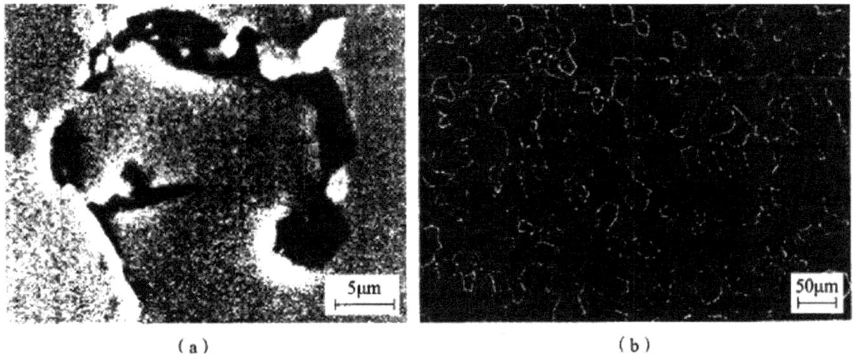

Fig. 4.32 Photo of pore growth

that stress concentration around hard particles cannot be relaxed, causing pore nucleation. This explanation is consistent with experimental results. Theoretical analysis indicates that the pore growth stage dominated by diffusions is controlled by the maximum principal stress, with little difference between stress states. Therefore, the pore growth rate under the two stress states is essentially equal at this stage. However, the main mechanism of pore growth is the power-law creep, and the growth stage where it plays a major role is controlled by the average stress. The average stress in the bidirectional equal-stress state is twice as high as that in the unidirectional tensile stress state, leading to a higher pore growth rate.

2. **7B04 aluminum alloy**

The 7B04 aluminum alloy used in this study has a grain size of 10μm (fine grain) and 20μm (coarse grain), as shown in Fig. 4.33. The banded microstructure of the coarse grain sheet is quite pronounced, with a large aspect ratio of grains and only a few fine grains present in the microstructure. In contrast, the fine grain sheet has an equiaxed microstructure, with grains distributed evenly.

Superplastic tensile tests were conducted on fine-grained 7B04 aluminum alloy plates under optimal deformation conditions (530 °C, $3 \times 10^{-4}s^{-1}$) with varying deformations. Positions subjected to large local deformations were observed, and the pore distribution is shown in Fig. 4.34. Figure 4.34a shows that when the deformation is 100%, a small number of pores are observed under optical microscope, the pores are dispersedly distributed and are small in size and volume fraction. At this stage, the pores have no significant impacts on the superplastic deformation process. However, as shown in Fig. 4.34b–f, as the deformation increases, the number and volume fraction of pores gradually increase, as more pores nucleate and grow. Due to the effect of tensile stress, grains rearrange along the tensile direction, making

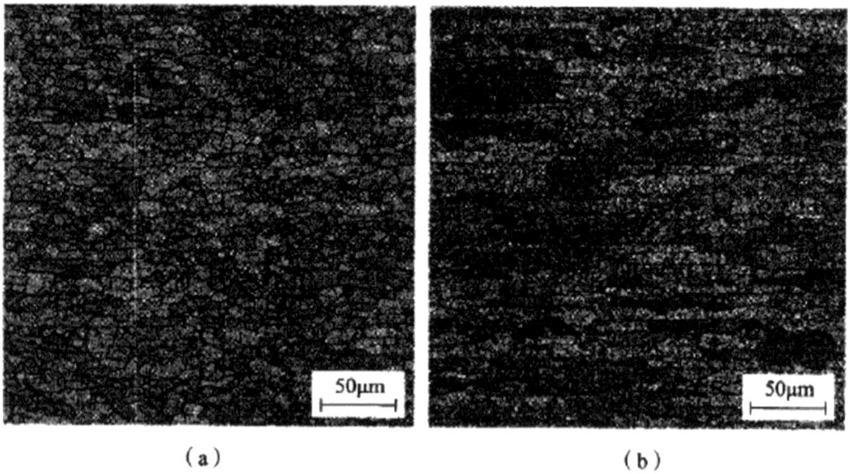

Fig. 4.33 Original microstructure of fine-grained and coarse-grained 7B04 aluminum alloy plate **a** Fine-grained plate; **b** Coarse-grained plate

the long axis of pores align with the tensile direction, and the bonding along this direction becomes more and more apparent with the increase of deformation. When the deformation reaches 1000%, as shown in Fig. 4.34e, large-sized pores (with a mean value of long and short axes greater than 200 μm) appear in the microstructure, and these pores may become the most vulnerable part of the specimen as they continue to grow. When the deformation reaches 1663%, as shown in Fig. 4.34f, the volume fraction of pores is large, and the number of large pores increases, with a more pronounced bonding state between pores. Due to the growth and bonding of pores, the deformation becomes unstable, leading to specimen fracture.

For coarse -grained 7B04 aluminum alloy plate, superplastic tensile tests with different deformation are conducted under the same deformation conditions (530 °C, $3 \times 10^{-4} s^{-1}$), and the distribution of pores is shown in Fig. 4.35. It can be seen from Fig. 4.35a that when the deformation is 20%, a small number of pores are observed under the optical microscope, as the deformation is obviously smaller than that when the pores are firstly observed in fine-grained plate, it can be concluded that the pores form at the early stage of tensile deformation. At this time, the small pores are distributed dispersedly with small size and volume fraction. As shown in Fig. 4.35b–d, with the increase of deformation, the number of pores and the average size also increase rapidly. When the deformation is 200%, there are many pores with a diameter of tens of micrometers. Compared with the case of fine-grained plate with the same deformation in Fig. 4.34b, the number and size of pores increase significantly. As shown in Fig. 4.35e, when the deformation reaches 310%, the size and number of pores continue to increase rapidly. The aspect ratio of pores is large, pore bonding is very obvious, and the pores are sharp in both ends. This indicates the presence of large stress concentration which leads to the rapid bonding and growth of pores, making the specimen fracture rapidly.

The diameter and volume fraction of pores in both coarse-grained and fine-grained 7B04 aluminum alloy sheets with different deformations are measured and calculated using Image-Pro Plus. Figure 4.36 compares the average pore diameter of both types of sheets against the true strain. The figure shows that the average pore diameter increases linearly, and fine-grained plates have a smaller average pore diameter under the same deformation. When fracture occurs, the average pore diameter in the fine-grained plate is 22.5μm, which is slightly smaller than that of the coarse-grained plate (24.7μm). The slope of the fine-grained plate is relatively small, indicating that the pores' growth rate is slower.

In Fig. 4.37, the pore volume fraction of coarse-grained and fine-grained sheets is compared with the true strain. The figure shows that pore volume fraction increases exponentially, with a slow increase in volume fraction at the early stage of pore growth, followed by a rapid increase at the later stage. The volume fraction of pores in fine-grained plates is smaller than that in coarse-grained plates for the same deformation, but there is little difference in pore volume fraction at the point of fracture (16.31% for fine-grained plates, and 16.49% for coarse-grained plates). The slope of the fine-grained curve is also smaller close to the point of fracture, indicating a slower growth rate of pore volume fraction.

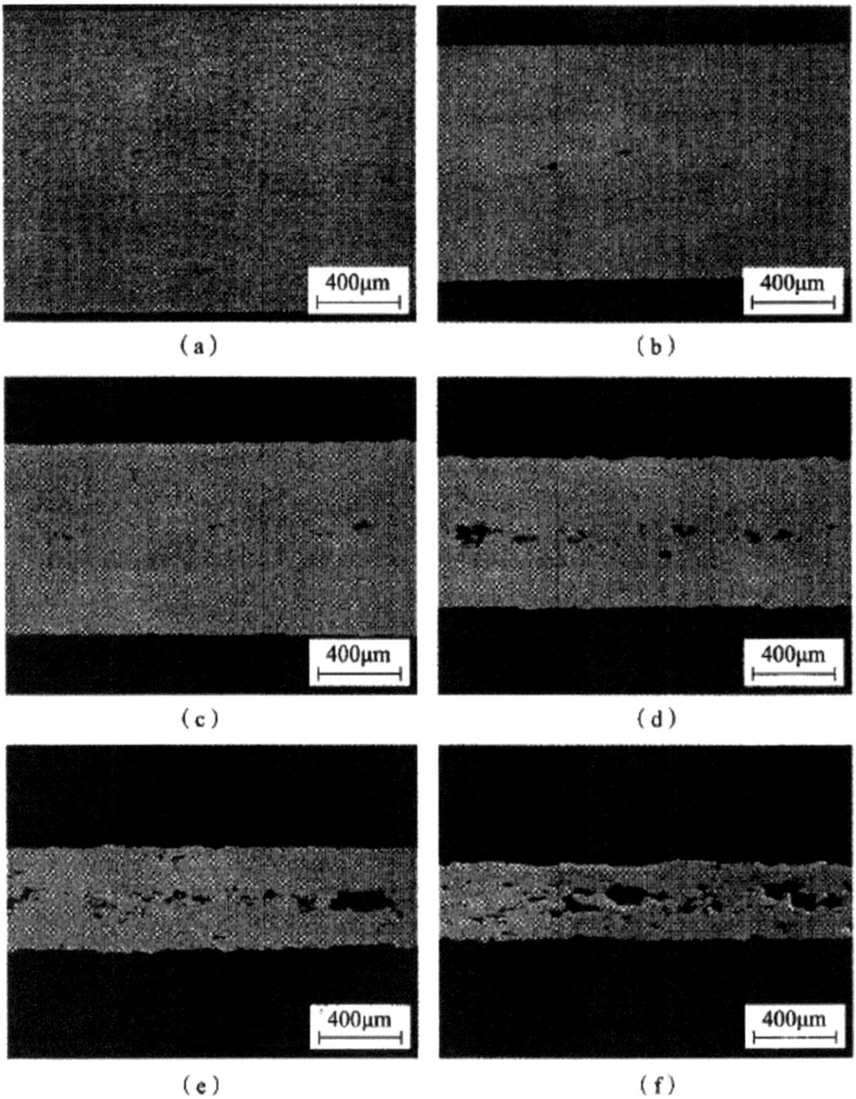

Fig. 4.34 Pore distribution of fine-grained plate with different deformations **a** 100%; **b** 200%; **c** 400%; **d** 600%; **e** 1000%; **f** 1663%

During superplastic tension, fine-grained 7B04 aluminum alloy has more grain boundaries, which favors grain boundary sliding and achieves coordinated deformation. At the early stage of deformation, if there is any tendency of defect initiation at a particular point, the deformation is promptly coordinated and the defect will be healed.. In this way, the growth process of the pores is delayed. The pore distribution in the deformed section shows that when the deformation reaches 100%, small

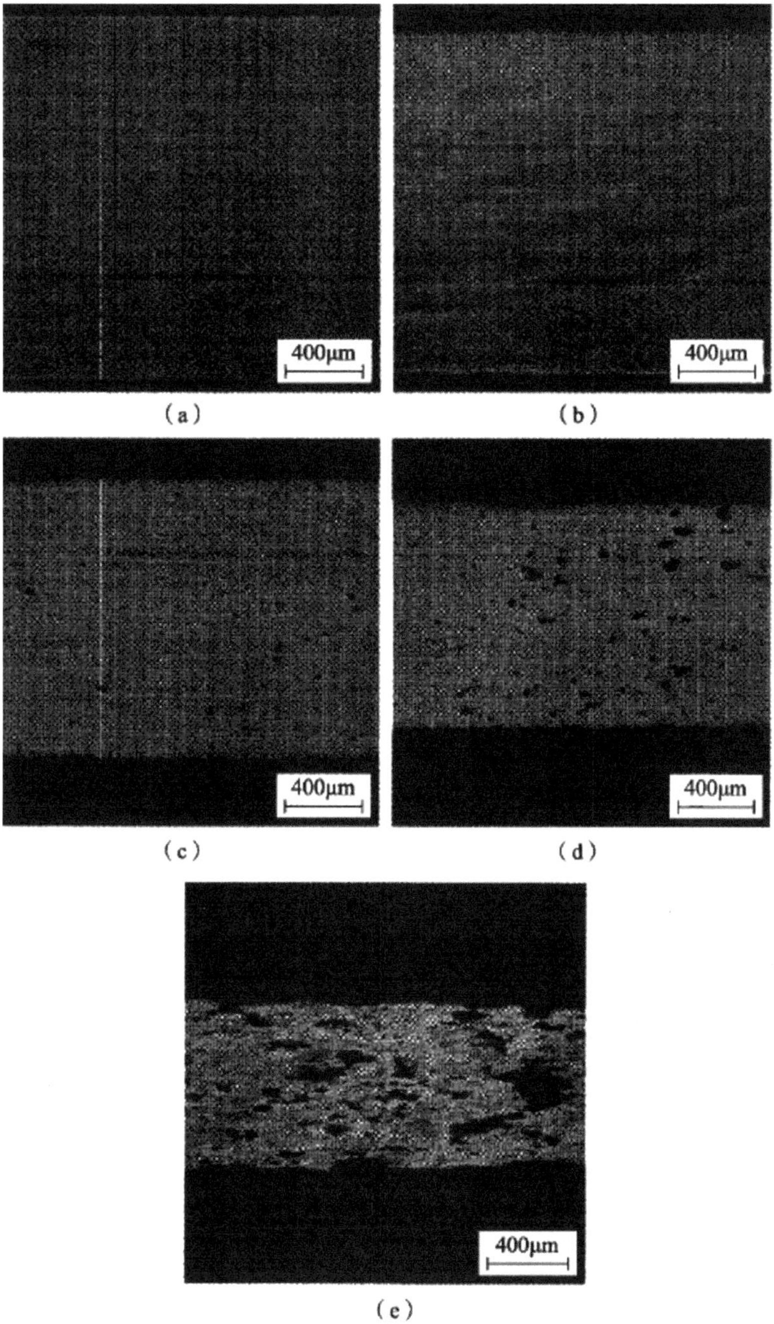

Fig. 4.35 Pore distribution of coarse-grained plate with different deformations **a** 20%; **b** 50%; **c** 100%; **d** 200%; **e** 310%

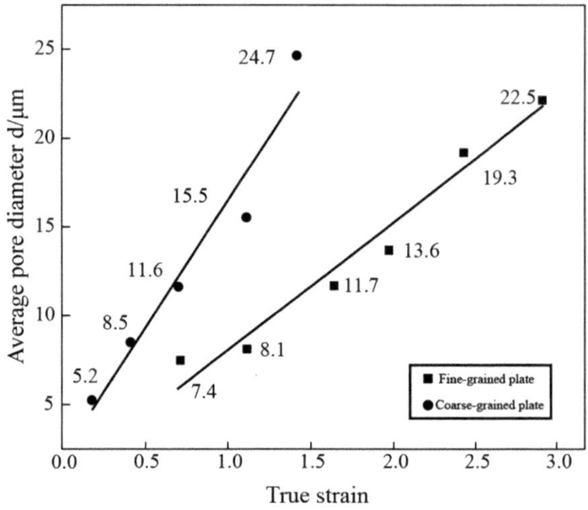

Fig. 4.36 Comparison of the average pore diameter of coarse-grained and fine-grained plates with the true strain

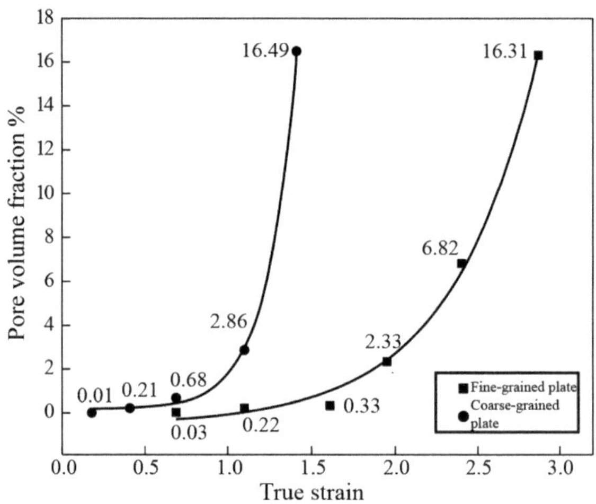

Fig. 4.37 Comparison of the pore volume fraction of coarse-grained and fine-grained plates with the true strain

pores appear, and the average diameter and volume fraction of the pores increase slowly. The deformation process is stable and less affected by defects, resulting in a large elongation. In contrast, the microstructure of the original coarse-grained 7B04 aluminum alloy plate is visibly banded, and the grain boundary sliding is difficult to occur during the deformation process. The stress concentration cannot be released in time, leading to early defect initiation, rapid increase in the size and number of pores, as well as quick bonding and aggregation. This results in rapid local instability of the material and significantly reduced elongation. While the size and growth rate of the pores in the fine-grained and coarse-grained plates differ, the calculation results of the pore volume fraction of pores upon fracture is similar, suggesting that the tolerance of 7B04 aluminum alloy to the pores is approximately 16%. In the early stage of growth, the pores are small and independently distributed, and their nucleation and growth facilitate the release of stress concentration at the grain boundary, which is beneficial to the coordination of deformation. However, as pores grow, they gradually bond and aggregate, leading to a rapid increase in the volume fraction of pores, which causes rapid instability and fracture of the material.

4.2.2 Diffusion Bonding Interface Defects

The presence of defects at the diffusion bonding interface can negatively affect the microstructure and mechanical properties of the joint. This section focuses on the typical distribution of defects and the distribution characteristics of first-order bending vibration stress at the diffusion-bonded interface of a TC4 three-layer hollow truss structure. Through numerical simulation based on stress response and process compliance, diffusion-bonded interface defects are designed in order to study the the evolution of diffusion-bonded interface defects, the effect of geometric parameters such as defect direction, defect size, and position on the damage of diffusion joints. Anti-welding coating is applied on the interface of two titanium alloy thick plates, defects are then obtained in the diffusion bonding process and characterized, preparing for study on the impact of interface defects on the mechanical properties of structural components.

Using the above method, we obtained a series of planar circular diffusion bonding interface defects at the designated location. We conducted ultrasonic testing of diffusion bonding interface to identify the defects and characterize the macro welding condition. We also used the metallographic microscopic analysis to observe the small area of close-fitting interface defects and characterize the microscopic welding rate at the diffusion bonding interface. We used a flat-bottom hole block as the ultrasonic reference block and placed the diffusion bonding block in the large-thickness ultrasonic testing system for C ultrasonic nondestructive testing. Results in Fig. 4.38 shows that the current ultrasonic testing method can be used for inspections of defects in the titanium alloy diffusion bonding block, and expected diffusion bonding interface defects have been obtained.

The wire cutting method is adopted to obtain the cross section specimen containing diffusion bonding interface defects for microstructure observation, as shown in

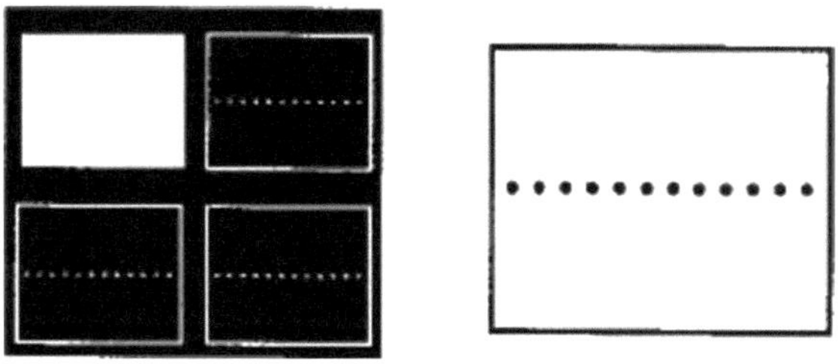

Fig. 4.38 Ultrasonic testing results of diffusion bonding interface defects

Fig. 4.39. The cross section of diffusion bonding interface defects is flat, and the height of interface defects is different, and the two ends of interface defects are herringbone. In addition, the microstructure of diffusion bonding interface area is observed, as shown in Fig. 4.40. It can be seen that the microstructure of titanium alloy plate after diffusion bonding is primary α phase, about 95% equiaxed microstructure, which is evenly distributed. Except the interface defects, the rest of the diffusion bonding interface is in metallurgical bonding state, with good diffusion bonding quality.

The evolution of TC4 alloy interface defects during superplastic forming is analyzed by numerical simulation using viscoelastic constitutive model including failure. The material constitutive model is a viscoelastic constitutive model based on Backofen superplastic equation and including K-R failure criterion. In this failure model, the failure at each point gradually accumulates and increases. The preset defect thickness is 50μm, the length is 100μm, and both ends of the defect are smoothly rounded. Figure 4.41 shows the schematic diagram of geometric model and local magnification.

In order to investigate the impact of defects on structural deformation, two points on the side of the rounding at the corner are chosen as the reference point, as illustrated

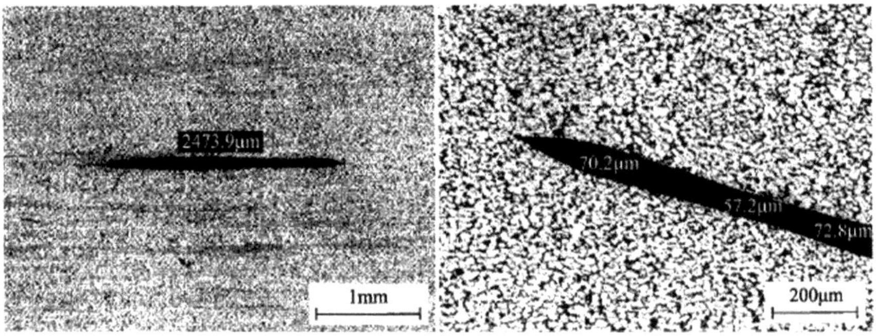

Fig. 4.39 Micro characterization of diffusion bonding interface defects

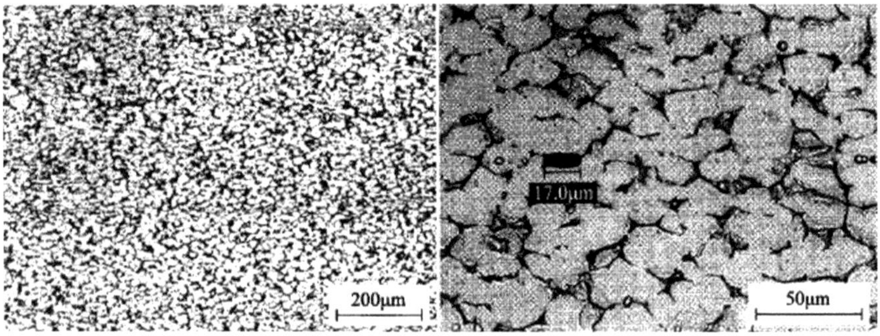

Fig. 4.40 Microstructure of diffusion bonding area

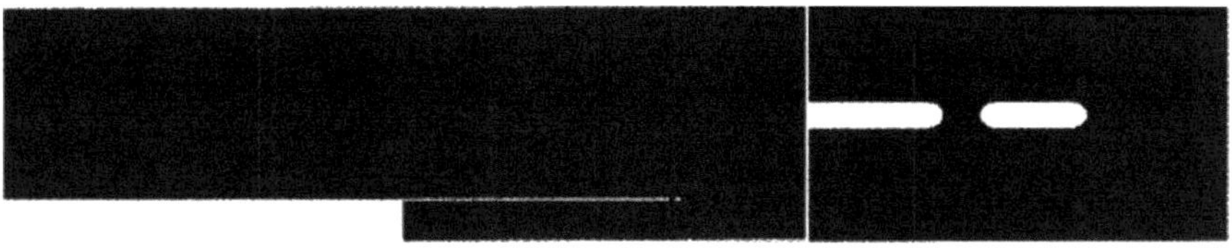

Fig. 4.41 Geometric model and local magnification for the internal defects of diffusion bonding interface

in Fig. 4.42. To quantify the deformation of the structure with defects, we used the relative displacement between these two reference points. The change in relative displacement over time varies, depending on the specific scenario. To conduct the numerical simulation, we set the distance between the center of the rounding on the left side of the defect and the center of the rounding at the corner point to 0.1 mm, 0.15 mm, 0.2 mm, and 0.3 mm, respectively.

Based on our numerical simulation results, the internal stress of the entire structure experiences an initial increase followed by a decrease (release) during superplastic forming. Additionally, the deformation and stress field near the corner point are affected to some extent by the presence of defects. Figure 4.43 demonstrates that, when the distance between the defect and corner point is 0.1mm, local failure occurs (as shown in A in Fig. 4.43a), and the effective diffusion bonding interface shrinks. Conversely, when the distance between the corner point and the defect is greater than 0.15 mm, no failure occurs during deformation, the defect thickness is smaller than that of the initial structure, and the defect tends to close. When the distance between the defect and corner point is greater than 0.3 mm, the defect consistently closes during the forming process, and has little effect on local deformation near the corner point. The change in the structure's deformation can be observed by examining the change curve of the different reference distances over time, as shown in Fig. 4.44. It is apparent from the figure that, when the distance between the defect and corner point is 0.1mm, failure occurs on the structure, and the reference distance exhibits the greatest difference from the case without a defect, indicating that the deformation change of the structure is significant. However, when the distance between the defect

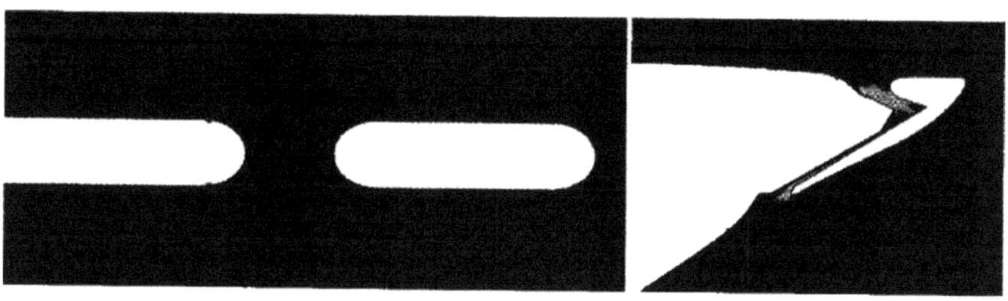

(a) (b)

Fig. 4.42 Definition of reference distance for structure deformation **a** Before deformation; **b** After deformation and corner point is 0.3mm, the reference distance has little difference from the case without a defect. Thus, it can be concluded that the impact of the defect on the corner point can be ignored when the distance between the defect and corner point is greater than 0.3mm.

In order to investigate whether the material at the defect will experience further failure after the initial failure at the corner point, we examined the effect of different defect sizes on the forming results. To model the defect, we simplified it to a rectangle in the middle and two semicircles at both ends. Therefore, we consider the length of the rectangle in the horizontal direction as the defect length when discussing the impact of defect size. Three cases are selected with defect lengths of 0 (no defect), 0.05mm, and 0.1mm, all located at a distance of 0.1mm from the corner point. Figure 4.45 shows the structure failure field obtained from numerical simulation. The black part indicates no failure, while the dark gray part (A, B) indicates full failure.

The simulation results demonstrate that when there is no defect or the defect has a short length, no failure occurs at the corner point. When the defect length is 0.1mm, failure occurs at the corner point, but no subsequent failure is observed at the defect location. These results suggest that when the defect is long, failure occurs at the corner point, but the end of the defect far from the corner point exhibits a tendency of closing instead of failure. When the defect is short, the material near the corner point will come into contact with the inner surface of the defect during

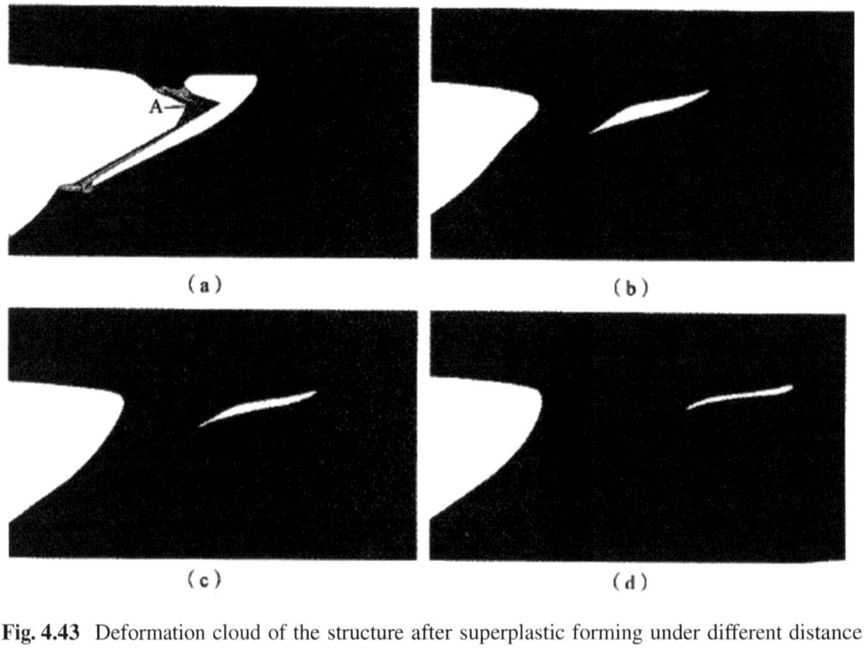

Fig. 4.43 Deformation cloud of the structure after superplastic forming under different distance between the defect and corner point The distance between the defect and corner point is **a** 0.1 mm; **b** 0.15 mm; **c** 0.2 mm; **d** 0.3 mm

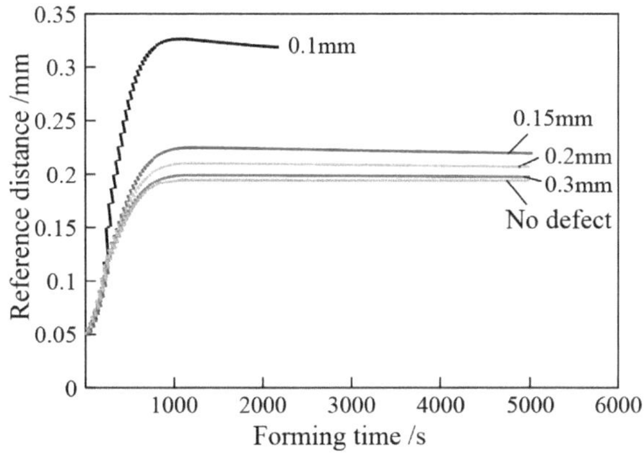

Fig. 4.44 Change of reference distance with forming time under different distance between defect and corner point

deformation, thus relieving local stress concentration. This prevents material failure near the corner point.

The simulation results also indicate that no subsequent failure occurs at the defect after the material failure at the corner point, suggesting that failure may only occur near the corner point. When K-R failure viscoelastic constitutive model is used for the material, the sharp corner area may experience failure during the forming process. If the distance between the defect and corner point is less than 0.3mm, the structure will fail; when the distance between the defect and corner point is more than 0.3 mm, the defect has little impact on the corner point.

To analyze the stress response near interface defects in titanium alloy discontinuous solids under alternating loads, and to study the effects of geometric parameters on the damage of diffusion bonding joints, we use the stress intensity factor at the tip of the interface defect as the primary evaluation index. Numerical simulations are performed to examine the effects of parameters such as interface defect plane, load direction, defect size, and location. In this study, we set the defect diameter to 2mm and apply a load of 600 MPa at the end. Angle between the non-welded interface defect inside the plate and the tensile direction is set from 0° to 90° in steps of 15°. The stress distribution near the defect on the structural symmetry plane and the distribution of stress intensity factor in the defect circumferential direction (taking the defect center as the origin of polar coordinates) are shown in Fig. 4.46 for each angle. Note that the loading direction is vertical in Fig. 4.46.

The stress intensity factor around the crack front near the defect is calculated and presented in Fig. 4.47. Since semi model is used, the defect surrounding angle is 180%. From Fig. 4.47, it can be observed that the stress intensity factor along the defect circumferential direction is very low, making it difficult for cracks to propagate in the structure. However, when the defect edge (taking the defect center as the origin of polar coordinates) is positioned at 0°, 90°, 180°, 270°, and 360°, the stress intensity factor reaches its maximum value, indicating that crack growth is more likely to occur at these positions.

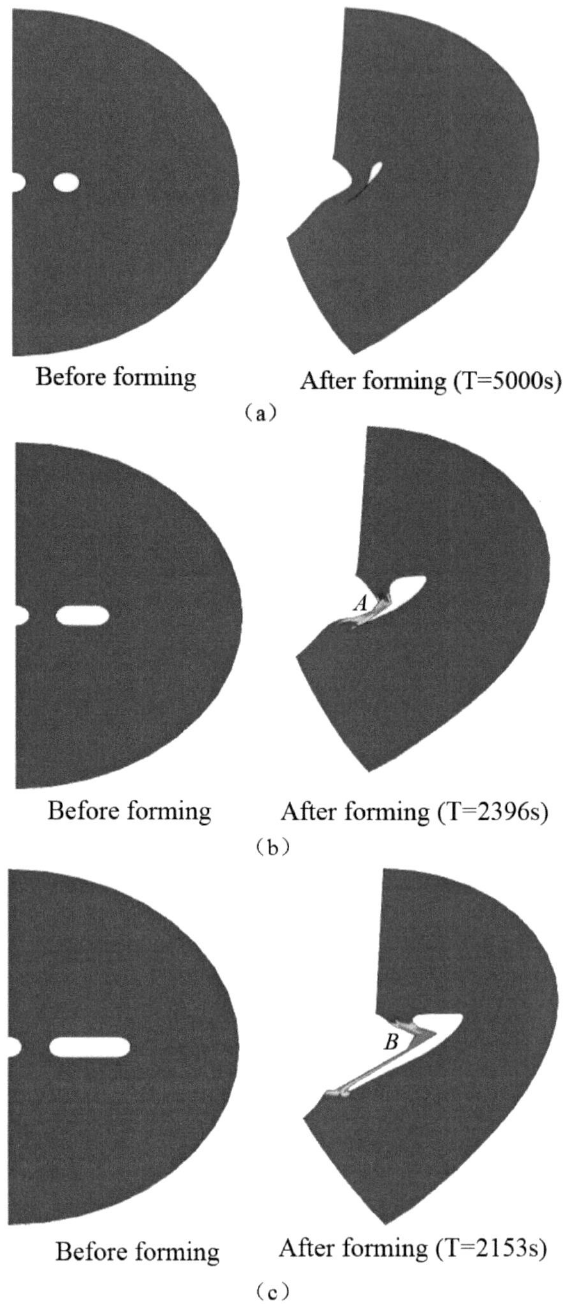

Fig. 4.45 Impact of defect size on superplastic forming failure field Failure field of the structure after forming when the internal rectangle length of defect is **a** 0 mm; **b** 0.05 mm; **c** 0.1 mm

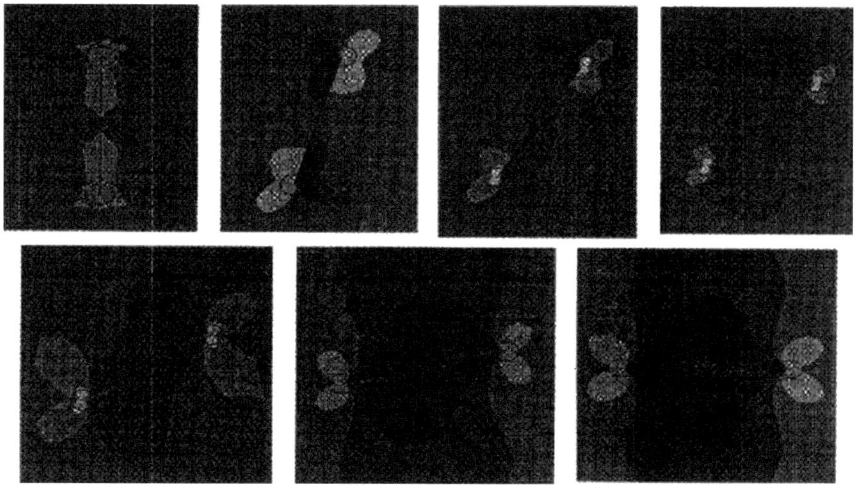

Fig. 4.46 Included angle between defect and tensile direction and stress distribution nearby

As shown in Fig. 4.48, the stress intensity factor along the circumferential direction of the defect increases as the angle between the defect and tensile direction increases from 0° to 90°. When the included angle is 0°, the stress intensity factor is minimal, and the likelihood of crack growth is low. However, as the angle increases, the maximum stress intensity factor on the crack surface rises sharply, and when the included angle is 90°, the probability of crack growth is highest. In this case, the corresponding crack type is mode I crack. With the exception of the 0°, the distribution of stress intensity factors on the defect surface is nearly uniform for all other angles, and crack growth happens almost simultaneously on the crack surface.

When discussing defects with a fixed angle, it is important to note that when the circumferential angle along the defect is 0°/180°, the stress intensity factor reaches its maximum value, leading to high crack growth rates in those positions. Additionally, if the included angle is 0°, the maximum stress intensity factor occurs at the side wall of non-welding defects. Conversely, when the angle is 90°, the tip of the defect experiences the maximum stress intensity factor. During the process of damage evolution, the crack initiation life dominates in the former case, while crack propagation life dominates in the latter.

Based on the numerical simulation results, the angle between the interface defect and the loading direction plays a significant role in the impact of interface defects on the life of diffusion bonding joints. When the load is parallel to the defect, structure failure is unlikely, and the crack initiation life is the primary concern. As the angle between the two increases, the stress intensity factor near the defect increases, resulting in an accelerated crack growth rate. At this point, the crack growth life gradually dominates. Interestingly, the above simulation results indicate that when the defect is parallel to the loading direction, the joint life is not sensitive to its size.

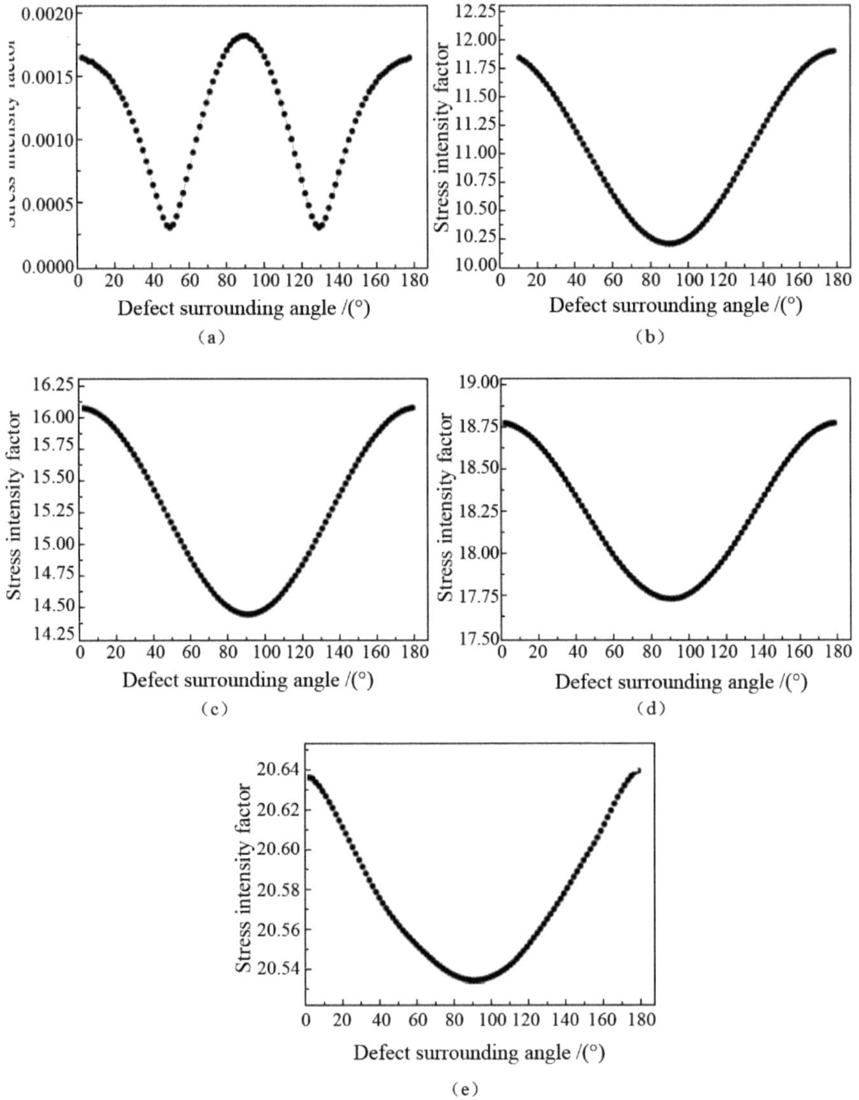

Fig. 4.47 Circumferential stress intensity factor at different included angles between defect and tensile direction **a** 0°; **b** 30°; **c** 45°; **d** 60°; **e** 90°

To further investigate the impact of the defect tip shape on the stress distribution and life of the joint, a longitudinal interface defect with the size of 1mm was used, and the same finite element model as above was established. A uniform load was applied to both ends, and the defect tip's geometric model was described by a five-point spline. Simulation results suggest that the crack life is sensitive to the defect's shape. The impact of the defect's location on the service life is related to the stress

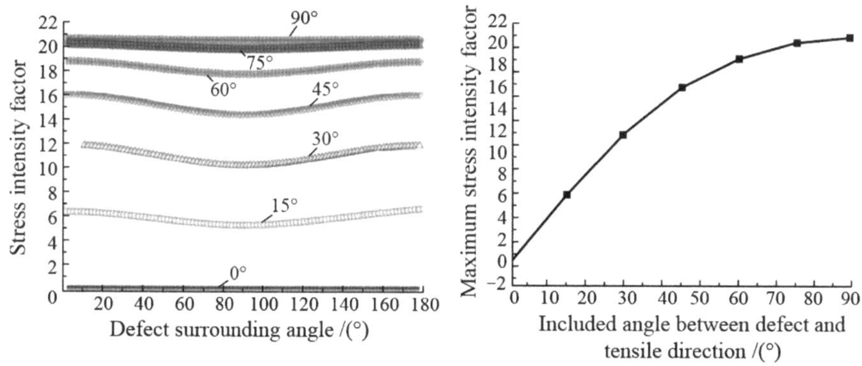

Fig. 4.48 Relationship between the included angle direction and the stress intensity factor and the maximum stress intensity factor in the circumferential direction of the defect

distribution in the structure. Moreover, when multiple defects are present, the stress level at the defect location is higher than that of a single defect, resulting in a shorter life.

The study examines the impact of defect length on the stress intensity factor when the defect is perpendicular to the loading direction. The defect radius in the model is 0.2mm, 0.5mm, and 1mm (as shown in Fig. 4.49), and the finite element model used is the same as the one mentioned earlier. Figure 4.50 illustrates the stress distribution near the defect on the structural symmetry plane and the stress intensity factor distribution in the circumferential direction of the defect for each condition. The simulation results indicate that when the diffusion surface is subjected to alternating loads, the stress intensity factors are distributed uniformly, even when there are transverse interface defects of varying sizes perpendicular to the loading direction. As depicted in Fig. 4.51, the size of the defect significantly impacts the crack life since the stress intensity factor around the defect increases as the interface defect size increases.

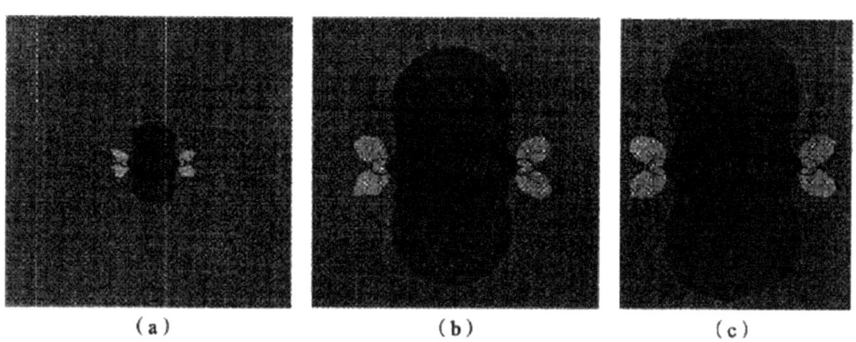

Fig. 4.49 Stress distribution for defect with the radius of **a** 0.2mm, **b** 0.5mm, **c** 1mm

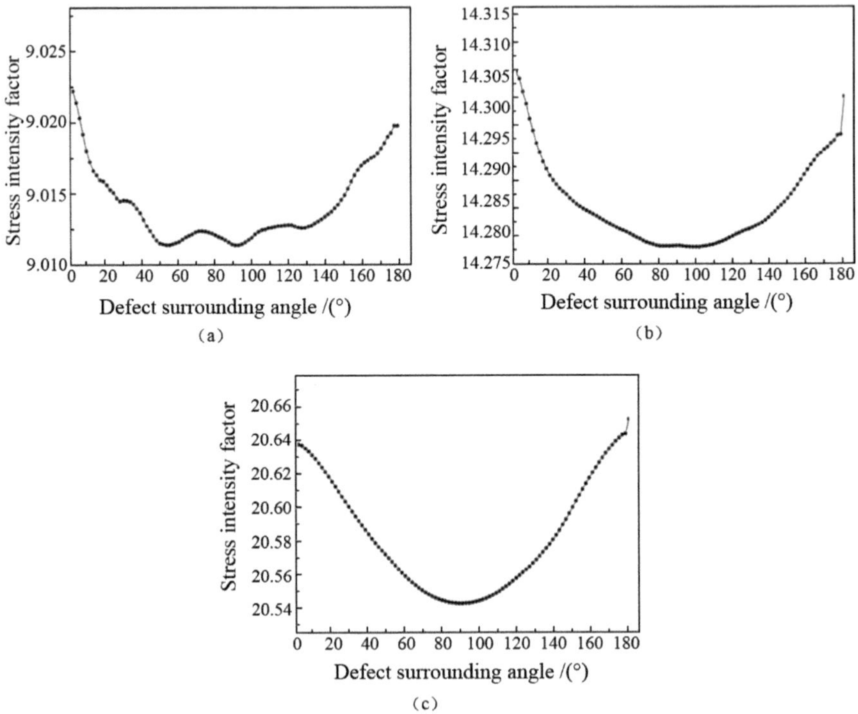

Fig. 4.50 Stress intensity factors in the circumferential direction of defect with the raidus of **a** 0.2mm, **b** 0.5mm, **c** 1mm

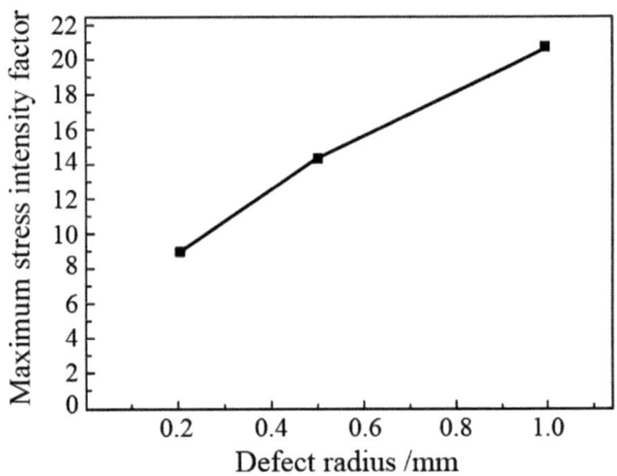

Fig. 4.51 Maximum stress intensity factor of crack front with different defect radius

4.3 Material Properties After Superplastic Forming/Diffusion Bonding

Superplastic forming/diffusion bonding of titanium alloys usually involves several physical processes, including heat generation, heat transfer, stress, strain, diffusion, and phase transformation. These processes cause various structural changes and defects, which can significantly affect the mechanical properties of the material. By studying the impact of these structural changes and defects on the mechanical properties, we can evaluate the effect of the superplastic forming/diffusion bonding process on the bearing capacity of the structure.

4.3.1 Material Properties Based on Superplastic Forming/Diffusion Bonding Thermal Cycle

Superplastic forming/diffusion bonding of titanium alloys is a complex process that requires "coordinative control of shape and property". This is difficult to achieve because many factors can influence the structure and properties, and these factors often interact with each other, making it challenging to determine and optimize the process parameters.

To study the effect of temperature on the TC4 alloy during superplastic forming/diffusion bonding, a thermal cycling test is conducted at three different temperatures (960 °C, 930 °C, and 900 °C) for four hours. Room temperature tensile specimens and funnel shaped smooth cylindrical high-cycle fatigue specimens are prepared after thermal cycling and air cooling. These specimens are then tested for room temperature tensile properties and axial tensile-tensile high-cycle fatigue properties (loading frequency of 100Hz, stress ratio $R = 0.1$, and the maximum stress level between 400 and 700MPa). In the fatigue test, the lifting method is used in the long life stage, and the grouping method in the middle and low life stage. At least 3 samples are tested at each stress level. If the fatigue life exceeds 10^7 cycles, the test is terminated.

Table 4.2 shows the room temperature tensile properties of TC4 alloy after thermal cycling in furnace and in the original state. The results indicate that the yield and tensile strength of the alloy decrease to some extent after the thermal cycling in furnace test, but the plasticity remains relatively unchanged. This is because the α-phase grains of the material coarsen after thermal cycling, leading to a decrease in material strength as per the Hall–Petch relationship. However, recovery and recrystallization can occur, resulting in an equiaxed grain and more uniform microstructure distribution, which improves plastic deformation and delays the initiation and growth of cracks.

Under the three thermal cycling conditions, the room temperature tensile properties of TC4 alloy remain similar. Although the increase in grain size has adverse effects on mechanical properties, the decrease in primary α phase proportion and more

Table 4.2 Tensile properties of TC4 alloy subjected to varying heat treatments at room temperature

Heat treatment system	Tensile strength R_m/MPa	Yield strength $R_{p0.2}$/MPa	Elongation A/%	Percentage reduction of area Z/%
Original state	1012	965	15.4	35.4
900 °C, 4h	1003	933	14.5	30.3
930 °C, 4h	997	929	14.5	32.0
960 °C, 4h	992	920	14.3	30.3

uniform distribution of microstructure help coordinate plastic deformation, which mitigates the adverse effects of grain size growth. Therefore, the room temperature tensile properties of TC4 alloy under the three thermal cycles are comparable.

Figure 4.52 shows the S–N curve of TC4 alloy after thermal cycling at 960 °C for 4 h. The arrow in the figure indicates that the fatigue life of the specimen exceeds 10^7 cycles at the stress level. The test data were paired using the up-and-down method, and the median fatigue limit of TC4 alloy in this state was calculated to be 445MPa. At a stress of 450MPa, the fatigue life of two specimens exceeded 10^7. When the maximum stress was 425MPa, the fatigue life of five consecutive specimens exceeded 10^7, indicating that the fatigue limit of TC4 alloy under 10^7 cycles in this state is not less than 425MPa. The S–N curve shows the average fatigue life at each stress level. It can be observed that the fatigue life increases with decreasing stress level, which is consistent with the typical fatigue behavior of titanium alloys. The fatigue life covers two stages, the fatigue crack initiation stage and the fatigue crack propagation stage. For high-cycle fatigue, the proportion of the fatigue crack initiation stage gradually increases with decreasing stress level, resulting in an overall prolongation of fatigue life. Furthermore, it can be seen from Fig. 4.52 that the scatter of fatigue life increases with decreasing stress level. Under a stress of 450MPa, the minimum life is only 429,915 cycles, and the maximum fatigue life of two specimens exceeds 10^7 cycles.

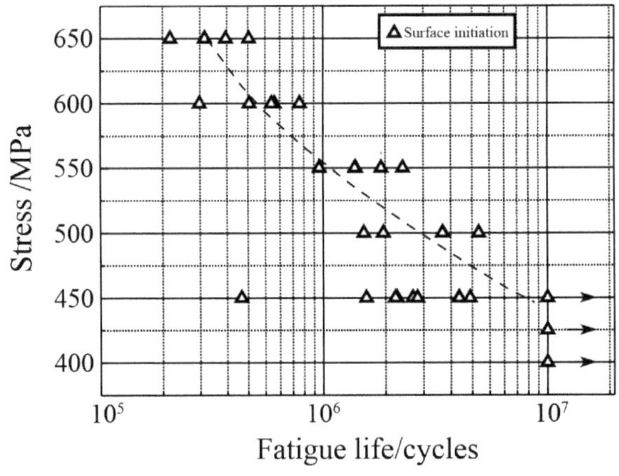

Fig. 4.52 S–N curve of high-cycle fatigue life of TC4 alloy after the thermal cycling in furnace at 960 °C for 4h

The S–N curve of TC4 alloy after thermal cycling at 930 °C for 4 h is shown in Fig. 4.53. The median fatigue limit of TC4 alloy in this state is calculated to be 455MPa. The average fatigue life curve in Fig. 4.53 shows that the fatigue life increases with a decrease in stress level, which is consistent with the case of 960 °C. Except for the 550MPa stress level, the fatigue life of specimens tested at all other stress levels exceeds 10^7 cycles. At the stress level of 500MPa, 8 specimens are tested, and one of which has a fatigue life exceeding 10^7 cycles, while the minimum fatigue life is 1255900 cycles. At the stress level of 475MPa, 8 specimens are tested, and 3 of which have a fatigue life exceeding 10^7 cycles, with the minimum fatigue life being 2,210,800 cycles. At the stress level of 450MPa, 10 specimens are tested, and 4 of which have a fatigue life exceeding 10^7 cycles, with the minimum fatigue life being 1,961,810 cycles. At the stress level of 425MPa, 10 fatigue specimens are tested, and 6 of which have a fatigue life exceeding 10^7 cycles, with the minimum fatigue life of the remaining r specimens being 3,571,800 cycles. It can be seen that the scatter of the fatigue life for TC4 alloy in this state is relatively large, and the law of change with the stress level is not obvious.

The graph in Fig. 4.54 shows the S–N curve of TC4 after thermal cycling at 900 °C for 4 h. The median fatigue limit of TC4 alloy in this state is calculated to be 430 MPa. The average fatigue life curve in the graph indicates that as the stress level decreases, the fatigue life increases. This trend is similar to the case of 960 °C and 930 °C. However, the graph also shows that the scatter of fatigue life of TC4 alloy at each stress level is relatively large. Some specimens exceed the fatigue life of 10^7 cycles at stress levels of 425 MPa and 450 MPa. For example, 8 specimens are tested at the stress level of 450 MPa, and 2 of them have a fatigue life exceeding 10^7 cycles, with a minimum fatigue life of 2,651,377 cycles. At the stress level of 425 MPa, 10 specimens are tested, and 6 of them have a fatigue life exceeding 10^7 cycles, while the minimum fatigue life of the remaining 4 specimens is 2,932,228 cycles. It is clear that the scatter of fatigue life of TC4 alloy is also large after thermal cycling in the furnace at 900 °C for 4 h, and its change with stress level is not apparent.

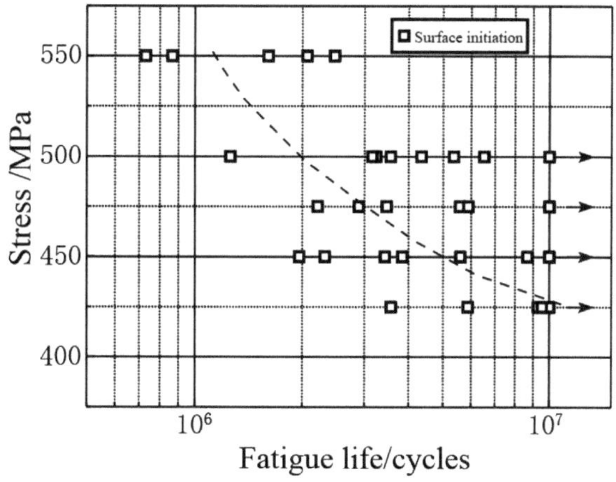

Fig. 4.53 S–N curve of TC4 alloy after thermal cycled at 930 °C for 4h

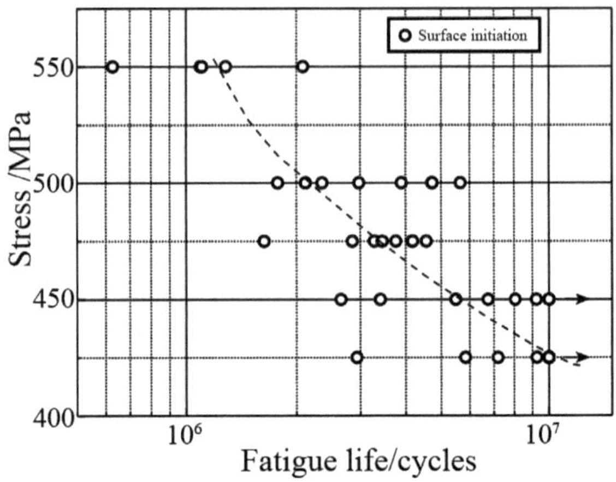

Fig. 4.54 S–N curve of TC4 alloy after the thermal cycling at 900 °C for 4h

The axial high-cycle fatigue properties of TC4 alloy undergo significant changes after thermal cycling under different conditions. The median fatigue limits after thermal cycling at 960 °C, 930 °C and 900 °C for 4 h are 445 MPa, 455 MPa, and 430 MPa, respectively. Comprehensive analysis of alloy fatigue life under various conditions (Fig. 4.55) indicates that for TC4 alloy subjected to thermal cycling, the best comprehensive fatigue performance is under the temperature of 930 °C, superior to that of 960 °C and 900 °C.

In the high-cycle fatigue damage process of titanium alloy, the crack initiation stage accounts for most of the fatigue life, which is mainly composed of dislocation slip and micro-crack formation. The influence of microstructure on fatigue crack initiation is mainly determined by lattice strength and dislocation slip path. The higher the lattice strength, the shorter the dislocation slip path, the more difficult the fatigue crack initiation, and the better the high-cycle fatigue performance. The lattice strength

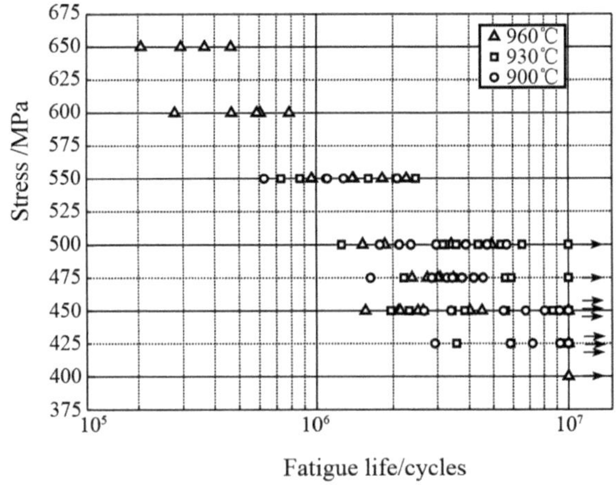

Fig. 4.55 Comparison of fatigue properties of TC4 alloy after thermal cycling at 960 °C, 930 °C and 900 °C for 4 h

of titanium alloy can be inferred from its yield strength. After different thermal cycles, the short-term mechanical properties of TC4 alloy have little difference, and the ultimate tensile strength and yield strength are relatively close, which indicates that the influence of lattice strength on the high-cycle fatigue properties of materials in different thermal cycles is weak.

The microstructure analysis shows that after thermal cycling at 900 °C for 4h, the average grain size of the primary α phase in the microstructure remains basically unchanged, the α plate thickness in the β transformation structure increases, and some elongated α phase grains with larger sizes remain. This uneven structure distribution results in a large scatter factor of high-cycle fatigue life and suboptimal fatigue performance. After thermal cycling at 930 °C for 4h, the microstructure is composed of short strip primary α phase grain and intergranular β transformation microstructure and basically achieves equiaxed. The average grain size of primary α phase has grown but the volume fraction has decreased, which is about 70% of the original condition. Although the grain size of primary α phase has grown in this state, the microstructure is basically equiaxed, and large long α grains no longer exist in the raw material, the microstructure distribution is relatively uniform, with the best high-cycle fatigue performance. The scatter of fatigue life is smaller compared with the fatigue life after thermal cycling at 900 °C for 4h. After thermal cycling at 960 °C for 4h, the microstructure is fully equiaxed, and the microstructure distribution is more uniform, which is conducive to reducing the scatter of fatigue life of the material. However, the average grain size of primary α phase grows significantly, and the phase volume fraction decreases significantly, about 60% of the original condition. The volume fraction of β transformation structure increases significantly compared to the raw material, and the secondary α phase lamellar and the remaining β lamellar are arranged alternately, with the thickness of the plates is increased, resulting in a decline in its high-cycle fatigue performance.

Ductile metal materials are generally assumed to follow a normal distribution for their fatigue failure distribution. When the probability of fatigue failure is expressed in the form of normal distribution, the cumulative probability distribution space is established. In engineering safety analysis and design, the fatigue life corresponding to a failure probability of POF = 0.1% is often used to determine the minimum safe life of materials. In this section, we plot the cumulative probability distribution of fatigue failure of TC4 alloy under different stress levels after thermal cycling at 960 °C, 930 °C and 900 °C for 4h using a normal distribution. Figure 4.56 shows the cumulative probability distribution of fatigue failure (failure probability-fatigue life). The normal distribution probability of fatigue failure for specimens under different stress levels in different states is linearly fitted, as represented by the dotted lines in Fig. 4.56. The results indicate that the fatigue failure probability of each specimen under each stress level has a good linear relationship. This shows that in the cumulative distribution space of fatigue life failure under different thermal cycles, the fatigue specimens under different stress levels have the same fatigue life distribution.

After thermal cycling at 960 °C for 4h, the slope of the fitting line tends to decrease as the stress level decreases, indicating that the scatter of fatigue life increases. However, after thermal cycling at 930 °C and 900 °C for 4h, the slope of the fitting

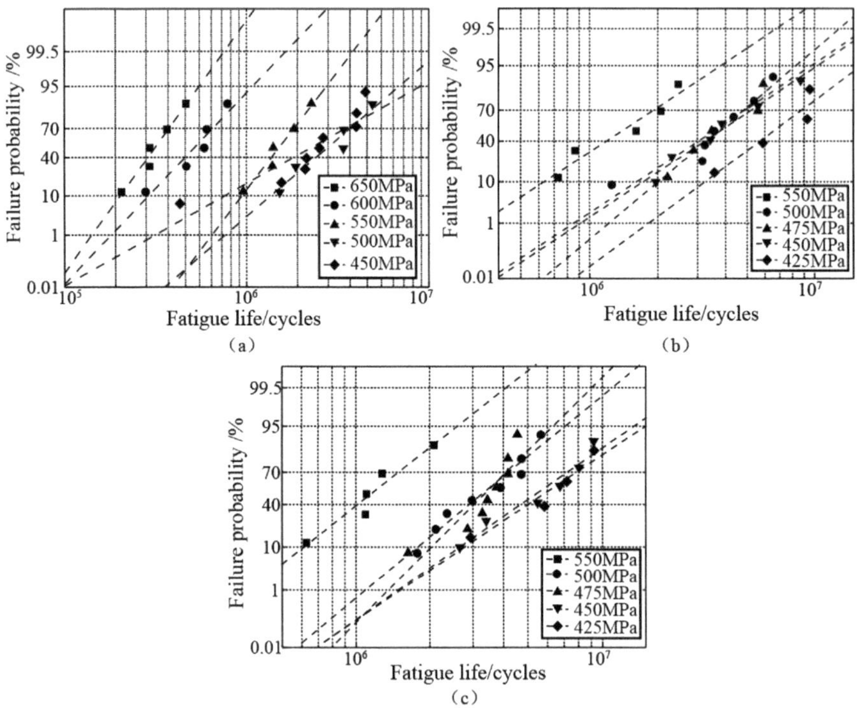

Fig. 4.56 Cumulative probability distribution of fatigue failure of TC4 alloy after different thermal cycling **a** 960 °C, 4h, **b** 930 °C, 4h; **c** 900 °C, 4h

lines are relatively close with no clear regularity found with decreasing stress level. This shows that the rule of fatigue life scatter changing with stress level is not obvious, which is consistent with the rule obtained in the above S–N curve analysis.

Figure 4.57 displays a comparison of fatigue life data for TC4 alloy under different stress levels after undergoing various thermal cycles. The fatigue life data includes the minimum fatigue life, average fatigue life, and POF = 0.1% life. The figure indicates that the average fatigue life of materials subjected to thermal cycling at 930 °C for 4h is better than those cycled at 960 °C for 4h and 900 °C for 4h, across different stress levels. However, the minimum fatigue life does not exhibit the same trend and is more random than the average fatigue life. Interestingly, POF = 0.1% life and minimum fatigue life show similar trends.

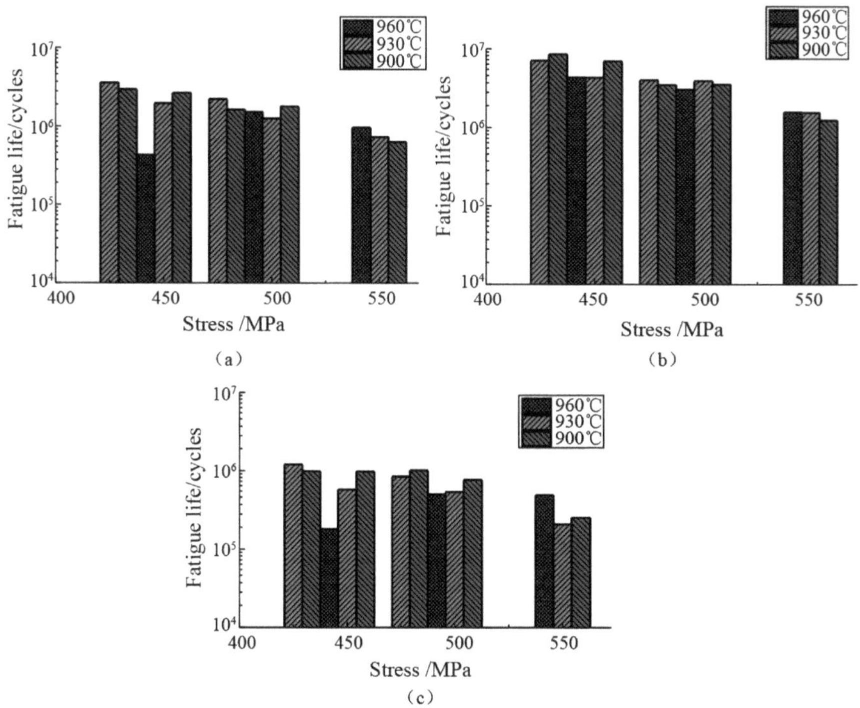

Fig. 4.57 Comparison of fatigue life data of TC4 alloy at each stress level after different thermal cycles **a** Minimum fatigue life; **b** Average fatigue life; **c** POF = 0.1% life

4.3.2 *Properties of Materials with Diffusion Bonding Interface Defects*

The application field of superplastic forming/diffusion bonding technology has expanded from non-bearing or secondary bearing structures to main bearing parts, rotating parts, and even hot-end rotating parts. Its mechanical properties, particularly its fatigue resistance, have become a crucial indicator of structural reliability. Despite achieving large area and high-precision bonding, current diffusion bonding technology may still have interface defects like holes, non-welding, or weak bonding, which can result in local stress concentration and impact part performance and reliability. Hence, studying the influence of interface defects on alloy properties and damage mechanisms is significant. Table 4.3 shows the tensile properties of TC4 alloy with diffusion bonding interface defects measured by a tensile test at room temperature. It also includes the performance of non-defective materials after diffusion for comparison. The table reveals that the average tensile strength of materials containing ϕ4mm diffusion bonding interface defects and those without interface defects are 995MPa and 980MPa, respectively, while the elongation is 15.7% and 15.1%, respectively, which are almost the same. Since the defect plane is parallel to the tensile direction and its thickness is less than 100μm, the area of the stressed section of the specimen remains nearly unchanged, making it have little impact on the static strength and elongation of the specimen. Moreover, a considerable number of dimples still appear near the defect of the tensile fracture, as shown in Fig. 4.58.

Figure 4.59 displays the S–N curve for the fatigue life of a TC4 alloy with a ϕ4mm diffusive interface defect. The test data is divided into two areas: area I, where the median fatigue limit calculated using the lifting method is 423MPa, and area II, where the median fatigue limit of the specimens that experience fatigue crack initiation at the defect is 393MPa. The median fatigue limit for all fatigue specimens is 402MPa. The scatter in the fatigue life of the titanium alloy specimen with the ϕ4mm internal interface defect is obvious in the figure. Meanwhile, the interface defect parallel to the fatigue load direction is not the only location for fatigue crack initiation. All specimens that crack from the surface are in the high-life area (area I), while all specimens that fail from the defect are in the low-life area (area II), except for one. It is important to note that due to processing errors, the diffusion bonding defects are not all located precisely in the center of the test bar. Therefore, the current test data cannot verify the impact of defect location on the service life.

Table 4.3 Tensile properties of TC4 alloy at room temperature

Material	R_m/MPa	Average tensile strength / MPa	A/%	Average elongation/%
Material with ϕ4mm defect	997	995	17	15.7
	996		15.6	
	993		14.4	
Materials without interface defect	976	980	15.1	15.1
	977		15.7	
	988		14.3	

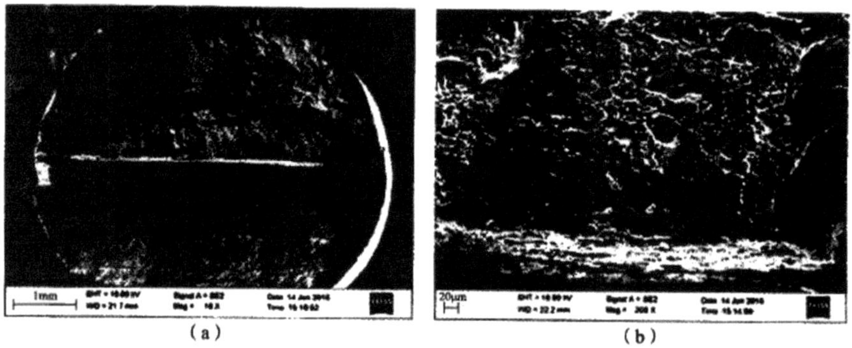

Fig. 4.58 Tensile fracture of TC4 alloy with ϕ4mm interface defect **a** Macro morphology; **b** Defect edge

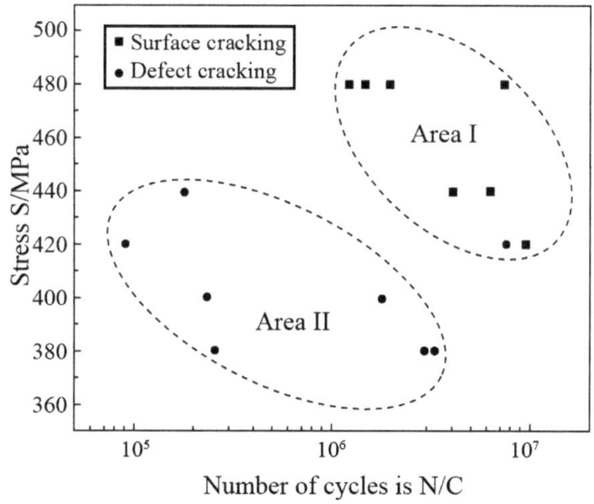

Fig. 4.59 S–N curve of TC4 alloy with ϕ4mm interface defect (stress ratio $R = -1$)

Figure 4.60 shows optical microscope and SEM graphs of three typical axial high-cycle fatigue fractures and the crack initiation area of TC4 alloy with ϕ4mm diffusion bonding interface defects. During the experiment, the fatigue specimen is subjected to cyclic tensile and compressive stress under the load of stress ratio R = -1, and fracture closure during the process results in wear in the initiation area. The fracture surface is divided into crack source area, growth area, and transient fracture area. Three typical fatigue fracture surfaces display clear river-like growth patterns, but the location of fatigue crack source of the specimens differs.

As shown in Fig. 4.60a, the fatigue crack of the specimen initiates on the outer surface of the specimen (the specimen is located in the area I of S–N distribution in Fig. 4.59). The crack nucleation and micro-crack growth occur due to the dislocation slip accumulation on the outer surface. When entering the steady-state growth area, the hard internal defects play a role in hindering the crack tip growth, and the fatigue cracks need to bypass the hard internal defects or interface defects.

In contrast, the fatigue crack of the specimen shown in Fig. 4.60b initiates at the tip of the internal defect (the specimen is located in the area II in the S–N distribution in Fig. 4.59), and the fracture presents a single "fish eye" crack source. The micro-crack propagates to the upper, lower, and right sides of the crack source to form a dark micro-crack growth area. Under the continuous action of the alternating cyclic load, the crack enters the steady-state growth area, and the hard interface defect filler will play a role in hindering the crack growth process.

In general, when a smooth and continuous specimen is subjected to axial fatigue loading, the crack source is usually located on the surface of the specimen due to crack nucleation caused by slip accumulation of surface dislocations. However, material discontinuities, such as defects, can cause different levels of stress concentration, which are closely related to the size, shape, and distribution of defects. To investigate this, the stress distribution of TC4 alloy specimens with ϕ4mm diffusion bonding interface defects was simulated using the finite element method. The Mises stress

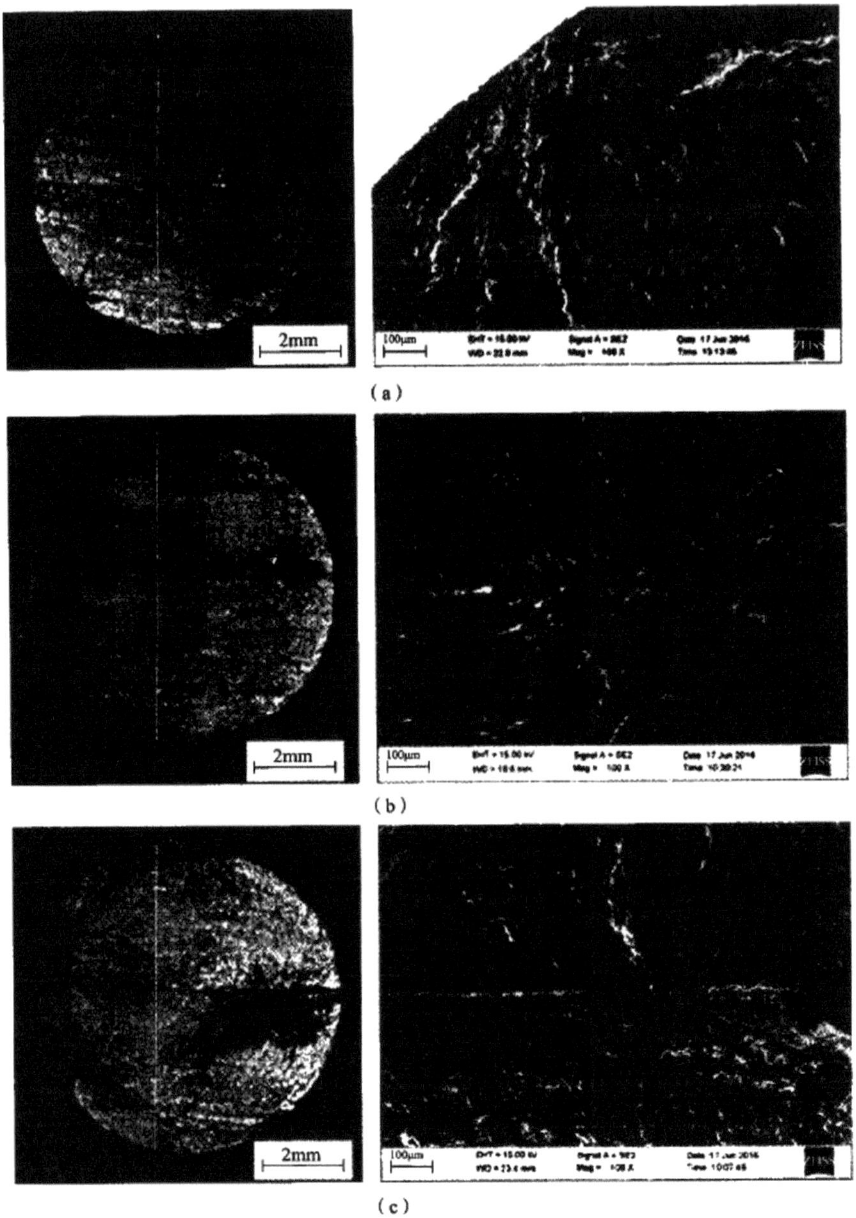

Fig. 4.60 Fatigue fracture morphology of TC4 alloy with ϕ4mm diffusion bonding interface defects (stress ratio $R = -1$) **a** Surface crack source, $\sigma_{max} = 440$MPa, $N = 4{,}074{,}900$ cycles; **b** The defect is located in the center and is the crack source, $\sigma_{max} = 400$MPa, $N = 1{,}826{,}100$ cycles; **c** The defect is off-center and is the crack source, $\sigma_{max} = 380$MPa, $N = 3{,}334{,}500$

distribution near the internal defects was obtained when the defects were located in the middle (Fig. 4.61a, b) and at the edge (Fig. 4.61c, d) under the maximum tensile and compressive stress.

When the interface defect at the center of the specimen is subjected to alternating loads parallel to its plane, a certain stress concentration is formed at the edge of the defect, but it is equivalent to the surface of the specimen. The stress gradient at the root of the fatigue specimen is small. Simulation results show that when the interface defect is at the off-center position (Fig. 4.61c, d), a greater stress concentration is formed at the surface of the specimen near the defect than at the center. The stress at the edge of the defect is 15 MPa higher than that on the surface of the specimen, and the stress gradient at the root of the fatigue specimen increases.

In both cases, the difference between the stress on the specimen surface and the stress at the defect tip is not obvious (about 3%), so the interface defect does not cause all crack sources to move to its tip. However, it should be noted that the degree of stress concentration caused by the internal defect (ϕ4mm, thickness of 0.05–0.1mm) whose plane is parallel to the axial load is far less than that caused by the through-hole (stress concentration factor Kt = 3).

According to the results of finite element simulation of stress distribution under alternating loads with interface defects, a certain stress concentration is formed near the boundary of interface defects, and the effect of stress concentration at the defect offset center is significant. When the interface defect is introduced into the continuous homogeneous material, high-cycle fatigue crack can initiate both from the outer and inner surfaces of the specimen. Therefore, the high-cycle fatigue life of the specimens with diffusion bonding interface defects are highly dispersed.

Compared with the traditional high-cycle fatigue test of titanium alloys with continuous internal and smooth external surfaces, the introduction and location of internal defects parallel to the stress axis brings internal interface as well as tip effects, and therefore causes redistribution of stress field under the fatigue load. However, this stress concentration effect is far less than that when the defect plane is perpendicular to the axis or in the case of through-hole, and is affected by boundary conditions such as defect size and location. Therefore, the high-cycle fatigue crack initiation mechanism and failure mechanism of titanium alloy with internal diffusion bonding interface defects are quite different, and results in the scattering of high-cycle fatigue life data. Further investigation is needed to fully understand the failure mechanism.

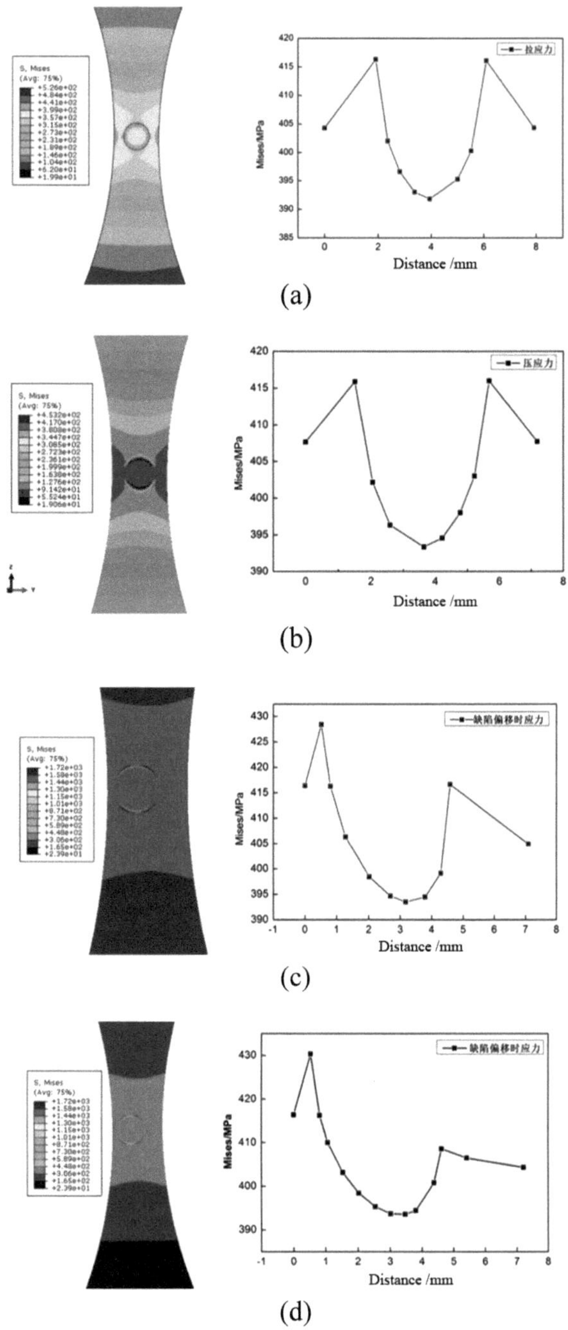

Fig. 4.61 Finite element simulation of stress distribution in defect region of TC4 alloy during fatigue (stress ratio $R = {}^{-}1$) **a** Stress diagram near the maximum tensile stress defect; **b** Stress diagram near the maximum compressive stress defect; **c** Stress diagram near the defect with maximum tensile stress when the defect is offset; **d** Stress diagram near the maximum compressive stress defect when the defect is offset

References

1. Wu Shidun. 1997. Theory of superplastic deformation of metals. Beijing: National Defense Industry Press.
2. Zhaorong, Lin. 1990. *Principle and Application of Metal Superplastic Forming*. Beijing: Aviation Industry Press.
3.] Nieh, T.G. 1997. Superplasticity in metals and ceramics[M]. Cambridge: Cambridge University Press.
4. Zhiqiang, Li., and Guo Heping. 2010. Application progress and development tendency of superplastic forming/diffusion bonding technology. *Aeronautical manufacturing technology* 8: 32–35.
5. Kaifeng, Zhang, and Wang Guofeng. 2012. *Advanced superplastic forming technology for materials*. Beijing: Science and Technology of China Press.
6. Kayeboshev, C.A. 1982. Plasticity and superplasticity of metals. Translated by Wang Yanwen. Beijing: China Machine Press.
7. Zhiqiang, Li., and Guo Heping. 2004. Application and development of superplastic forming/ diffusion bonding technology. *Aeronautical Manufacturing Technology* 11: 50–52.
8. Wen Jiuba, Yunlin Yang, and Yongshun Yang, et al. 2005. Superplastic application technology. Beijing: China Machine Press.
9. Wenjuan, Zhao, Ding Hua, Cao Furong, et al. 2007. Microstructure evolution and deformation mechanism of Ti-6Al-4V alloy during superplastic deformation. *The Chinese Journal of Nonferrous Metals* 17 (12): 1973–1980.
10. Lixia, M.A., L.X. Wan, M. Weidong, W.D. Li, et al. 2022. On the superplastic deformation mechanisms of near-a TNW700 titanium alloy. *Journal of Materials Science and Technology* 108: 173–185.
11. Ganghua, Huang. 2009. *Numerical simulation and process research on superplastic forming/ diffusion bonding of titanium alloy [D]*. Nanjing: Nanjing University of Aeronautics and Astronautics.
12. Jinsheng, Pan. 2000. *Fundamentals of material science*. Beijing: Tsinghua University Press.
13. Xiangming, Wang, and Liu Wenting. 2010. *Structural design and application of aircraft titanium alloy*. Beijing: National Defense Industry Press.
14. Zhao, B., Z.Q. Li, H.L. Hou, et al. 2010. Three dimensional fem simulation of titanium hollow blade forming process]J]. *Rare Metal Materials and Engineering* 39 (6): 963–968.
15. Guo Heping, Yuansong Zeng, and Xiuquan Han, et al. 2008. Superplastic forming/welding combination technology for titanium alloy integral structure of aircraft. *Welding* 11:41–45.
16. Qiang, G.G., L.D. Sheng, L.X. Qiang, et al. 2018. Simulation and experiment investigation on superplastic forming/diffusion bonding process of a Ti-6A1-4V alloy rear fuselage part. *Defect and Diffusion Fo-rum* 385: 407–412.
17. Bing, Zhao, Li. Zhiqiang, Han Xiuquan, et al. 2010. Three-dimensional finite element analysis of forming process of titanium alloy hollow monolithic structure based on rigid viscoplastic constitutive relation. *The Chinese Journal of Nonferrous Metals* 04: 14–21.
18. Yoon, J.H., H.S. Lee, Y.M. Yi, et al. 2007. finite element analysis on superplastic blow forming of Ti6A14V multi-sheets. *Materials Science Forum* 546–549: 1361–1366.
19. Zhang Jiuhai, and Peng He. 2000. Development of numerical simulation of diffusion joint behavior. *Transactions of The China Welding Institution*, 04: 84–91, 101.
20. Shen Junjun. Numerical Simulation of Interface Characteristics and Hole Closure of TC4 Titanium Alloy Diffusion Bonding [D]. Harbin: Harbin Institute of Technology, 2007.
21. Shun, Zhang. 2018. *Study on diffusion bonding process and mechanical properties of titanium alloy based on molecular dynamics [D]*. Nanjing: Nanjing University of Aeronautics and Astronautics.
22. Liu, Y.X., W. Chen, Z.Q. Li, et al. 2016. The HCF behavior and life variability of a Ti-6A1-4V alloy with transverse texture. *International Journal of Fatigue* 23 (97): 79–87.
23. Jha, S.K., C.J. Szczepanski, R. John, et al. 2015. Deformation heterogeneities and their role in life-limiting fatigue failures in a two-phase titanium alloy. *Acta Materialia* 82: 378–395.

24. Schuve, J. 2001. *Fatigue of structures and materials*. Dordrecht: Kluwer Academic.
25. Wu, G.Q., C.L. Shi, W. Sha, et al. 2013. Effect of microstructure on the fatigue properties of Ti—6Al—4V titanium alloys. *Materials and Design* 46 (2): 668–674.
26. Zhentong, Gao, and Xiong Junjiang. 2000. *Fatigue reliability*. Beijing: Beijing University of Aeronautics and Astronautics Press.
27. Safiullin, R.V., O.A. Kaibyshev, R.Y. Lutfullin, et al. 1994. Solid state joint formation under conditions of the process of superplastic forming and diffusion bonding. *Materials Science Forum* 170–172: 639–644.
28. Wujing, Deng, Shao Jie, Chen Wei, et al. 2020. Influence of interface defects on high-cycle fatigue behavior of titanium alloy diffusion welded joints. *Aeronautical Manufacturing Technology* 63 (22): 78–83.
29. Wujing, Deng, Chen Wei, Li. Zhiqiang, et al. 2017. Influence of diffusion bonding interface defects on mechanical properties of Ti-6A1-4V alloy. *Aeronautical Manufacturing Technology* 18: 74–78.

Chapter 5
Test Method of Superplastic Forming/Diffusion Bonding Structure

The single-layer plate structure and two-layer plate structure manufactured by the superplastic forming/diffusion bonding (SPF/DB) process are commonly used for secondary load-bearing or non load-bearing structures, and usually only require the shape accuracy of parts. Three-layer plate structure and four-layer plate structure are commonly used in the main load-bearing structure, which requires not only high accuracy of part shape, but also good design manufacturing compliance of internal structure, and reasonable distribution of internal and external residual stress. The manufacturing process of SPF/DB structure is long, and the quality control of some manufacturing links depends on the process and its stability. It is difficult to control the consistency of welding interface, internal and external geometric accuracy, and internal residual stress. The inspection process is set in the workpiece manufacturing process to avoid bringing defects into subsequent manufacturing links and ensure the final part performance. This chapter focuses on the testing methods for shape, internal structure, diffusion bonding interface quality, residual stress, etc.

5.1 Shape Inspection Method

The inspection tasks of SPF/DB structure are divided into final inspection and inspection during manufacturing. The oxidation of workpiece surface materials caused by high temperature environment and the gradual removal of workpiece materials in the manufacturing process will destroy the initial benchmark set on the blank, making it difficult to machine and measure the profile in the subsequent process. In view of the characteristic that the initial processing datum is destroyed, it is necessary to reconstruct the datum origin and measurement coordinate system before the profile inspection. After the measurement is completed, the transformation relationship between the reconstructed coordinate system and the theoretical coordinate system is used to optimally match the measurement data and the theoretical model, and then the geometric and contour accuracy of the workpiece or part is evaluated.

Z. Li, *Superplastic Forming/Diffusion Bonding Technology of Titanium Alloys*,
https://doi.org/10.1007/978-981-99-3909-1_5

When measuring engine static and rotor blades and other parts, they are also faced with the characteristics of large curvature difference in different areas of the outer surface of the workpiece and large shape deviation between parts, which makes contact measurement method more difficult. The measurement parameters need to be set according to the characteristics of the parts to be measured, and the testing efficiency of a single part is low. Therefore, when selecting the shape testing method, it is necessary to select the appropriate testing method from the standard template method, coordinate testing method, optical scanner method and other measurement methods according to the measurement accuracy and cycle requirements, combined with the characteristics of the workpiece at different stages of the manufacturing process.

5.1.1 Coordinate Measuring Method

Coordinate measurement is a contact measurement method. Typical testing equipment is Coordinate measuring machine (CMM Fig. 5.1). During the measurement, the sensor obtains the point position information through physical contact with the parts. The advantages of this measuring equipment are mature technology and high measurement accuracy, and the geometric measurement accuracy for parts with complex shape can reach 1.0 μm, which meets the requirements of geometric dimension inspection of common aerospace parts. However, the preparation work of this method is heavy, and it needs to plan the measurement path and prepare the measurement program based on the characteristics of the parts and measurement requirements, which takes a long time. In addition, this method requires strict equipment installation conditions, resulting in high use and maintenance costs, and the size of the measured parts is usually limited by the equipment measurement space.

The main process of coordinate measurement is as follows: place the measured part in the measuring space of coordinate measuring equipment, plan the number of measuring points and measuring path according to the testing requirements and theoretical model, and calculate the coordinate position of each point on the measured part. The geometric dimension, shape and position features of the measured part are solved through mathematical operation after obtaining the coordinate information of the point through testing. Equal altitude method is a common measurement method: select a series of contour planes according to the shape characteristics of the measured object; select multiple measuring points on each contour plane, calculate the normal direction of the measuring point according to the theoretical model, and set the return point and positioning point of the measuring head of sensor along the normal direction; when testing, move the measuring head to the positioning point of the measuring position, the measuring head moves slowly along the normal direction until it contacts the part, then the measuring head slowly moves back to the return point, and the coordinate data information of this measuring point is obtained through this process; next, move the measuring head to the next measuring point, repeat the

Fig. 5.1 Coordinate measuring machine (CMM)

above process to complete the measurement of the next coordinate, and repeat the process until all coordinate testing tasks are completed.

When the shape of the part is complex or the datum is missing, the first step is to find the datum plane of the part to be measured, establish the measurement coordinate system, and use coordinate transformation to make the measurement coordinate system coincide with the coordinate system of the theoretical model, which is usually the first step of the shape measurement of the superplastic forming diffusion bonding part. If the measuring object is a workpiece, the common practice is to use the contour data of the near free area of the workpiece to establish the reference coordinate system. If the measuring object is the final part, the datum coordinate system is established through the contour data of the machining area with good rigidity and regular geometric shape, and then, plan the measurement track, generate the measurement code, and the computer completes the measurement according to the program. Finally, compare the measured data with the theoretical model to obtain the deviation between the measured part shape data and the theoretical model. The coordinate measurement process is shown in Fig. 5.2, where A_i is the return point, B_i is the positioning point, and C_i is the theoretical measurement position.

The coordinate method has a long preparation time, and the measuring head measures point by point, so that the efficiency is low, but if these two shortcomings are avoided, the measurement efficiency can be significantly improved. Under the guidance of this idea, optical measurement technology has been developed, such as binocular vision measurement, laser joint arm measurement method, etc. The measuring head of the optical measuring equipment will not have physical contact with the parts, avoiding the risk of mutual collision and equipment damage, reducing the preparation workload before measurement and shortening the preparation period. The optical measurement method uses image or line scanning to obtain the geometric

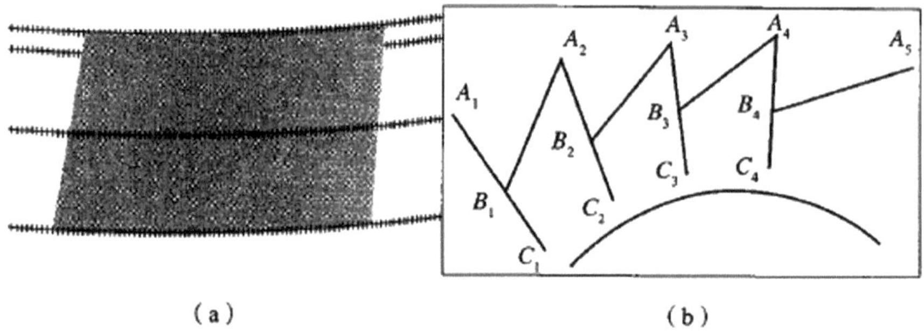

Fig. 5.2 Measurement process of coordinate testing method. **a** Generate measurement points; **b** Plan measurement path

data of the measured part, and the data acquisition cycle is short. This method generally has no limit on the size of the measured parts, and has good adaptability to the environment. These characteristics endow optical testing methods with high testing efficiency and flexibility. In SPF/DB structure manufacturing, this method is widely used in the process of tooling, dies, workpieces and other parts that have relatively low requirements on the accuracy of shape testing.

5.1.2 Optical Testing Method

The binocular vision measurement method is characterized by projecting the measuring grating onto the surface of the workpiece to be measured, recording the grating image by the camera, and calculating the point cloud coordinates of the workpiece to be measured based on the optical triangle principle and data processing by the digital image processor. When the shape of the measured part is complex and the surface state is poor, the measurement accuracy is improved by evenly coating the developer on the surface of the measured object. If the size of the measured part is large, which causes the measurement area to exceed the image capture range of the camera, the measurement can be conducted in different areas first, and then the data of multiple areas can be spliced to obtain the overall information.

The binocular vision method has high measurement efficiency and lower measurement accuracy than the coordinate measurement method. When the measured object has sharp shape features, it is highly likely to lose data points, causing measurement data holes, reducing the measurement accuracy in these areas, which should be avoided in engineering applications.

The binocular vision measurement method has been applied in the SPF/DB structure shape measurement (Fig. 5.3a). Use a high pixel SLR camera (Fig. 5.3b) to take pictures of all identification points of the tested part (Fig. 5.4a). The captured image data is processed by software, the measurement coordinate system is constructed, and the coordinate information of the identification point is calculated, and the part shape data is spliced under the global coordinate system. The measurement results are shown in Fig. 5.4b, c. The measurement time of a single workpiece is less than 10 min.

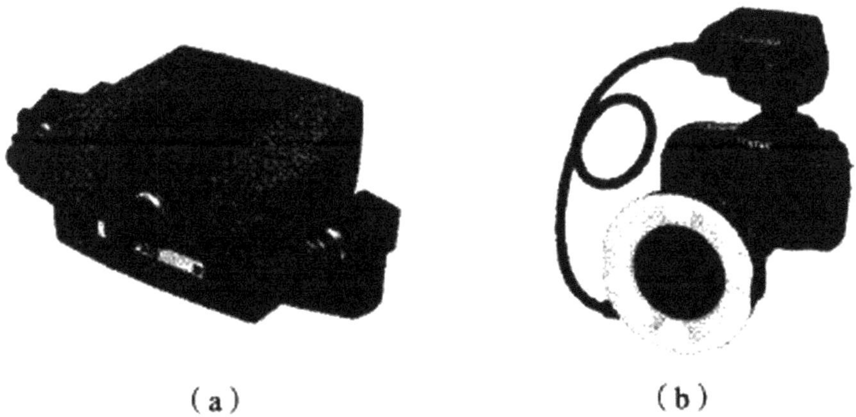

Fig. 5.3 Optical testing equipment. **a** Binocular vision measurement equipment; **b** Single-lens reflex camera

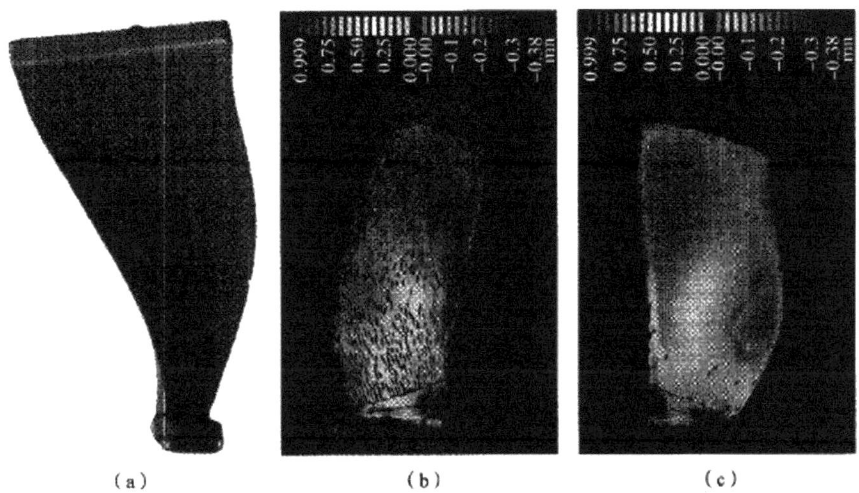

Fig. 5.4 Binocular vision measurement results of workpieces with complex bending and twisting profiles. **a** Tested workpiece; **b** Concave measurement data and deviation structure; **c** Convex measurement data and deviation structure

5.2 Internal Structure Inspection Method

There are mainly 6 types of internal structural defects of SPF/DB structure: deviation of skin thickness, partial thinning of truss in thickness, distortion of truss, deviation of welding interface size, deviation of welding position, deviation of included angle between truss and skin, etc. The above deviations affect the bearing capacity of the structure and are key testing items. Hollow sandwich structures are usually closed, which is the main difficulty in testing. For such structures, it is difficult to observe the geometric characteristics of internal structures and measure the internal shape accuracy by visual inspection or endoscope. Therefore, high-energy ray, ultrasonic and other methods are mainly used. Common nondestructive testing methods include CT testing, X-ray testing, etc.

5.2.1 X-Ray Testing

When the X-ray passes through the tested part, the intensity of the ray signal is affected by the thickness of the material. After the signal is received by the sensor, a gray image is formed, and the gray value of the image reflects the thickness characteristics of the tested part. X-ray testing methods are mainly divided into two categories: X-ray radiographic testing technology and X-ray real-time imaging testing technology. X-ray radiographic testing technology is mainly used for static imaging testing of structural members, including film radiography and digital radiography; X-ray real-time imaging method is mainly used for real-time or near real-time dynamic imaging testing of structural parts, including image intensifier and digital imaging plate real-time imaging technology, with fast testing speed.

X-ray real-time imaging method is mainly used for real-time or near real-time dynamic imaging testing of structural parts, including image intensifier and digital imaging plate real-time imaging technology, with fast testing speed. This testing method is sensitive to the size of the part wall thickness, and is suitable for the thickness consistency testing of materials inside the closed hollow sandwich structure. In addition, the preparation of this method is simple, the testing efficiency is high, and there are many engineering applications. Generally, the X-ray testing is set after the superplastic forming process to qualitatively judge the forming quality of the internal structure of the workpiece.

X-ray testing method is commonly used to detect the local thinning and distortion of the truss inside SPF/DB structures. Figures 5.5 and 5.6 show the X-ray images of a hollow sandwich structure. The images show several high brightness lines parallel to the length direction of the part, reflecting that some trusses have local thinning near the location their intersection with the skin.

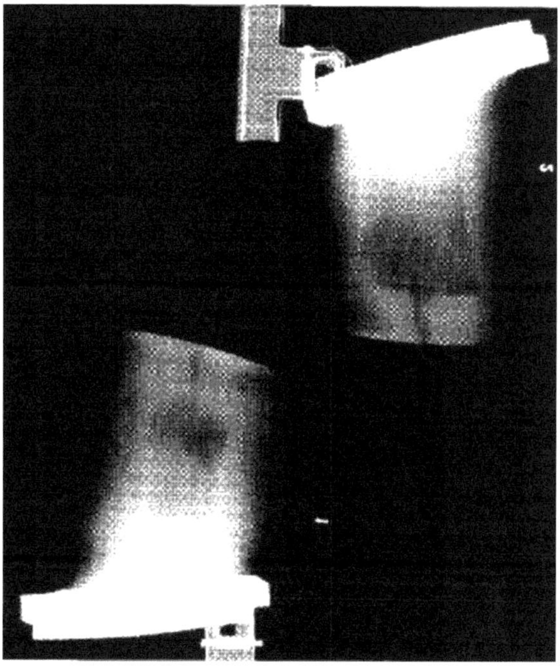

Fig. 5.5 X-ray image of hollow sandwich structure (the white area is a solid area, and the inside of the white area is a cavity area)

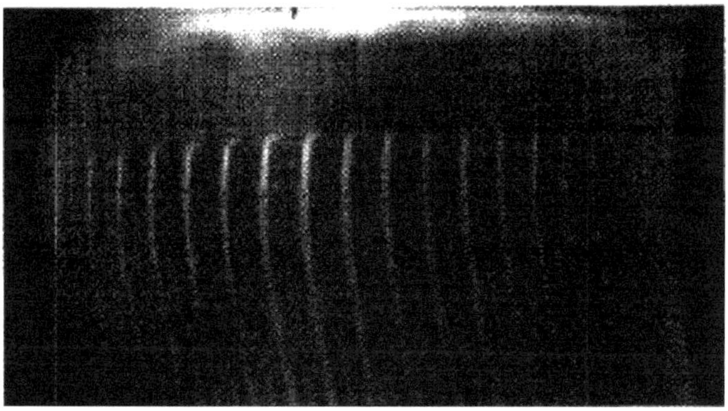

Fig. 5.6 X-ray image of hollow sandwich structure (some areas have bright curve characteristics, which is caused by the truss necking near the truss skin intersection)

5.2.2 CT Testing Method

CT testing method provides a way for accurate testing of the internal structure of hollow sandwich, but such equipment is expensive to build and use, and a measurement process takes a long time, so it is mostly used in the initial research and development stage of SPF/DB structure, supporting structure design and process optimization. This method is mainly used for sampling verification when the parts are manufactured in batch production, and it is also used to assist the analysis of faulty parts.

CT testing method provides a way to visualize the internal geometric characteristics of structures. This method uses the influence of material thickness on the ray intensity to generate a gray image to display the distribution of materials on the measurement plane. The dark area in the image indicates large material thickness dimensions. The large difference in the thickness of the component materials in the measurement plane and the adjustment of the image display parameters will affect the determination of the material boundary. Calibration blocks with the same geometric material and similar thickness as the tested object can be prepared, and CT testing parameters and image display parameters can be optimized to ensure the measurement accuracy of geometric quantities such as length and thickness. Calibration is not required for measuring relative position geometric characteristics such as angle. Figure 5.7 shows the contour section image of titanium alloy three-layer plate structure. The variation rule of skin thickness in the three areas and the included angle between skin and truss can be intuitively obtained from the figure.

Figure 5.8 shows the crack characteristics of a hollow sandwich structure after fatigue experiment detected by CT. Figure 5.8a shows the image of the section perpendicular to the thickness direction of the specimen. The black thin lines in the figure are cracks, from which information on the number, location and length of cracks can be obtained. Figure 5.8b shows the section information perpendicular to the height direction, from which it can be seen that the titanium alloy superplastic forming/diffusion bonding technology and application cracks have penetrated the skin, and there is a black rectangular area perpendicular to the thickness direction on the internal surface of the structure, indicating that this area may be the crack initiation location. CT testing technology provides the geometric characteristics of

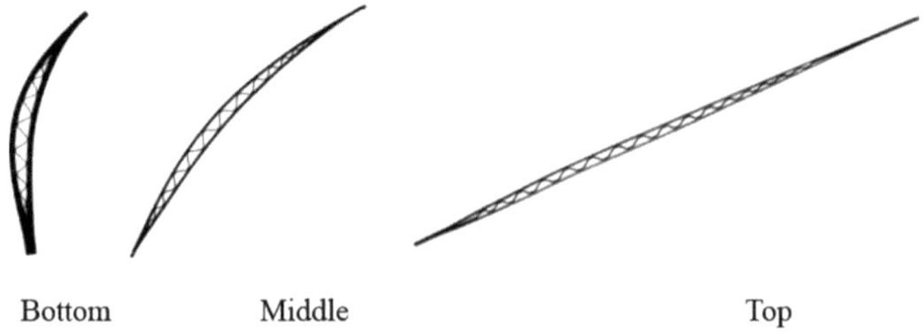

Fig. 5.7 Image of hollow sandwich structure with equal height cross section

the final stage of the failure of the specimen, and this information has an important reference role for the subsequent failure analysis.

In recent years, three-dimensional CT measurement method has been developed. This method relies on a series of plane data obtained by CT measurement to build a three-dimensional digital model of the measured part, which improves the visualization level of CT testing data, and can quickly obtain geometric features on any interface of the part through the three-dimensional digital model, greatly improving the flexibility of analysis.

3D CT measuring equipment generally provides three scanning functions: 3D cone-beam scanning, large-field scanning and spiral-cone-beam scanning. The three-dimensional cone-beam scanning method is the most commonly used method in engineering due to its short testing time and low cost. The maximum thickness that can be measured depends on the energy level of the ray. For the hollow sandwich structure made of titanium alloy materials with a width of about 300 mm, the maximum thickness of the outer contour of about 15 mm, the wall thickness of one side of about 2.5 mm, and the maximum height of the internal cavity of 10 mm, the industrial CT system with 450 kV conventional energy array and 6 and 9 meV high-energy accelerator array can be used for measurement to obtain high-quality sectional image features. If the length direction dimension parameters of the measured parts exceed the measuring range of the equipment, the principle of geometric similarity can be used to splice multiple measured data, and the software will automatically establish a three-dimensional digital model based on the measured data, as shown in Fig. 5.9, so that the geometric features of any contour section can be obtained. In addition, it can also analyze the geometric characteristics of sections in any direction.

5.3 Diffusion Bonding Interface Quality Testing Method

The area of diffusion bonding interface in common hollow sandwich structures varies from hundreds of square millimeters to several square meters. From the perspective of structural design and strength, it is hoped that all diffusion bonding areas can reach the metallurgical bonding strength, and that the quality of diffusion bonding interfaces is consistent between different parts, however, which is difficult to achieve in engineering. Under the action of multiple factors, interface defects with random location and size may appear in the diffusion bonding area designated by the design. The defect height is 0.01–0.1 mm when viewed from the direction perpendicular to the normal direction of the diffusion bonding interface. Therefore, the geometric characteristics of interface defects are similar to those of initial cracks.

Optimizing the position of diffusion bonding interface and controlling the size of interface defects in structural design, manufacturing and testing can prevent the interface defects from expanding and forming cracks when parts are subjected to alternating loads, and ensure the safety of structure in service. In the structural design stage, the interface defects that may occur in the structure is considered, and the diffusion bonding position is designed and the allowable value of defect size is given

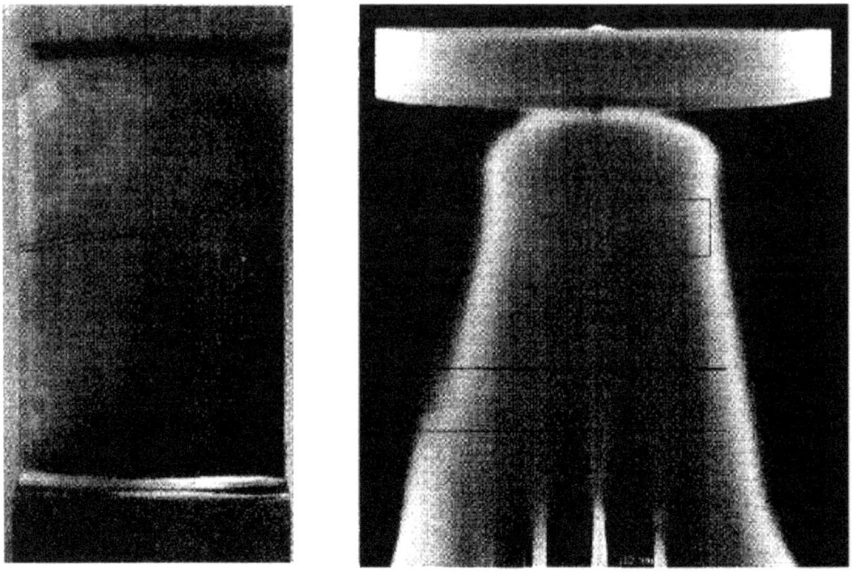

(a)

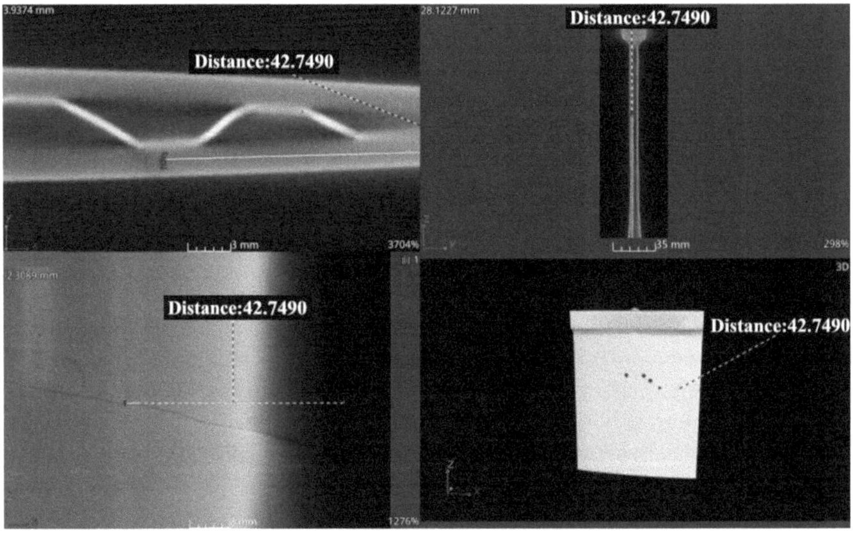

(b)

Fig. 5.8 Crack characteristics of hollow sandwich structure after fatigue experiment detected by CT. **a** Macro failure characteristics and crack cross section characteristics; **b** Crack characteristics in longitudinal section

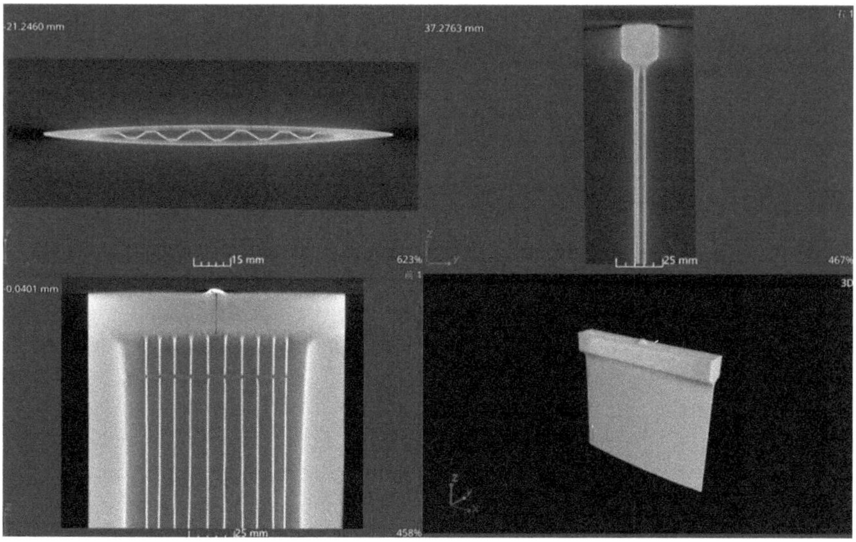

Fig. 5.9 CT testing results of hollow sandwich structure and 3D digital model reconstruction model

under the constraints of the structure weight and bearing demand. For the area with high bearing stress in the structure, the diffusion bonding interface can be set inside the structure or deep away from the structure surface to reduce the stress level on the diffusion bonding interface. Even if there is a certain size defect, it will not expand to form cracks during service. In the manufacturing process, the quality of diffusion bonding interface can meet the requirements and ensure its stability and consistency by optimizing the process parameters and diffusion bonding process. The quality of diffusion bonding interface can meet the requirements and ensure its stability and consistency by optimizing the process parameters and diffusion bonding process. In the three links, "testing" is the basis and core. In the development stage, its data will support the development process of defect threshold; in the batch production phase, the data is the basis for judging whether the parts are qualified; and in the process of field service, its data is the criterion for the safety of components and whether they can continue to serve.

Diffusion bonding interface is generally in the hollow sandwich structure, which is difficult to measure directly. In engineering, the geometrical discontinuity of diffusion bonded interface defects in the direction perpendicular to the interface is used to determine the location and size of defects by measuring thickness changes. The common methods include immersion ultrasonic testing, micro CT testing and metallographic testing. The first two methods are nondestructive testing methods.

5.3.1 Non-destructive Testing Method

Ultrasonic testing method is widely used in engineering to evaluate the welding interface quality of SPF/DB parts. Water immersion ultrasound is one of those methods. It uses water as the coupling agent. When testing, the parts are completely immersed into water, the probe sends ultrasonic signals, and the signals enter the parts along the direction perpendicular to the diffusion bonding interface. After the ultrasonic signals are reflected by the defects or the internal surface of the structure, the signals between the defects and the welded area will be different, so as to identify the internal characteristics of the structure. This testing method is also called longitudinal wave vertical reflection method. There are two ways to improve the ability of this method to identify defects: one is to improve the gain. In theory, this approach can improve the ability to identify small defects, but it also leads to an increase in the noise level and tends to annihilate the signals of small defects; the other way is to optimize the ultrasonic frequency and use prefabricated defect standard specimens for calibration. The equipment and typical acoustic signals for immersion ultrasonic testing are shown in Fig. 5.10.

The shape of the prefabricated defect standard specimen is usually a cylinder, with a cylindrical blind hole at the bottom, and the bottom of the blind hole is a plane. This feature is consistent with the defect characteristics. Therefore, the specimen is also called a flat-bottom hole standard specimen. The diameter of the flat-bottom hole and the height of the cylinder depend on the size and location characteristics of the defects to be detected, but the size is usually one order of magnitude smaller than the size required to be detected. The scanning test data of the standard specimen of the prefabricated flat-bottom hole can be used to optimize the test parameters such as the probe frequency to obtain a high-precision testing method for diffusion bonding interface defects. A certain type of structural material is TC4 alloy. The testing accuracy can be evaluated to reach 0.8 mm^2 by using the standard specimen which is 70 mm in height, the diameter and the height of the flat-bottom cylinder hole is 0.8 mm and respectively. Figure 5.11 shows the ultrasonic signal of the standard

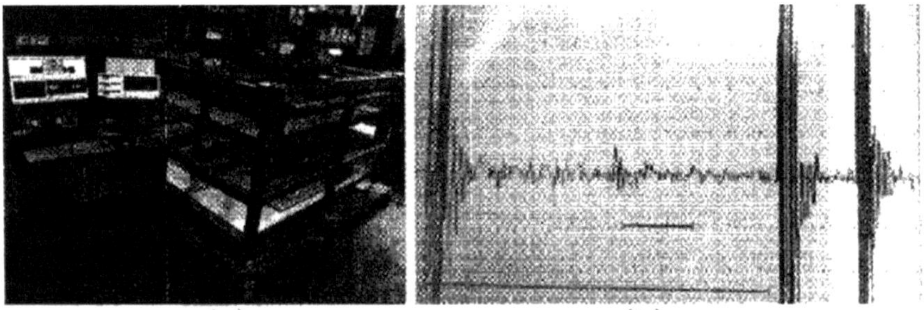

(a) (b)

Fig. 5.10 Equipment and typical acoustic signals for immersion ultrasonic testing. **a** Testing equipment; **b** typical acoustic signal specimen of prefabricated flat-bottom hole. The image of the standard specimen tested by ultrasonic C-scan imaging is shown in Fig. 5.12.

The longitudinal wave vertical reflection method requires that the incident direction of sound wave is perpendicular to the diffusion bonding interface, but the workpiece shape is complex, and the diffusion bonding interface is located at the midpoint of the structure thickness, resulting in an included angle between the workpiece surface and the diffusion bonding interface, which is difficult to strictly meet the requirements. To solve this problem, the included angle between the workpiece surface and the diffusion bonding interface can be reduced by optimizing the workpiece shape. After the testing is completed, the added materials can be removed by machining to obtain the final shape of the part, as shown in Fig. 5.13. When the above methods cannot be used, bevel blocks can be prepared based on the shape characteristics of the workpiece to optimize the testing parameters when the ultrasonic wave

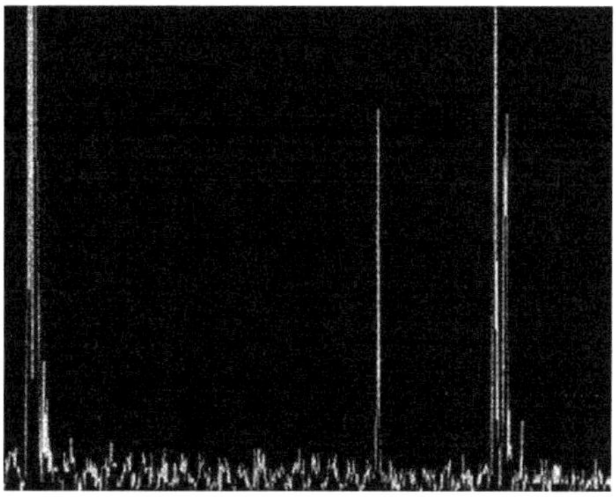

Fig. 5.11 Ultrasonic signal with flat-bottom hole diameter of 0.8 mm and cylinder length of 70 mm standard specimen

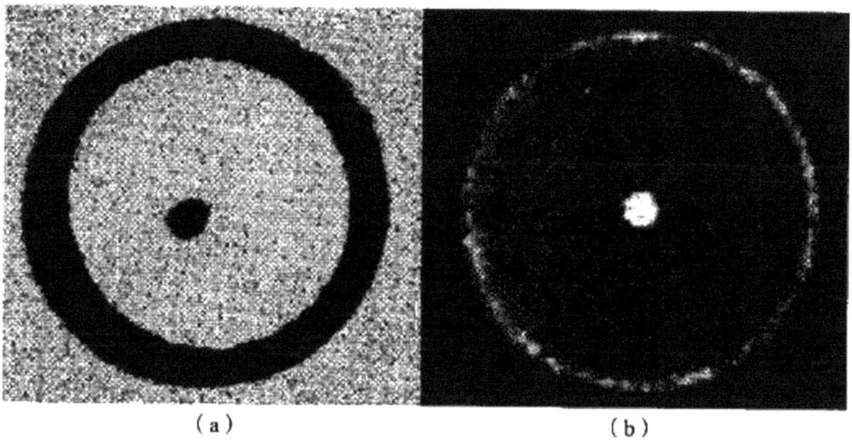

Fig. 5.12 Ultrasonic C-scan imaging results of TC4 alloy flat-bottom hole standard specimen. **a** The diameter of the flat-bottom hole is 0.8 mm, and the length of the cylinder is 5 mm; **b** The C-scan image of flat-bottom hole standard specimen

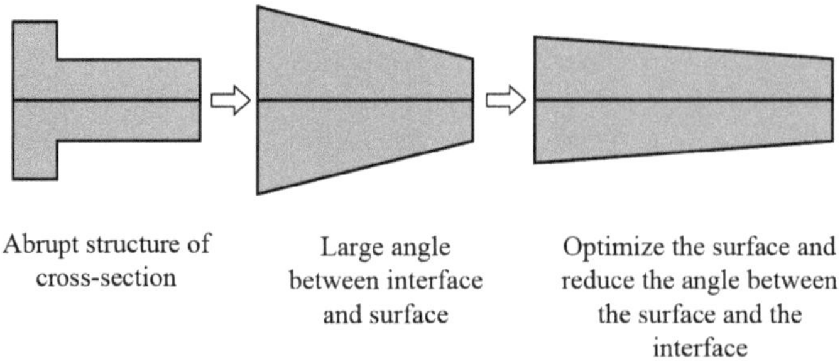

Fig. 5.13 Optimization of workpiece shape—reducing the angle between the surface and diffusion bonding interface

is not vertically incident. The shape of the bevel block is shown in Fig. 5.14. Typical ultrasonic C-scan image features are shown in Fig. 5.15.

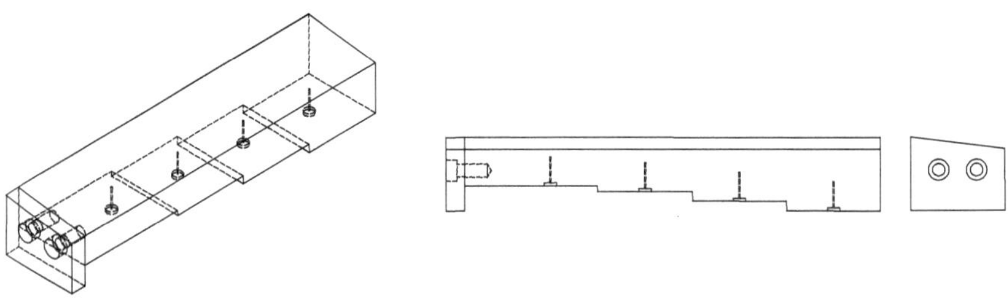

Fig. 5.14 Bevel specimen

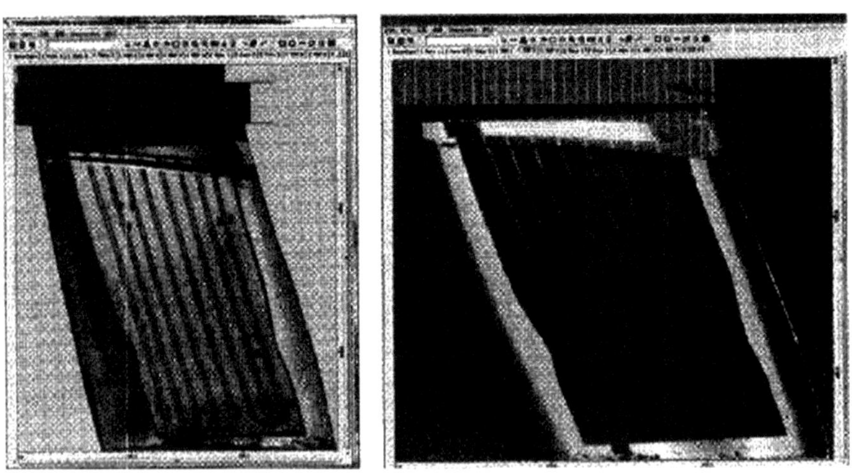

Fig. 5.15 Ultrasonic C-scan imaging results

In recent years, the ultrasonic phased array testing method has been developed, which uses a new type of coupling agent. It is no longer necessary to immerse all the tested parts in water, significantly improving the testing flexibility. The new testing method uses a phased array ultrasonic probe to improve the testing speed (Fig. 5.16a). The testing accuracy of this method is equivalent to that of the water immersion ultrasonic testing method. Through the oblique angle probe, the incident direction of the acoustic wave is made to form a certain angle with the normal direction of the surface of the part to be tested, and the cracks that originate on the internal surface of the part and whose extension direction is perpendicular to the bearing direction can be detected (Fig. 5.16b), thus expanding the testing range, which is of great significance to improve the reliability of the hollow sandwich part.

Micro CT testing is a method to detect diffusion bonding interface defects under laboratory conditions. This method can obtain the three-dimensional image of defects with high measurement accuracy. Combining the three-dimensional imaging data with CAD/FE modeling method, the three-dimensional characteristics of defects can be obtained, which is convenient for further establishing the analysis model. It

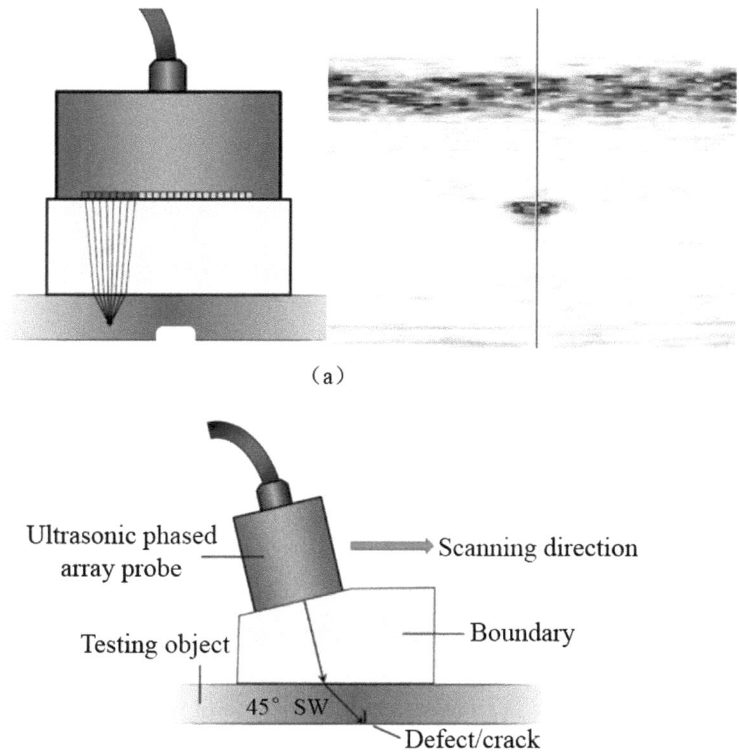

Fig. 5.16 Ultrasonic phased array testing method. **a** Test diffusion bonding interface defects; **b** Test cracks that originate from the internal surface and propagate along the thickness direction

Fig. 5.17 Micro CT testing results (obtaining the characteristics of diffusion bonding interface defects)

has great advantages in studying the three-dimensional characteristics and mechanical properties of diffusion bonding joint interface defects. The disadvantage of this method is that the size of the tested specimen is strictly limited, which generally cannot exceed $\phi 10$ mm $\times$ 100 mm. The diffusion bonding joint interface defect image obtained by micro CT testing method is shown in Fig. 5.17.

5.3.2 *Metallographic Testing Method*

Metallographic testing method is often used to analyze the microstructure of materials, and can also be used to evaluate the quality of diffusion bonding interface, which is a destructive testing method. This method is used to support the optimization of process parameters in the part development stage, to spot check and verify the quality conformance and consistency of the workpiece in the batch production stage, and to evaluate the quality of the local welding interface of the part in the failure analysis stage. As shown in Fig. 5.18, the metallographic specimen is made by cutting parts and taking local specimens, and then tested with a metallographic microscope. This method can directly obtain the geometric characteristics of defects and quantitatively measure their two-dimensional dimensions. Diffusion bonding joint defects are shown in Fig. 5.19, where three common types of defects are illustrated:

(1) Isolated defect. The section of some defects is approximately circular, with a diameter of about 10 μm. Some defects are strip, with a height of about 10 μm and a width of about 50 μm;
(2) Diffuse defect. There are several isolated point defects in the field of view of the microscope, some of which are round and some of which are nearly strip;
(3) Large size defect. The height is 10 μm, the length is more than 100 μm.

The metallographic inspection method can obtain the two-dimensional dimension parameters of defects, that is, the length and height of defects. After obtaining the length and size data of defects, the welding rate is often used to characterize the

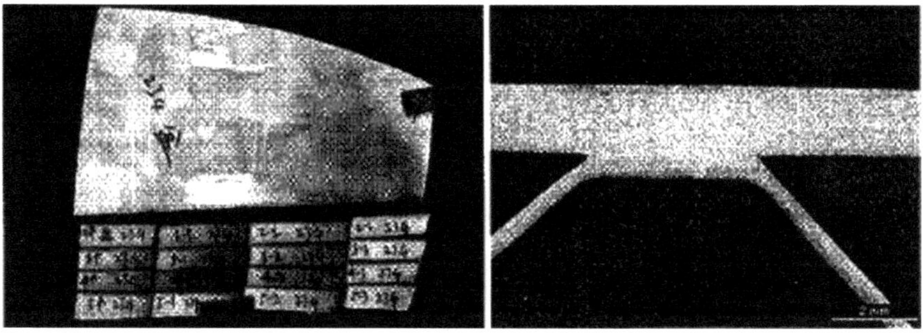

Fig. 5.18 Preparation of metallographic specimens by sampling hollow sandwich structures using wire cutting

quality of the welding interface. The calculation method is as follows

$$\text{Welding rate} = \frac{\text{Line length of diffusion bonding interface} - \text{line length of unwelded area}}{\text{Line length of diffusion bonding interface}} \times 100\% \tag{5.1}$$

5.4 Testing and Prediction of Structural Residual Stress

The residual stress in the hollow sandwich structure is may generated in the hot working and surface strengthening processes. Due to the weakness in the rigidity of this type of structure. Residual stress has a significant effect on the shape accuracy and the performance of parts under periodic load. However, it is difficult to accurately measure and control the magnitude and distribution of residual stress.

During hot working, the blank is placed in a high temperature environment and produces plastic deformation under gas pressure loading. Creep deformation occurs in the process of gas pressure holding, which relaxes the internal stress and theoretically does not produce residual stress. After forming, the parts are cooled and discharged from the high temperature condition. Due to the cooling speed difference between the mould and the workpiece and the difference in material expansion coefficient, the mould holds and extrudes the workpiece, resulting in residual stress. The residual stress during hot working can be controlled by increasing the holding time and slowing down the cooling rate. The residual stress in the workpiece can be effectively reduced by discharging the workpiece at room temperature discharge instead of high temperature, or by two-stage or multi-stage cooling rate.

In the process of surface strengthening, the materials on the surface and subsurface of the part undergo plastic deformation under the action of external forces, and the main extension direction of the material is perpendicular to the normal direction of the

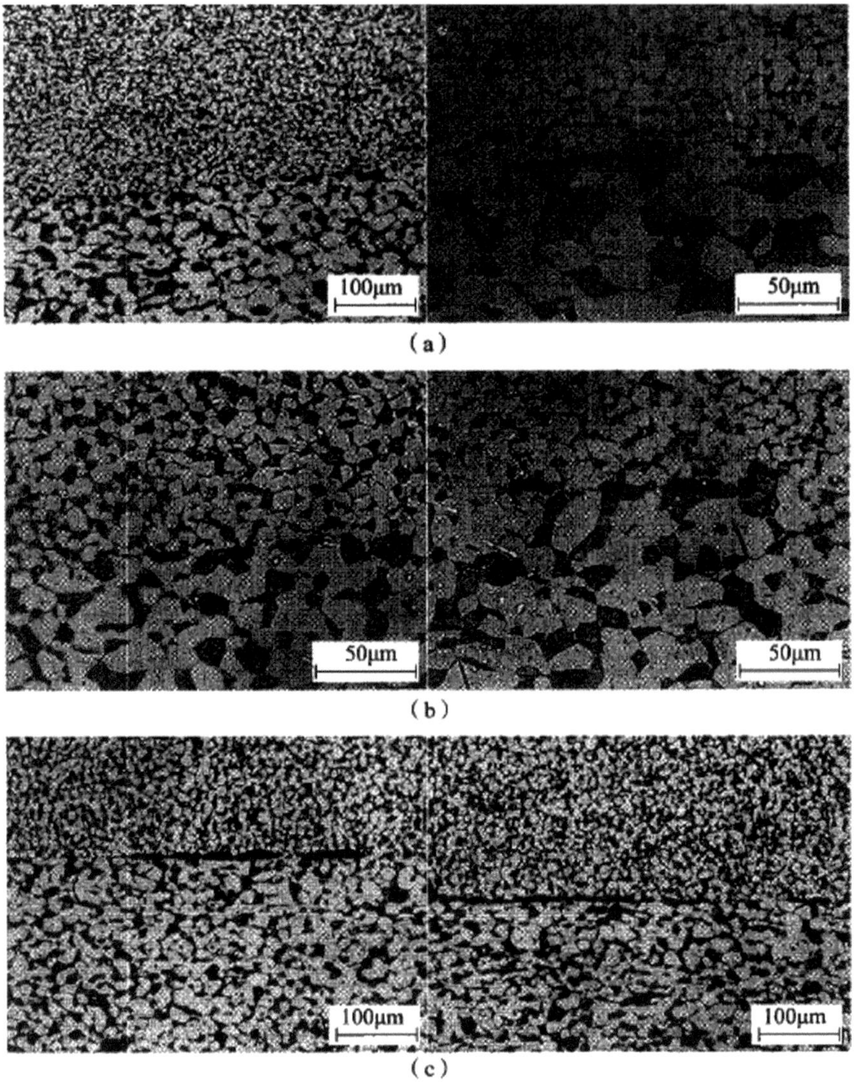

Fig. 5.19 Defect characteristics of diffusion bonding interface. **a** Isolated defect; **b** Diffuse defect; **c** Large size defect

surface. As the material below the sub surface does not produce plastic deformation, but it will undergo tensile deformation under the traction of the extension deformation of the surface material and exert reaction on the surface material, therefore, when the strengthening process is completed, residual stress will be generated in the structure, in which the compressive residual stress is between the surface and the subsurface of out side surface of parts, and the tensile stress is in the internal area of the part.

The internal area of the SPF/DB structure is closed, and the accessibility of the surface strengthening process is poor. Generally, only the external surface area can be strengthened. The internal surface of the structure is not strengthened and is affected by the tensile residual stress. Such characteristics reduce the ability of the internal surface to resist fatigue loading, which has a negative impact on the anti fatigue performance of the structure. Therefore, the surface strengthen process suitable for surface strengthening of SPF/DB structures should meet the geometric requirements. First, the process should be able to generate sufficient compressive residual stress on the out-surface to improve the fatigue resistance of the part; second, the depth of compressive stress generated by out-surface strengthening should not be too large to reduce the tensile residual stress on the internal surface.

The main difference between SPF/DB structure surface strengthening and traditional solid structure is to avoid high tensile residual stress caused by surface strengthening, properly reduce the surface strengthening level, and give consideration to the internal and external surface residual stress levels, so as to obtain the optimal fatigue resistance of the structure. Based on the analysis of residual stress field characteristics generated by various processes such as rolling strengthening, laser shock strengthening, dry shot peening and wet shot peening, the wet shot peening process meets the above requirements and is suitable for surface strengthening of SPF/DB structures. For a specific part, the residual stress after surface strengthening is determined by the strengthening process parameters and structural parameters. The optimization process of part structure and strengthening parameters is carried out based on the requirements and residual stress testing results.

The residual stress measurement of SPF/DB structure involves the external and internal surfaces of the structure, and it is difficult to measure the residual stress of the internal surface. The reason is that this kind of structure is usually an internal cavity closed structure, which is formed in a high temperature environment:

(1) The sensor cannot be embedded inside the blank, as it will be destroyed in the forming process.
(2) It is difficult to directly deliver the measured energy into the structure.
(3) The accuracy of the destructive measurement method is low. For the destructive measurement, it is required to cut the part locally for sampling, which leads to the release of residual stress, and the accurate information of structural residual stress cannot be obtained.
(4) The residual stress in the structure is relatively low, which requires high measurement accuracy.

Due to the measurement requirements and geometric characteristics of the structure, it is difficult to find a method at the current technical level, which is suitable for testing the internal and external residual stresses of SPF/DB structures. Generally, a variety of methods are combined according to the test requirements, such as the residual stress testing method of X-ray diffraction (XRD) is adopted to test the residual stress on the structure out-surface. The residual stress between the out-surface and subsurface of the structure is measured by XRD and stripping method.

The residual stress in the structure is measured by the residual stress measurement method of neutron, synchrotron radiation and other high-energy rays.

5.4.1 X-Ray Diffraction Testing Method

X-ray diffraction (XRD) is a common residual stress testing method. The basic principle is that when there is residual stress in the specimen, the lattice plane spacing will change. When Bragg diffraction occurs, the diffraction peak will also move, and the size of the moving distance is related to the size of the stress. The X-ray of the wavelength is used to shoot on the specimen several times at different incidence angles, measure the corresponding diffraction angle 2θ, and calculate the slope M of 2θ to sin2ψ, then the stress σ_ψ can be calculated (where, 2θ is the included angle between the incident X-ray and the diffracted X-ray, ψ is the included angle between the normal line of the specimen surface and the normal line of the diffracted crystal face, and σ_ψ is the normal stress component with the included angle of Φ direction between the X axis and the measurement coordinate system).

The energy of the light source of the commonly used XRD residual stress testing equipment is relatively low. The depth of the X-ray entering the metal material is between several microns and tens of microns. The measured data reflect the average value of the residual stress within this depth range. The basis for testing the residual stress of titanium alloy materials by XRD is EN15305:2008. {213} is usually selected as the diffraction lattice plane cluster, and the diffraction angle is 142°. The intensity of X-ray diffraction signal (diffraction peak intensity) has a significant influence on the measurement accuracy of residual stress; Diffraction signal intensity is affected by collimator diameter, exposure time and other parameters.

The collimator is similar to the shutter of a camera. The ray generated by the X-ray source first passes through the collimator and then shines on the surface of the workpiece to be tested. The collimator determines the size and shape of the light spot on the workpiece surface. There are two kinds of common collimators: circular and rectangular. The larger the collimator diameter is, the larger the diffraction peak intensity is. The intensity of the diffraction peak can also be increased by increasing the exposure time under the condition that the size of the collimator is unchanged. The diameter of the collimator and counting time remain unchanged, but the counting times increase. For example, when the counting times increase from 10 to 20, the diffraction peak intensity only increases slightly.

The influence of collimator diameter, exposure time and counting times on the diffraction peak intensity of TC4 alloy is shown in Fig. 5.20. Generally speaking, selecting a collimator with larger size can effectively shorten the exposure time of the measurement process, and is suitable for parts with larger size and more uniform surface residual stress distribution.

For the measurement of TC4 alloy zero stress standard specimen, the collimator diameter is 3 mm, the exposure time is 1s, 2s and 3s respectively, and the exposure times are 10. Each parameter combination is tested 4 times respectively, and the

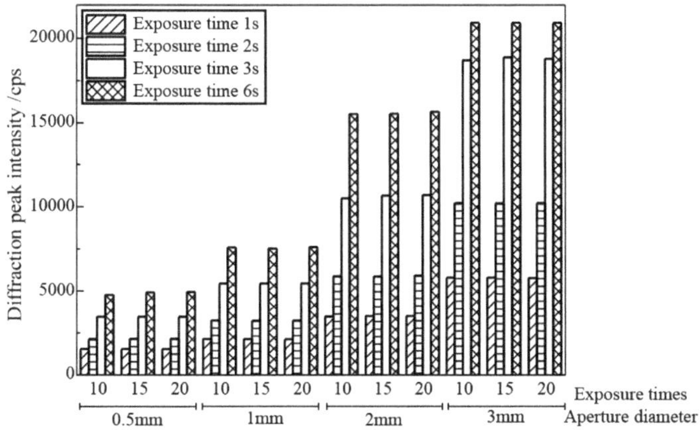

Fig. 5.20 Influence of test parameters on the intensity of diffraction peak of standard zero-stress specimen

results are shown in Fig. 5.21. It can be found that the influence of exposure time on the diffraction peak intensity is only inferior to the collimator diameter.

Satisfactory measurement results can be obtained under the conditions of collimator diameter of 1.0 mm, exposure time of 2 s, 11 diffraction angles, and collimator diameter of 2.0 mm, exposure time of 2 s, 11 diffraction angles. Combined with the influence of test parameters on the standard deviation of test results of standard zero stress specimen, the following parameter combinations are selected for TC4 alloy test: collimator diameter $\geq$ 2 mm, exposure time $\geq$ 2 s, diffraction angle $\geq$ 11, the number of exposures is 10.

The residual stress is determined by fitting the diffraction signal. Generally, XRD residual stress measuring equipment provides Gaussian function parabolic function, Cauchy function, Pearson function and other fitting functions by default. The fitting results of the four functions on the residual stress test signals of TC4 alloy are shown in Fig. 5.22.

For the zero-stress standard specimen, the calculated data $\geq$80% I_{max} (I_{max} is the maximum value of the diffraction peak), and the residual stress data given by the Gaussian function fitting method is within the standard value (0 $\pm$ 14) MPa; for the high residual stress standard specimen, when the calculated data $\geq$ 40% I_{max}, the residual stress data given by the Gaussian function fitting method is within the standard value (-662 ± 35) MPa. The fitting results of parabolic function for zero stress and high stress standard specimen data are poor, and the error of residual stress data given is large. The fitting curve of Cauchy function declines slowly to zero, which is suitable for the diffraction peak curve with flat and wide shape. When the residual stress is small, there is error in the results of the given residual stress, and when the residual stress is large, the calculation range is expanded, the fitting degree of this method is increased, and the accuracy of test results is improved. The change trend of Pearson function fitting results is the same for low stress specimens and high stress specimens. With the continuous expansion of the calculation data

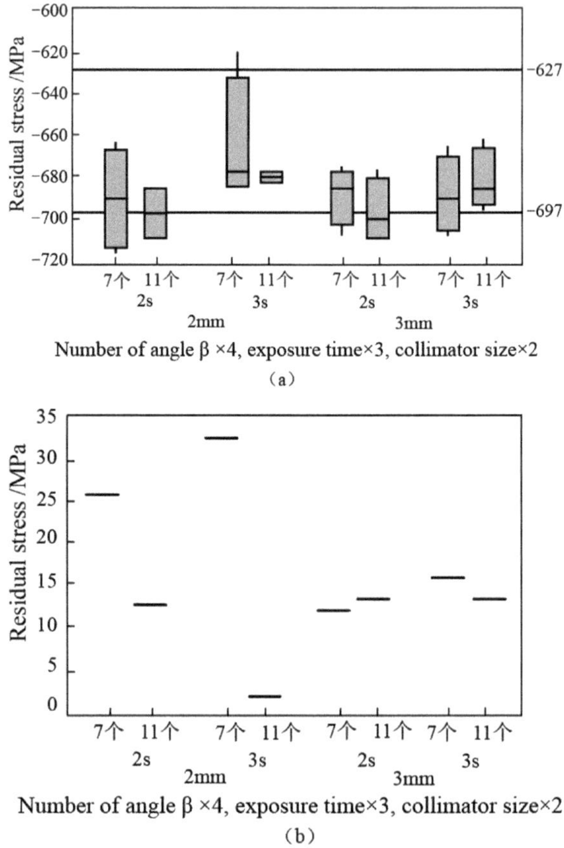

Fig. 5.21 Influence of exposure time on test results of zero-stress TC4 alloy standard specimen. **a** Box plot of residual stress; **b** Box plot of standard deviation of residual stress

range, the better the fitting degree of this method is, the more accurate the residual stress data is given.

If the grade of metal material remains unchanged, fine tune the proportion of alloy elements or change the phase composition through heat treatment, the micromechanical properties of the material will change, thus affecting the measurement accuracy of residual stress. At this time, it is necessary to calibrate the XRD test parameters. Taking TC4 alloy for an example, oxygen element is its interstitial strengthening element, which changes the macroscopic physical properties of titanium alloy by affecting the properties of primary α phase. In general, the oxygen content in TC4 alloy is 0.13% (mass fraction), but some parts for special purposes require that the oxygen content in TC4 alloy be increased to about 0.20% (mass fraction). At this time, the measurement accuracy is improved by calibrating the residual stress test parameters. Calibration equipment (such as in-situ loading testing machine) compares the difference between theoretical stress and test stress through four-point bending loading, corrects the elastic constant of specific oxygen content, and then obtains the influence rule of oxygen content on the elastic constant of TC4 alloy

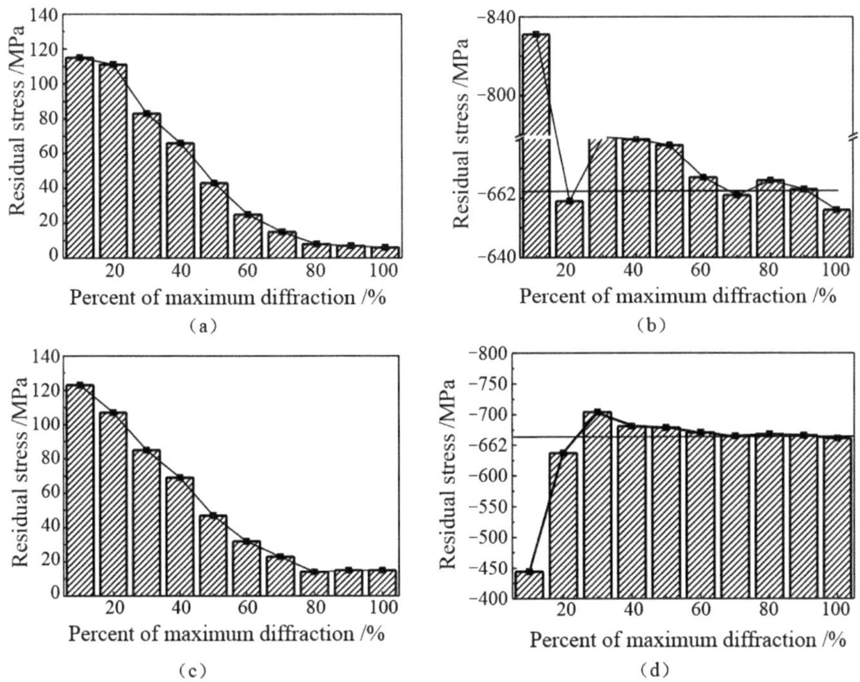

Fig. 5.22 Influence of fitting function on stress test results. **a** Gauss function zero stress standard specimen; **b** Gauss function high stress standard specimen; **c** Pearson function zero stress standard specimen; **d** Pearson function high stress standard specimen

residual stress test. The characteristics of the four-point bending loading tester are shown in Fig. 5.23.

The elastic constant correction process is as follows: the stress generated by the applied load by the in-situ loading testing machine is recorded as σ_{loading}; the initial residual stress on the surface of the specimen is denoted as σ_0 by XRD stress meter, the stress state of the specimen surface is tested with a stress meter under loading, and the test result is recorded as σ_{XRD}. The three have the following equation relationship:

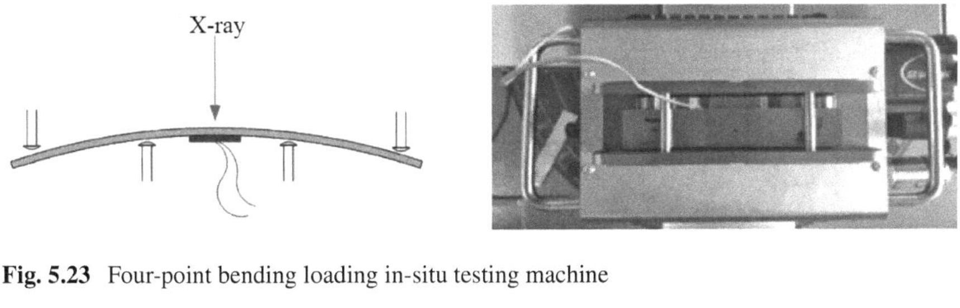

Fig. 5.23 Four-point bending loading in-situ testing machine

$$\sigma_0 + \sigma_{\text{Loading}} = \sigma_{XRD} \tag{5.2}$$

The following relationship exists under different loading loads:

$$\begin{aligned}
\sigma_0 + \sigma_{\text{Loading}_1} &= \sigma_{XRD_1} \\
\sigma_0 + \sigma_{\text{Loading}_2} &= \sigma_{XRD_2} \\
\sigma_0 + \sigma_{\text{Loading}_3} &= \sigma_{XRD_3} \\
&\vdots \\
\sigma_0 + \sigma_{\text{Loading}_n} &= \sigma_{XRD_n}
\end{aligned}$$

Calculate the difference value of Eq. (5.2), then

$$\begin{aligned}
\sigma_{\text{Loading}_2} - \sigma_{\text{Loading}_1} &= \sigma_{XRD_2} - \sigma_{XRD_1} \\
\sigma_{\text{Loading}_3} - \sigma_{\text{Loading}_1} &= \sigma_{XRD_3} - \sigma_{XRD_1} \\
&\vdots \\
\sigma_{\text{Loading}_n} - \sigma_{\text{Loading}_1} &= \sigma_{XRD_n} - \sigma_{XRD_1}
\end{aligned} \tag{5.3}$$

Equation (5.3) is summarized as

$$\Delta\sigma_{\text{Loading}} = \Delta\sigma_{\text{XRD}} \tag{5.4}$$

Substitute Eq. (5.2) into Eq. (5.4), then

$$\begin{cases}
\Delta\sigma_{\text{Loading}} = 1\Big/\frac{1}{2}S_{2,\text{Software}}^{\{\text{hkl}\}} \times \Delta\frac{\partial\varepsilon_{\psi,\varphi}^{\{\text{hkl}\}}}{\partial\sin^2\psi} \\
\Delta\sigma_{XRD} = 1\Big/\frac{1}{2}S_{2,\text{Calibration}}^{\{\text{hkl}\}} \times \Delta\frac{\partial\varepsilon_{\psi,\varphi}^{\{\text{hkl}\}}}{\partial\sin^2\psi}
\end{cases} \tag{5.5}$$

where: $\frac{1}{2}S_{2,\text{Software}}^{\{\text{hkl}\}}$ is the X-ray elastic constant of crystal face {hkl} given in the stress instrument test software; $\frac{1}{2}S_{2,\text{Calibration}}^{\{\text{hkl}\}}$ is the X-ray elastic constant of the calibrated crystal face {hkl}; $\varepsilon_{\psi,\varphi}^{\{\text{hkl}\}}$ is the strain of crystal face {hkl} in the direction of angle Φ and ψ; ψ is the angle between the specimen normal and the diffraction crystal face normal; $\Delta\sigma_{\text{loading}}$ and $\Delta\sigma_{\text{XRD}}$ are respectively applied stress and XRD test stress increment; Φ is the included angle with the X axis of the measurement coordinate system.

Simplify Eq. (5.5), then

$$\frac{\Delta\sigma_{XRD}}{\Delta\sigma_{\text{upload}}} = \frac{\frac{1}{2}S_{2,\text{Calibration}}^{\{\text{hkl}\}}}{\frac{1}{2}S_{2,\text{Software}}^{\{\text{hkl}\}}} \tag{5.6}$$

Then, the calibrated X-ray elastic constant:

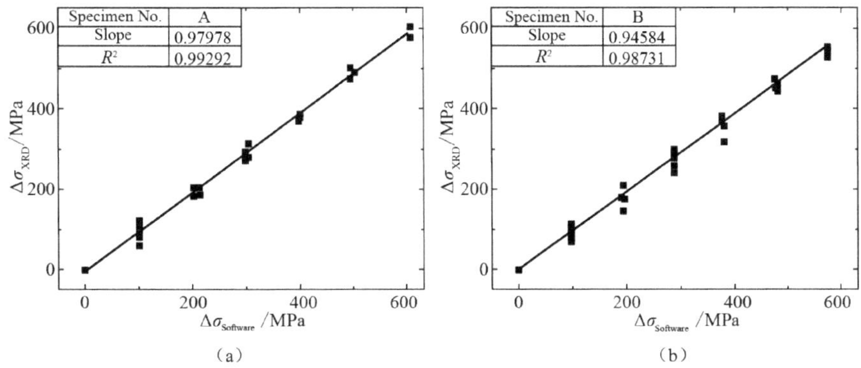

Fig. 5.24 Straight line fitting diagram of $\Delta\sigma_{\text{XRD}}$ to $\Delta\sigma_{\text{loading}}$. **a** 0.13% (mass fraction); **b** 0.20% (mass fraction)

$$1/2 \times S_{2,\text{Calibration}}^{\{\text{hkl}\}} = 1/2 \times S_{2,\text{ software}}^{\{\text{hkl}\}} \times \Delta\sigma_{\text{XRD}}/\Delta\sigma_{\text{Loading}}$$

where: $\Delta\sigma_{\text{XRD}}/\Delta\sigma_{\text{loading}}$ is the linear slope M of $\Delta\sigma_{\text{XRD}}$'s linear fitting of $\Delta\sigma_{\text{loading}}$.

When the oxygen content of TC4 alloy is 0.13% (mass fraction), the X-ray elastic constant after calibration is 2.02% less than that set by the test software, and its value is 11.6474×10^{-1} MPa^{-1}, as shown in Fig. 5.24a; when the oxygen content of TC4 alloy is 0.20% (mass fraction), the X-ray elastic constant after calibration is 5.42% lower than that set by the test software, and its value is 11.2440×10^{-6} MPa^{-1}, as shown in Fig. 5.24b.

According to JJF 1059.1-2012 (Evaluation and Expression of Measurement Uncertainty), the uncertainty of residual stress measurement results of TC4 alloy materials is analyzed, in consideration of the following factors: ① Repeatability of measurement; ② Stress constant K; ③ Slope fitting of stress factor M. The uncertainty introduced by the repeatability of measurement results is evaluated by statistical analysis of independent repeated test results. The error of stress constant is evaluated according to the uncertainty of Class B; the stress factor M involves the fitting of diffraction peaks and diffraction angles (2θ) and sin2ψ, which are completed by commercial testing and analysis software. The resulting error appears after the stress test results in the form of "±statistical error", and this part of error is also evaluated according to Class B uncertainty. The stress is

$$\sigma = KM \tag{5.7}$$

where: K is stress constant, and

$$\text{K} = \frac{\text{E}}{2(1+\mu)} \times \cot\theta_0 \times \frac{\pi}{180}$$

where: μ is Poisson's ratio; θ_0 is Bragg angle associated with d_0; E is modulus of elasticity.

M is stress factor, and

$$M = \frac{\partial(2\theta)}{\partial(\sin^2\psi)}$$

Repeat the measurement for 10 times for TC4 alloy material standard specimen (Table 5.1), and calculate the measurement repeatability uncertainty component. The uncertainty component introduced by the measurement result repeatability is 3.2 MPa:

$$u(\sigma) = \frac{s(x)}{\sqrt{n}} = \frac{10}{\sqrt{10}} = 3.2 \tag{5.8}$$

Calculation of uncertainty components generated by elastic constants: assuming that the elastic constant follows uniform distribution, the confidence factor k = 3, and the elastic constant of the high stress standard specimen is the true value, the error between the estimated value and the true value is within ± 3%, then the uncertainty component is 11.0 MPa:

$$u(K) = \frac{0.03 \times 637}{\sqrt{3}} = 11.0 \tag{5.9}$$

Calculation of uncertainty components generated by stress factor M: assuming that the stress factor is constant and the confidence factor k = 3, the uncertainty component is 5.2 MPa:

$$u(M) = \frac{9}{\sqrt{3}} = 5.2 \tag{5.10}$$

According to the international standard, the combined uncertainty is 95%. If k = 2 is adopted, then

Table 5.1 Residual stress test results of high stress standard specimen of titanium alloy

Test point	1	2	3	4	5
Stress /MPa	−634 ± 11	−632 ± 11	−649 ± 9	−647 ± 9	−644 ± 8
Test point	6	7	8	9	10
Stress /MPa	−623 ± 9	−634 ± 11	−649 ± 9	−623 ± 8	−633 ± 9
Average stress /MPa		Average value of statistical error /MPa		Standard deviation of stress	
−637		9		10	

$$u_c(\sigma) = \sqrt{u^2(\sigma) + u^2(K) + u^2(M)} = \sqrt{(3.2)^2 + (11.0)^2 + (5.2)^2} = 12 \quad (5.11)$$

$$U = ku_c = 2 \times 12 = 24 \quad (5.12)$$

According to the results of Eq. (5.12), the uncertainty of XRD residual stress measurement method is 24 MPa.

The combination of stripping method and XRD method is an effective way to measure the residual stress from the workpiece surface to the subsurface area. "Stripping" is to remove materials one time by one time with equal thickness by electrochemical corrosion method, and the accuracy of material thickness removal is controlled at about 0.01–0.015 mm. Figure 5.25 shows the specimen after measuring the residual stress with this method. Figure 5.26 shows the influence of the wet shot peening surface strengthening process parameters and structure thickness on the residual stress from the surface to the subsurface.

Note: mmN, mmA and mmC are the symbols of shot strength, which respectively represent the shot strength measured with N, A and C type Almen test pieces. Example: the shot strength is 0.4 mmN, which means that under certain shot peening

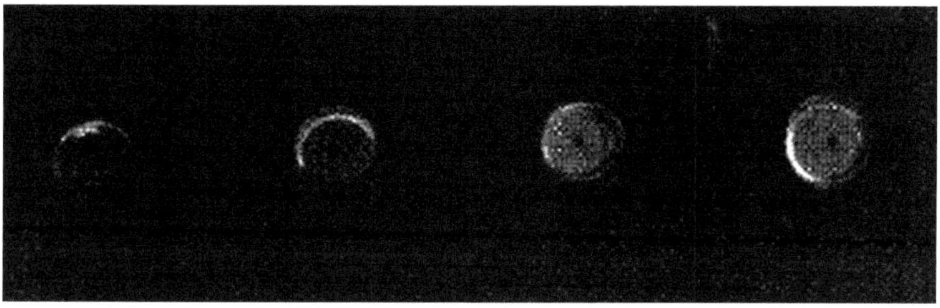

Fig. 5.25 Specimen after residual stress measurement by stripping method and XRD

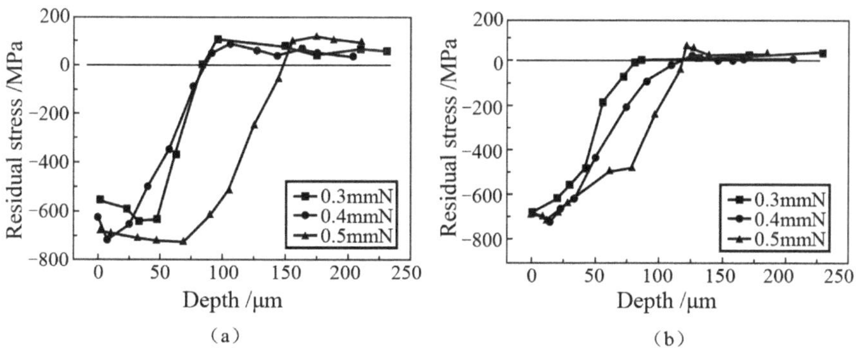

Fig. 5.26 Residual stress distribution of tc4 alloy after wet shot peening. **a** The plate thickness is 1.5 mm; **b** The plate thickness is 6.0 mm

process parameters, the arc height of the saturation point measured with N-type Almen test piece is 0.4 mm.

5.4.2 Neutron Diffraction Testing Method

The neutron diffraction testing method provides higher ray energy than XRD, and the ray can enter deeper into the material, so as to obtain the crystallographic information of materials in these areas. In the residual stress measurement, the change of the lattice plane spacing is obtained through high-energy ray, and the residual stress information of the material is calculated.

CMRR, the US SNS, Japan J-PARC, and the UK ISIS experimental reactor can all provide neutron sources for residual stress testing. CMRR uses a silicon single crystal monochromator to select neutrons of a specific wavelength from the polychrome neutron beam, and the generated neutron wavelength is 0.12–0.28 nm. The experimental device is shown in Fig. 5.27. When used to measure TC4 alloy, the neutron wavelength $\lambda = 0.1592$ mn matched with the diffraction lattice plane can be used, and the neutron fluence rate can be $4.7 \times 10^6 \text{n/(cm}^2 \bullet \text{s)}$. The resolution of the spectrometer can reach 0.25% under this condition.

The proportion of α phase in the bimodal alloy TC4 is relatively high. The crystal structure of α phase is a close packed hexagonal structure. The theoretical neutron diffraction full spectrum ($\lambda = 1.592$ Å) is shown in Fig. 5.28.

The theoretical structure information and theoretical diffraction spectrum of the α-Ti alloy are dense and complex, and there are multiple lattice plane near the same diffraction angle. In residual stress measurement, it is required that the shape of

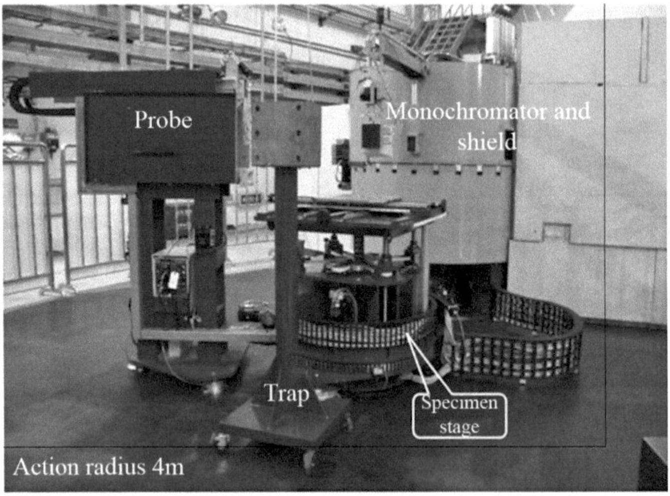

Fig. 5.27 Neutron stress analysis spectrometer

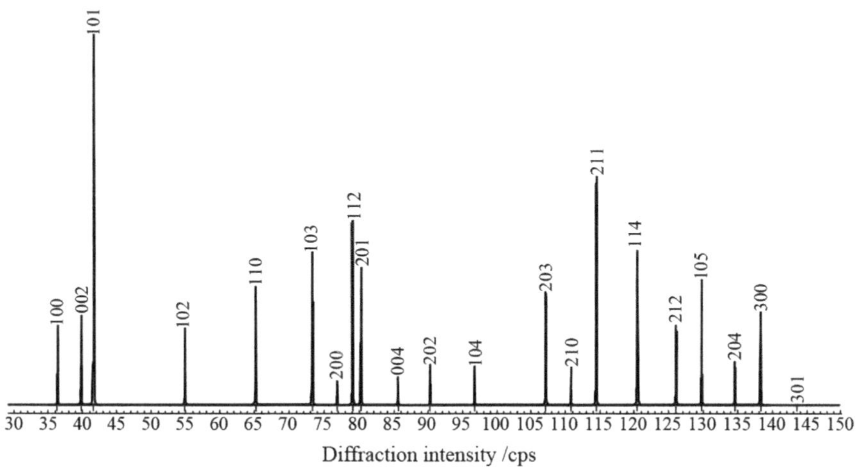

Fig. 5.28 Theoretical structure of α-Ti alloy—diffraction intensity and neutron spectrum

neutron diffraction signal should be single, the shape of diffraction area should be cubic, and the diffraction angle should be in the range of 60° ~ 110°. Ti (10–13) and the diffraction angle of 73.5° can be selected to test the lattice plane. The characteristics of the diffraction peak of the lattice plane are shown in Fig. 5.29.

When the neutron diffraction method is used to measure the internal residual stress of the hollow sandwich structure, the step measurement method is generally used. The theodolite is used to provide the information of the measurement space and control the position accuracy of the measurement area. Two theodolites are used in the measurement as shown in Fig. 5.30: one theodolite controls verticality and symmetry in horizontal plane; the other theodolite is used to check the positioning accuracy of the structure. The two theodolites measure data to control the spatial position of the diffractive body, so that the diffractive body gradually goes deep into

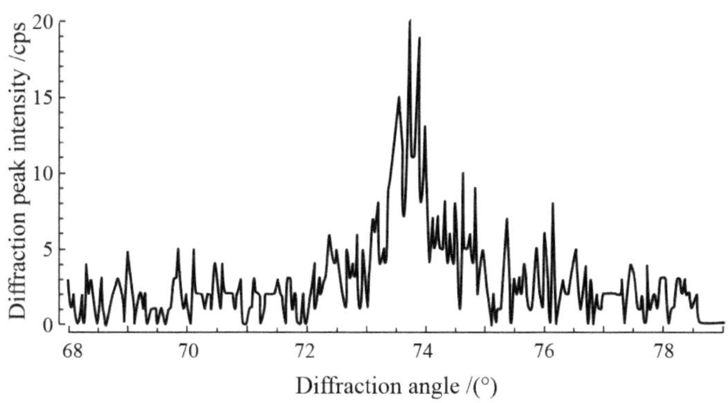

Fig. 5.29 Results of Ti(10–13) diffraction pre-scanning

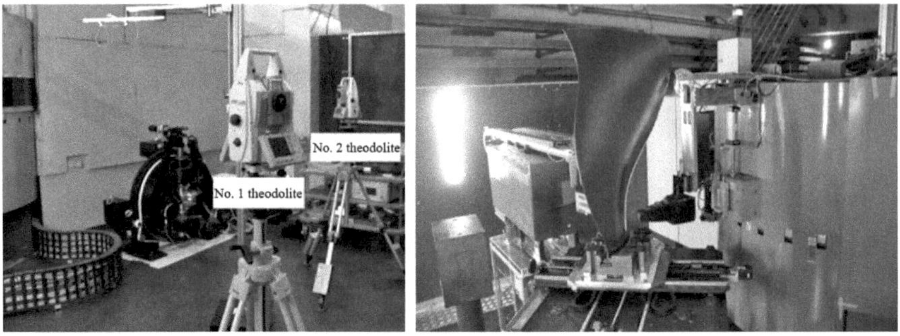

Fig. 5.30 Clamping and positioning of hollow sandwich structure

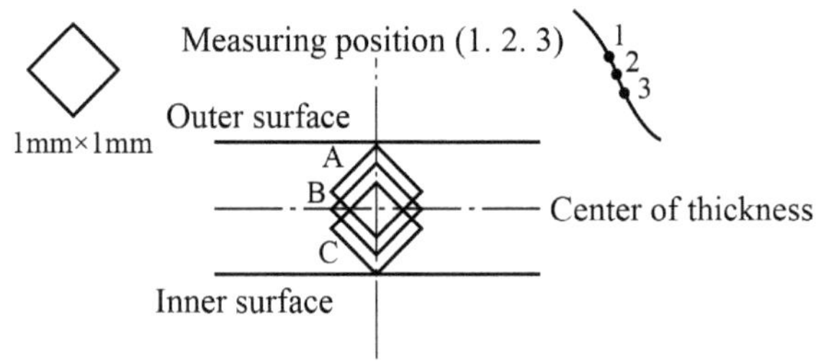

Fig. 5.31 Measurement of residual stress variation along thickness by step method

the structure along the normal direction of the structure surface, and the residual stress distribution inside the structure is obtained (Fig. 5.31).

When the residual stress of the structure is measured by the step method, the spatial position of the measurement area is judged by the information of the diffraction signal, as shown in Fig. 5.32. When the diffractive body does not contact the surface of the structure (position A), the integral intensity of the diffraction peak is 0; when the diffractive body just enters the tested structure completely (position C), the diffraction signal strength is the largest; and when half of the diffractive body is in the structure (position B), the integral intensity is just half of the maximum.

When measuring the residual stress by neutron diffraction, use the same state materials to prepare the specimen with the shape similar to the comb as the zero-stress standard specimen to calibrate the test parameter d_0. The shape of the standard specimen is shown in Fig. 5.33. The d_0 data of TC4 alloy material obtained by this method are shown in Table 5.2. After obtaining the data in three directions, the peak position, peak intensity and other parameters of the diffraction peak are output through background fitting, background deduction and peak shape Gaussian fitting, and the three-dimensional stress value of the structure is obtained through the following stress calculation formula:

$$\sigma_{xx} = \frac{E_{hkl}}{(1 + \nu_{hkl})(1 - 2\nu_{hkl})}\left[(1 - \nu_{hkl})\varepsilon_{xx} + \nu_{hkl}\left(\varepsilon_{yy} + \varepsilon_{zz}\right)\right] \tag{5.13}$$

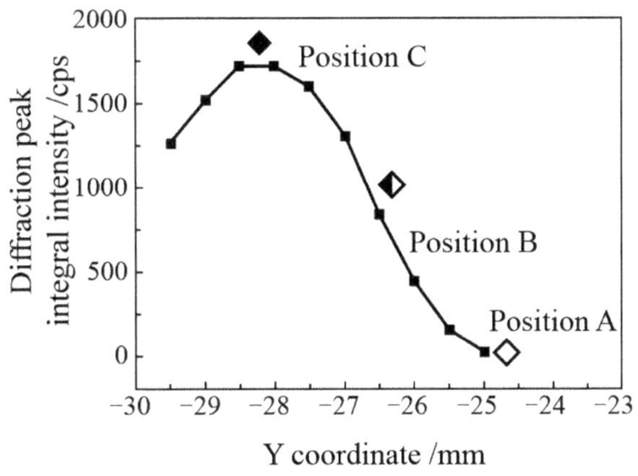

Fig. 5.32 Variation of diffraction peak intensity with measurement position

Fig. 5.33 Neutron diffraction zero-stress standard specimen

Table 5.2 Core data of TC4 alloy material

Testing direction	2θ/(°)	d_0/Å
Chordwise	73.6461	1. 3273
Normal	73.6494	1. 3272
Span-wise	73.6457	1. 3273

$$\sigma_{yy} = \frac{E_{hkl}}{(1+\nu_{hkl})(1-2\nu_{hkl})}\left[(1-\nu_{hkl})\varepsilon_{yy} + \nu_{hkl}(\varepsilon_{zz}+\varepsilon_{xx})\right]. \tag{5.14}$$

$$\sigma_{zz} = \frac{E_{hkl}}{(1+\nu_{hkl})(1-2\nu_{hkl})}\left[(1-\nu_{hkl})\varepsilon_{zz} + \nu_{hkl}\left(\varepsilon_{xx}+\varepsilon_{yy}\right)\right] \tag{5.15}$$

$$u(\sigma_{xx}) = \frac{E_{hkl}}{(1+\nu_{hkl})(1-2\nu_{hkl})}\sqrt{(1-\nu_{hkl})^2 u^2(\varepsilon_{xx}) + \nu_{hkl}^2\left(u^2\left(\varepsilon_{yy}\right)+u^2(\varepsilon_{zz})\right)} \tag{5.16}$$

$$u\left(\sigma_{yy}\right) = \frac{E_{hkl}}{(1+\nu_{hkl})(1-2\nu_{hkl})}\sqrt{(1-\nu_{hkl})^2 u^2(\varepsilon_{yy}) + \nu_{hkl}^2\left(u^2(\varepsilon_{zz})+u^2(\varepsilon_{xx})\right)}. \tag{5.17}$$

$$u(\sigma_{zz}) = \frac{E_{hkl}}{(1+\nu_{hkl})(1-2\nu_{hkl})}\sqrt{(1-\nu_{hkl})^2 u^2(\varepsilon_{zz}) + \nu_{hkl}^2\left(u^2(\varepsilon_{xx})+u^2(\varepsilon_{yy})\right)} \tag{5.18}$$

5.4.3 *Numerical Prediction Method of Structural Residual Stress*

There are some shortcomings in the way of physical testing SPF/DB structure residual stress: first, the data obtained from physical tests are discontinuous; Secondly, it is difficult to accurately measure the data of the internal surface of the structure, especially the data of the joint between the internal supporting structures such as trusses, webs and skin. Stress concentration exists in these areas, and the coupling effect of residual stress and alternating stress controls the crack initiation, which is crucial to the prediction of structural life. Combining the numerical method with the test data to predict the residual stress distribution of the structure, can more directly support the optimization of the surface strengthening process, and lay a foundation for the integration of the design and process of SPF/DB structures. In the long-term research, corresponding numerical prediction methods of structural residual stress have been formed based on two ideas.

One is direct simulation. That is, a numerical model reflecting the real physical process is established, and the evolution law of stress, strain and deformation of materials in the strengthening process is obtained through nonlinear dynamic calculation. In some studies, constitutive models containing strain rates are used to describe the behavior of materials under impulse excitation. In others, new numerical methods, such as smooth particle dynamics (SPH), are used to try to improve the prediction accuracy of stress and strain in the dynamic process. From the above approaches, it is easy to see that the direct simulation method needs to solve a large number of nonlinear equations, which is constrained by hardware, and the size of the numerical model is small. For example, for dry shot peening strengthening process simulation, the area of the established strengthening model is usually hundreds of square millimeters, and the number of shots is between tens and hundreds. For some nonlinear, multi physical field coupling surface processes, such as laser shock strengthening and wet shot peening strengthening. The numerical simulation method is still in the exploration stage, and its characteristics make it mainly used in the research of residual stress formation process.

Another simulation approach is based on the equivalence principle. In this method, the residual stress formed by surface strengthening is added to the finite element model as the initial boundary condition. For example, temperature load and linear expansion coefficient are used to generate initial stress or initial plastic strain to form equivalent stress that is the same as the resultant force and moment of residual stress generated under a specific surface strengthening process and a certain process parameter, so as to obtain the deformation data of the structure and the residual stress/strain data far away from the applied boundary conditions. It has few restrictions on finite element modeling. The model can be either a solid element or a shell element. The material constitutive model is generally a linear elastic model, and most of it does not involve the solution of material nonlinear problems, so it allows a large scale of the model. If this method is used in the shot peen forming research of aircraft panel parts, shell element can be used to model the thin plate structure with little difference

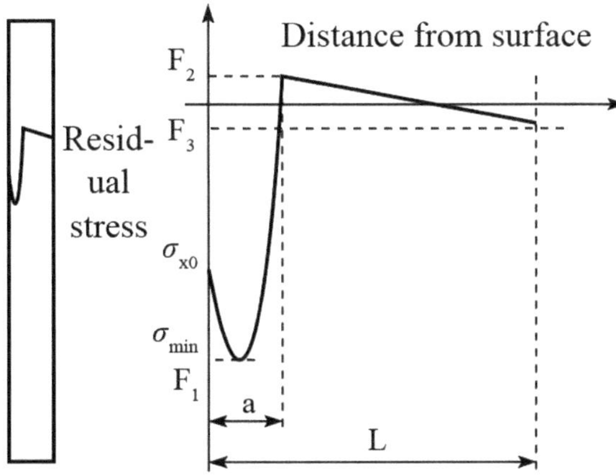

Fig. 5.34 Schematic diagram of residual stress distribution along the thickness of thin plate after unilateral strengthening

in thickness. For structures with large thickness difference or mixed solid and shell, solid element modeling can be used. It is required to precisely control the thickness of the unit, which is difficult for parts with complex geometry.

According to the existing data, it is assumed that after one side of the plate is strengthened by wet shot peening, the distribution of residual stress along the thickness direction of the thin plate is shown in Fig. 5.34, with the following characteristics:

(1) The strengthened side is the compressive residual stress, its thickness is α, and the compressive stress follows the normal distribution law, which is described by Eq. (5.19). The constants in Eq. (5.19) can be obtained by Eqs. (5.20)–(5.22);
(2) At the point where the thickness is α, the residual stress is positive, and then the residual stress changes according to the linear rule until the unreinforced surface L of the thin plate.

$$\sigma = A\sin(\omega x + \theta) \tag{5.19}$$

where: x is thick plate; constant A is the maximum compressive residual stress; the constants ω and θ are related to the compression stress distribution.

$$A = \min(\sigma) \tag{5.20}$$

$$\omega = \frac{\pi}{2(x_{\sigma\,\min} - x_{\sigma 0})} \tag{5.21}$$

$$\theta = (\frac{3\pi}{2} - x_{\sigma\,\max}\omega) \tag{5.22}$$

where $x_{\sigma min}$ is the location coordinate of the minimum residual stress; $x_{\sigma min}$ is the location coordinate of the maximum residual stress; $x_{\sigma 0}$ is the location coordinate of surface residual stress. σ_{min}, σ_{max} and $x_{\sigma 0}$ are the minimum residual stress, maximum residual stress and surface residual stress respectively. The σ_{x0}, $x_{\sigma min}$ and $x_{\sigma 0}$ data can be obtained by X-ray diffraction residual stress tester (XRD) combined with delamination method.

In order to verify the prediction results, first, the distribution law of compressive stress of TC4 alloy material after wet shot peening is obtained by using XRD combined with electrochemical corrosion delamination method, and the compression stress data under 0.3 and 0.5 mmN shot strength are fitted based on Eq. (5.19), as shown in Fig. 5.35. The calculation shows that the mean square deviation between the predicted value and the measured value is 33.3 MPa and 15.5 MPa respectively. This shows that the compression stress field produced by wet shot peening can be accurately characterized by using the sine function.

Equations (5.24) and (5.26) are obtained according to Eqs. (5.23) and (5.25) of force and moment balance principle. Where θ, a and L are as previously mentioned, F_1 is the residual stress on the surface. Therefore, there are only two unknowns in the two equations, and the tensile residual stress F_2 and the residual stress F_3 of the unreinforced surface can be obtained by calculation.

$$\sum Fi = 0$$

$$\Rightarrow \int_0^a f_1(x)dx + \int_a^L f_2(x)dx = 0 \tag{5.23}$$

$$F_1 \frac{4a}{5\pi + 4\theta}\left[\cos\left(\frac{3}{4}\pi - \theta\right) - 1\right] + \frac{F_3 - F_2}{2}(L - 1) + F_2(L - a) = 0. \tag{5.24}$$

$$\sum Mi = 0$$

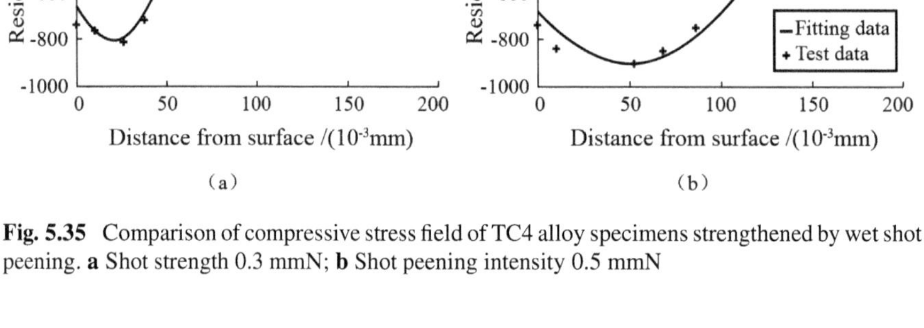

Fig. 5.35 Comparison of compressive stress field of TC4 alloy specimens strengthened by wet shot peening. **a** Shot strength 0.3 mmN; **b** Shot peening intensity 0.5 mmN

$$\Rightarrow \int_0^a f_1(x)xdx + \int_a^L f_2(x)xdx = 0 \tag{5.25}$$

$$\begin{aligned}
&- 2\pi\left(\frac{4a}{5\pi + 4\theta}\right)^2 F_1 \\
&- \left(\frac{4a}{5\pi + 4\theta}\right)^2 F_1\left[\sin\left(\frac{3}{4}\pi - \theta\right) - \left(\frac{3}{4}\pi - \theta\right)\cos\left(\left(\frac{3}{4}\pi - \theta\right)\right)\right] \\
&+ \left(\frac{4a}{5\pi + 4\theta}\right)^2\left(\frac{3}{4}\pi - \theta\right)F_1\left[1 - \cos\left(\left(\frac{3}{4}\pi - \theta\right)\right)\right] \\
&+ \frac{1}{3}\frac{F_3 - F_2}{L - a}(L^3 - a^3) - \frac{F_3 - F_2}{L - a}\frac{a}{2}(L^2 - a^2) \\
&+ \frac{1}{2}F_2(L^2 - a^2) = 0
\end{aligned} \tag{5.26}$$

In order to predict the strengthening deformation and residual stress of complex parts, the second idea mentioned above is adopted, that is, the residual stress field data obtained from the test on the standard test block is applied to the corresponding area of the finite element model (Fig. 5.36). In order to give consideration to the calculation efficiency and element quality, the element thickness is usually mm, which is much greater than the thickness of the compressive stress layer. Therefore, based on the principle of force and moment equivalence, and using Eqs. (5.24) and (5.26) to calculate the stress values applied on the surface nodes of the finite element, according to the basic principles of elasticity, when using this method to predict structural deformation and residual stress far away from the applied boundary conditions, the prediction results are accurate enough. In addition, we also require the finite element model established to meet the following requirements:

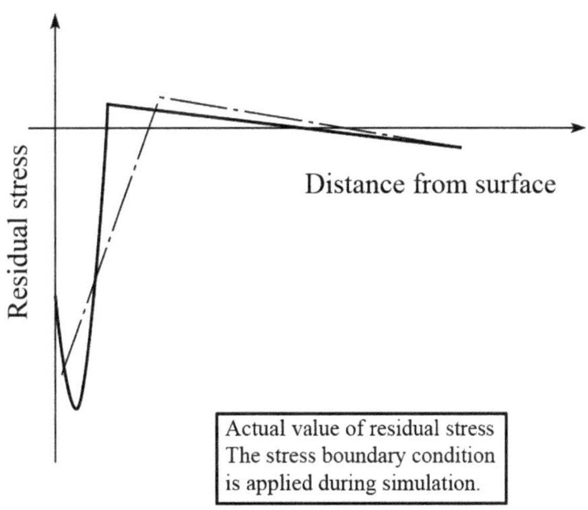

Fig. 5.36 Prediction method of strengthening deformation and residual stress based on equivalent principle

(1) The finite element of model should be 6-faceted 8-node element;
(2) The thickness of the surface layer element must be consistent and its thickness cannot be greater than 5 times of the compressive stress layer;
(3) The number of elements in the part thickness direction should not be less than 3 layers.

Based on this method, the deformation and residual stress distribution of thin plates and thin-walled parts after wet shot peening are predicted respectively. During finite element modeling, the thickness of surface element should be controlled to 0.2 mm. There are four layers of elements in the thickness direction of the strengthening area. The temperature load on the node on the strengthened surface is applied to generate thermal stress so as to simulate the residual stress equivalently. The model parameters are shown in Table 5.3.

For 40 mm × 15 mm × 1.6 mm TC4 alloy plate specimen (Fig. 5.37a), the residual stress of the unstrengthened surface and the distribution of residual stress along the thickness direction are predicted when one side of the surface is strengthened and the shot peening intensity is 0.3 and 0.5 mmN. It is possible to obtain the surface strengthening stress cloud diagram with shot peening intensity of 0.3 mmN (as shown in Fig. 5.37b) after all the degrees of freedom of nodes at one vertex of the finite element model are constrained during modeling, and the temperatures of 341 °C and 525.1 °C are applied to one side of the surface node.

With the shotpeening intensity of 0.3 mmN, the residual stress variation in the specimen thickness direction is predicted as shown in Fig. 5.38. The Figure shows that there is high compressive residual stress on the shotpeened surface, and the residual stress along the thickness direction changes from compressive stresses to tensile stress first, and then the tensile stress presents a linear change and decreases gradually until there is compressive residual stresses on other side of the specimen.

Furthermore, in order to obtain the residual stress on the surface of the unstrenghtened side of the plate, TC4 alloy material is used to manufacture a specimen of the same dimensions (Fig. 5.39a), and strain gauges are installed on the unstrengthened surface before the test (Fig. 5.39b). Shotpeening intensity is 0.3 and 0.5 mmN. For the residual stress on the unstrengthened surface calculated by using the data and Eqs. (5.25) and (5.26), the predicted value and measured result are shown in Table 5.4. It shows that the predicted strain capacity is slightly lower than the tested value, with the errors of -1–1% and −3.8%, respectively.

According to the equivalent modeling method of compressive residual stress field and the validation study results, the finite element modeling requirements of residual stress analysis are coupled into the finite element modeling software, as shown in

Table 5.3 Calculation parameters for surface strengthening simulation of characteristic specimen

Density/($\times 10^3$kg/m^3)	Modulus of elasticity /GPa	Poisson's ratio	Thermal expansion coefficient
4.44	109	0.3	1×10^{-6}

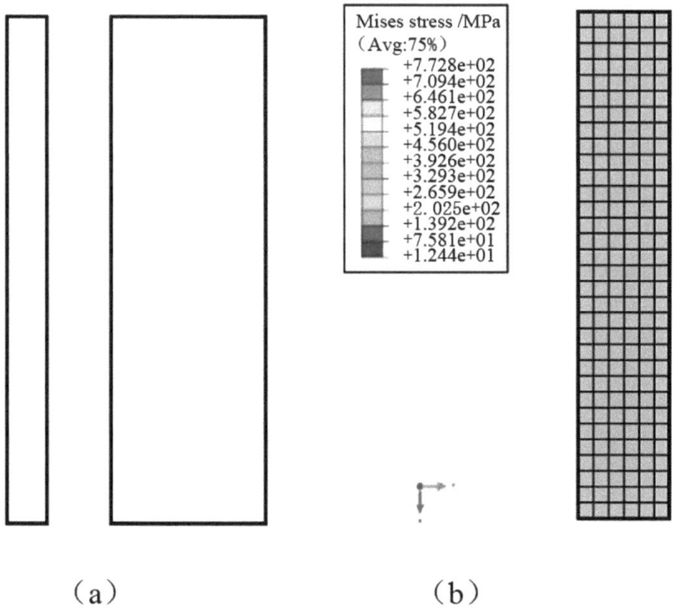

Fig. 5.37 Shape and surface strengthening stress cloud diagram of the plate specimen. **a** Shape of the specimen; **b** Residual stress distribution

Fig. 5.38 Residual stress variation in the thickness direction after strengthening the single-side surface of the plate specimen

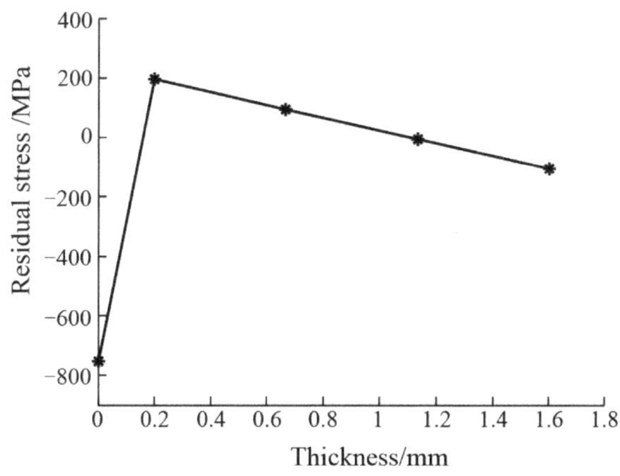

Fig. 5.39 Plate specimen. **a** Strengthened surface; **b** Rear surface

Table 5.4 Deformation and strain during surface strengthening

Shotpeening intensity /mmN	0.3	0.5
Measured strain value	-6.486×10^{-5}	-8.51×10^{-5}
Predicted strain value	-6.417×10^{-5}	-8.188×10^{-5}
Error/%	−1.1	−3.8

Fig. 5.40. The software can apply temperature loads to the surface of hollow structures based on the specified path and moving speed.

The finite element model of hollow sandwich structure is adopted and applied with three boundary conditions: ① Apply the surface load concurrently at a time (Fig. 5.41a); ② Simulate the surface strengthening process, and load step by step in sequence (Fig. 5.41b); ③ Simulate the surface strengthening process, load in sequence step by step, and increase the spanwise spacing (Fig. 5.41c). The path is shown in Fig. 5.41.

The residual stress distribution and deformation law of the hollow sandwich structures under the three boundary condition application strategies are shown in Figs. 5.42 and 5.43. From the Figures, it is concluded that the deformation are similar under the three boundary condition application strategies, and there is a slight different in stress distribution. For specific, the surface compressive stress field is uniform when the load is concurrently and evenly applied; in the consideration of surface strengthening path, the residual stress fluctuates along the spanwise direction; when the parts are strengthened, it is necessary to reduce the difference in residual stress at different parts through reasonable control of the path.

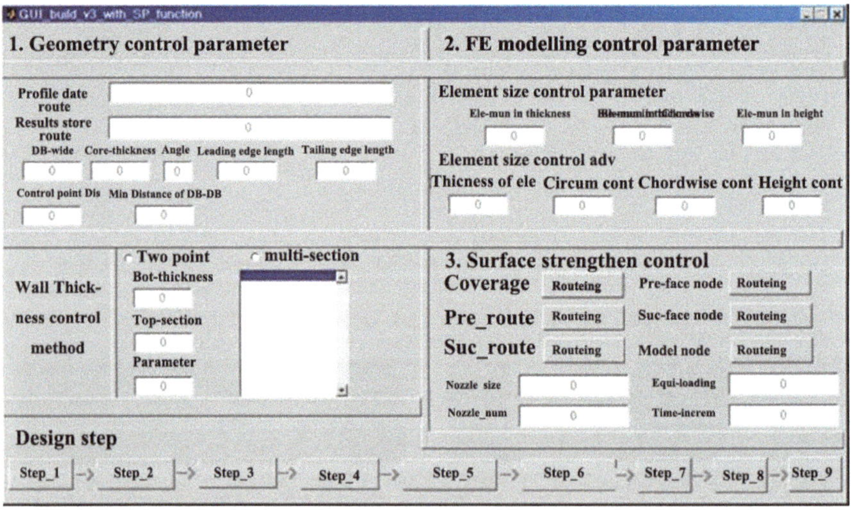

Fig. 5.40 Hollow structure modeling software with active grid size control function

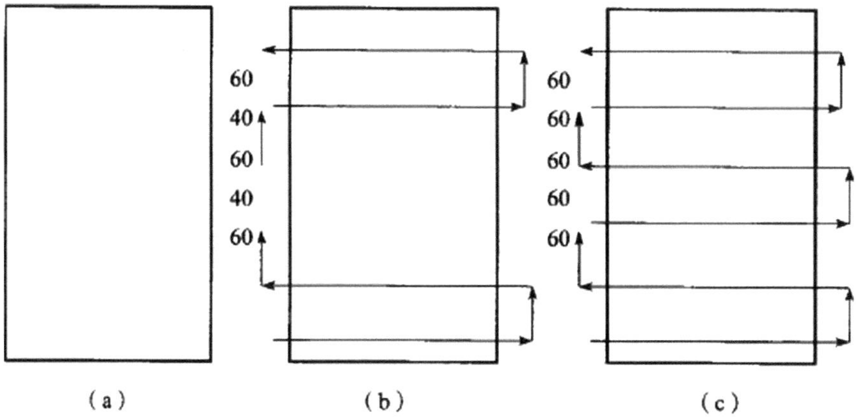

Fig. 5.41 Surface strengthening path. **a** Boundary condition application strategy 1; **b** Boundary condition application strategy 2; **c** Boundary condition application strategy 3

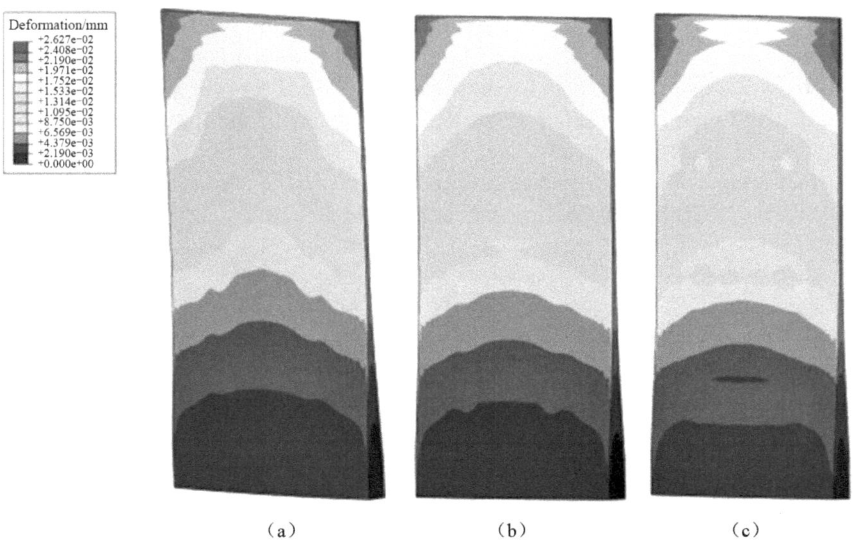

Fig. 5.42 Deformation contour diagram under three application strategies of boundary conditions. **a** Boundary condition application strategy 1; **b** Boundary condition application strategy 2; **c** Boundary condition application strategy 3

The verification of the residual stress prediction process of the hollow structure was completed by hollow sandwich structure, and on this basis the residual stress distribution of the hollow sandwich parts was analyzed. The finite element model of residual stress simulation of characteristic specimen was established, as shown in Fig. 5.44. Under the condition of applying the boundary conditions according to the

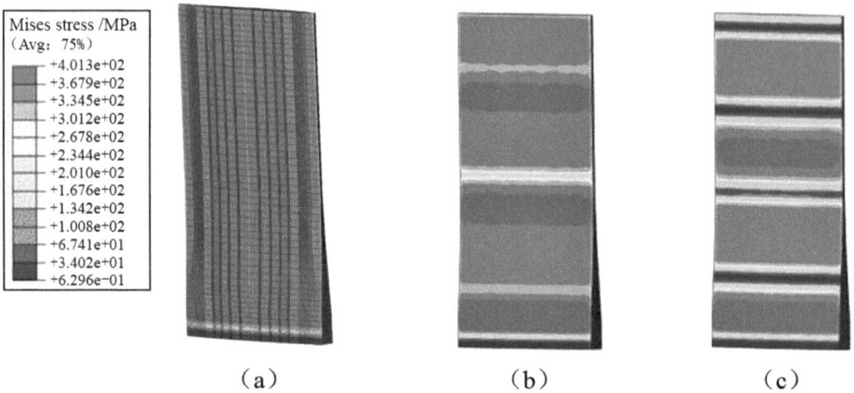

Fig. 5.43 Stress contour diagram of the hollow sandwich structure under three application strategies of boundary conditions. **a** Boundary condition application strategy 1; **b** Boundary condition application strategy 2; **c** Boundary condition application strategy 3

manufacturing parameters, the deformation cloud diagram after surface strengthening can be obtained, as shown in Fig. 5.45. The comparison of the predicted numerical results and test data as shown in Fig. 5.46. From the Figure, it can be seen that the predicted result follows the same law with the measured data, and the predicted results are conservative. The internal residual stress distribution of the structure is controlled by structure deformation law, so the predicted residual stress field is accurate if the deformation data is accurate.

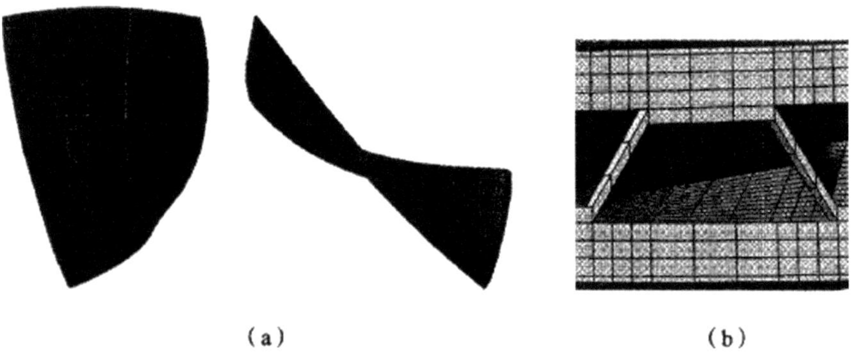

Fig. 5.44 Finite element model for residual stress prediction of certain hollow sandwich structure. **a** Characteristics of the model; **b** Characteristics of cross-sectional elements

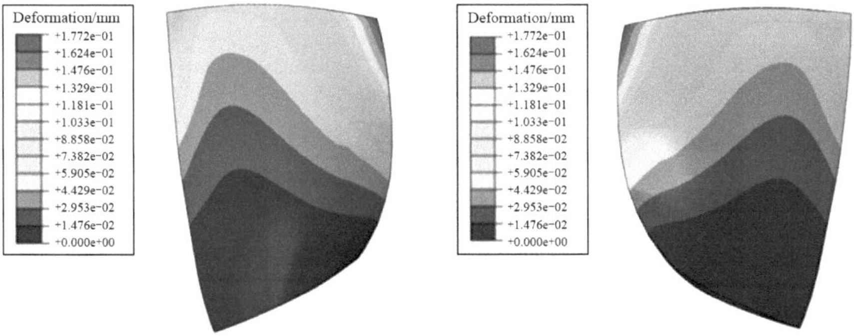

Fig. 5.45 Deformation contour diagram of the hollow sandwich structure after surface strengthening

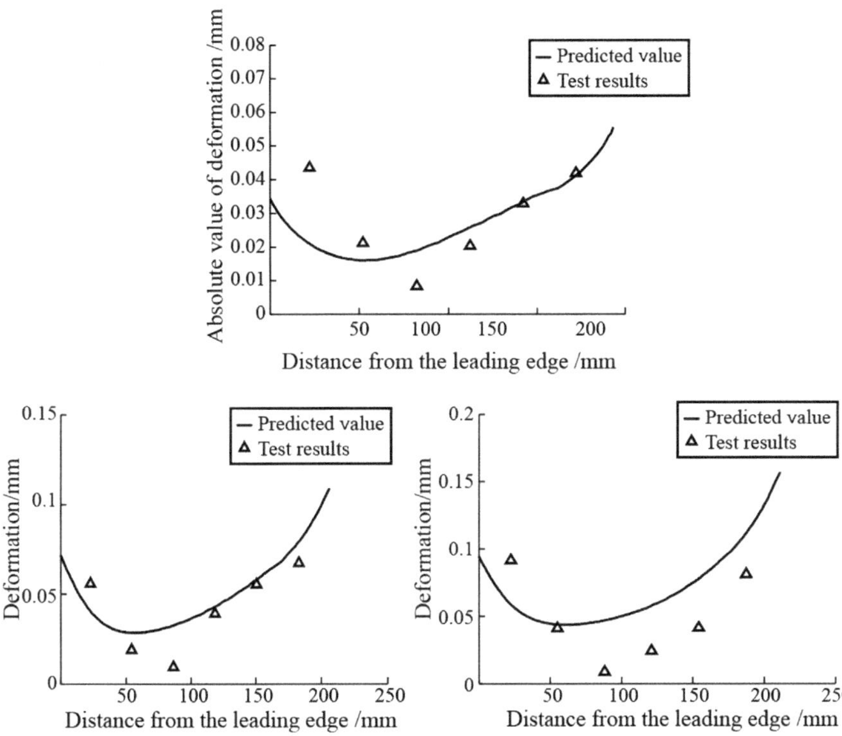

Fig. 5.46 Comparison of deformation at different heights of the hollow sandwich structure after surface strengthening

References

1. David Serra. 2008. Superplastic forming applications on aero engines: A review of ITP manufacturing processes. In *6th EUROSPF Conference*.
2. General Administration of Quality Supervision, Inspection and Quarantine of the People's Republic of China. 2013. *Evaluation and expression of uncertainty measurement: JJF 1059.1—2012*. Beijing: China Quality and Standards Publishing & Media Co., Ltd.

Chapter 6
Design and Evaluation Method of Superplastic Forming/Diffusion Bonding Structure

Superplastic forming/diffusion bonding (SPF/DB) process is intended to optimize the internal geometrical characteristics and reduce the stress concentration while delivering flexibility in structural design and response to the diversified requirements. For the geometric design involving high difficulty and multiple iterations, the structural design must be closely combined with the process design, and the corresponding performance evaluation methods are also developed to achieve integration of design, manufacture and evaluation. This Chapter addresses the load carrying characteristics, design methods, performance evaluation and failure analysis of typical structures.

6.1 Analysis on Bearing Characteristics of Common Structures

The single-layer plate structure usually matches with the simplest geometrical shape. Blanks are mostly sheet plates, which are commonly found in the non-bearing parts with regular shapes, either rotary type of plane shell. The geometrical characteristics are as shown in Fig. 6.1, the parts generally have a wall thickness of about several millimeter. By adopting stiffening rib geometry at the local area of the structure reinforcement geometrical characteristics, it is possible to improve the rigidity and vibration characteristics of parts. The single-layer plate structure is designed with civil engine inlet lip (Fig. 6.2), inlet, corrugated plate, etc. Such parts are easy to manufacture, and the area can be over several square meters. It is possible to produce larger-size parts by combination of the manufacturing process of the single-layer plates and other connecting processes. There is no inherent defect in the single-layer plate structure, so the traditional methods can be used in terms of its structure design, strength assessment, etc. The single-layer plate structure has low difficulty in design, relatively low rigidity and is easy to deform, so it is usually combined with skeleton and reinforcing rib structure, applicable for the aircraft skin. The single-layer plate

Z. Li, *Superplastic Forming/Diffusion Bonding Technology of Titanium Alloys*,
https://doi.org/10.1007/978-981-99-3909-1_6

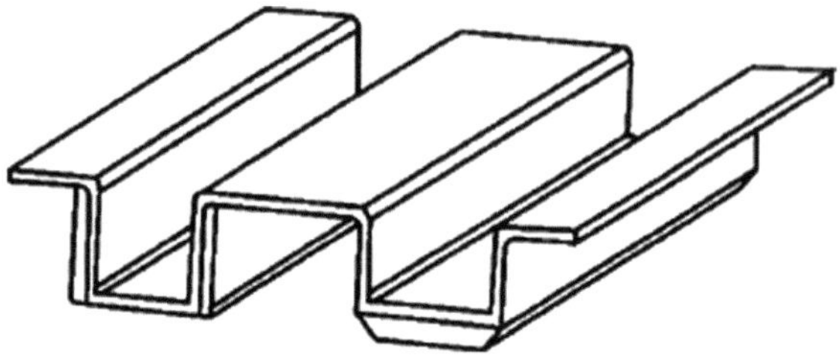

Fig. 6.1 Geometrical characteristics of single-layer plate structure

Fig. 6.2 Inlet made by single-layer plate structure

structure with rotor characteristics can be adopted for pressure container, such as fuel tank, cylinders, etc.

There are generally 2 blanks to manufacture two-layer plate structures, the blank shape is usually obtained by calculation after spitting the part into two halves along the middle surface. When the complexity of the inner and outer shapes of the part is increased, it is not possible to a pure shell structure, but also form a solid-shell mixed structure by changing in the blank thickness and combining various blanks.

The two-layer plate structure of pure shell can be seen as a structure with smooth surface on one side and with stiffening ribs on the other side. Such structure is normally made of two sheet plates, with the geometrical characteristics as shown in Fig. 6.3a. With flexible selection of skin thickness, it allows to manufacture larger-size parts fearing rigid structure, light weight, low cost and high reliability, which can be used to replace the skin frame or skin bonding structure in aircraft and engines. As with relatively thin skin, this structure has limited bearing capacity, and is often used in the manufacture of the bearing structure such as various flaps and hatch doors on the aircraft and engines. For two-layer plate structure such as hatch doors, local reinforcement can be achieved by adding the materials in a specific location, in order to improve the structural strength of these areas; and the reinforcing position are often used to connect other parts, such as actuating cylinder, hinge, as shown in Fig. 6.3b.

The shell-solid mixed two-layer plate structure has smooth surface, hollow interior and without reinforcing ribs in the hollow area. The internal and external shape characteristics of the structure are as shown in Fig. 6.4. The weight reduction efficiency of this type of parts are improved with the increases in its dimensions, the size in the thickness direction has the most significant effect on the weight reduction

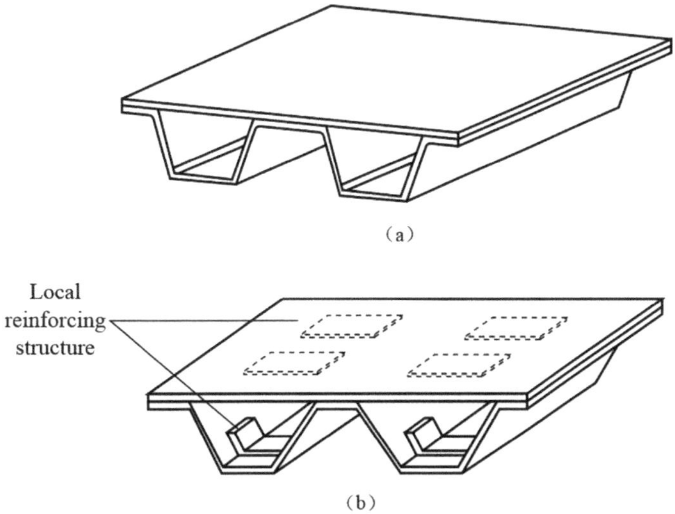

Fig. 6.3 Geometrical characteristics of pure shell two-layer plate structure. **a** Two-layer plate structure; **b** two-layer plate structure with local reinforcement

efficiency. The engineering practice has demonstrated that the weight of shell-solid mixed two-layer plate structure can be reduced by 15–60% as compared with the same shape of the solid structure. Furthermore, this type of structure has excellent performance of bearing quasi-static load, and theoretically there is no presence of inherent defect. When it is deployed for bearing the static or quasi-static load, simple structure design and strength check are applied; in the engineering practice, the welding interface is normally arranged in the symmetry plane to allow welding interface position to overlap with the neutral layer position when parts are under bending, and make the welding interface be vertical to the principal stress direction when the parts are under tensile load, it is therefore that the welding interface does not bear high stress, even if there are less welding interface defects on the surface, its effect on the bearing capacity of the structure may also be ignored; the shell-solid mixed two-layer plate structure is lack of support in the hollow area, the partial rigidity of skin is not good, and there are a variety of local vibration modes. Figure 6.5 shown the comparison of the first six order vibration mode of the two-layer plate hollow structure and the solid structure. As seen from the Figures, the internal characteristics of the hollow structure are similar to those shown in Fig. 6.4, the hollow and solid structures have the same low-order vibration mode; and starting from the 4th order vibration mode, the hollow structure shows local deformation, and at the same order, the natural frequency of the hollow structure is much lower than that of the solid structure. Table 6.1 shows the comparison of the first 6-order inherent frequencies between the solid structure and the two-layer plate hollow structure.

In consideration of these characteristics of two-layer plate hollow structure, the solid-shell mixed two-layer plate structure is suitable for manufacturing secondary

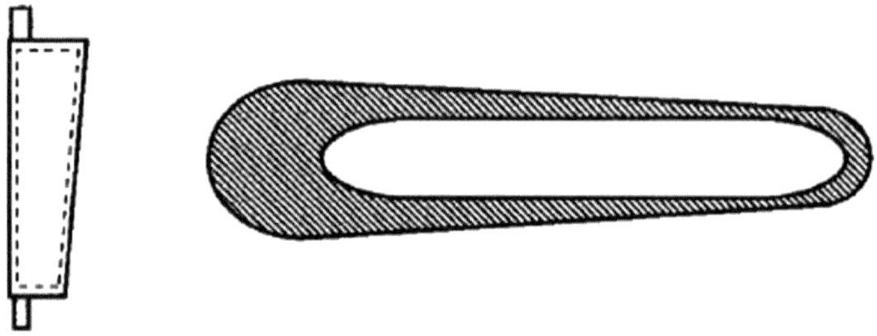

Fig. 6.4 Geometrical characteristics of shell-solid mixed two-layer plate structure

load carrying structure and static load carrying structure such as the inlet and outlet guide vans of aero engines, adjustable blades, and the rudder, wing surface structure of the aircraft.

The three-layer plate structure can be divided into shell structure (Fig. 6.6a) and shell-solid mixed structure (Fig. 6.6b). The three-layer plate structure is characterized by its skin and the W-shaped truss reinforcement structure in the middle. With free change of skin thickness in the design process, the structure has strong load carrying capacity and good overall rigidity. In the common load frequency domain of aircraft and engine (about 10–2000 Hz or the first 10-order natural frequencies of the structure), the natural frequencies of the three-layer plate structure are close to those of the solid structure (Fig. 6.7), and the two have the same vibration mode. From a theoretical view, there is no inherent defect in the structure, and the structural design and strength evaluation methods of the traditional solid structure can be followed to enable the three-layer plate structure to be used in manufacturing various type of critical, static and vibration load bearing structures, such as engine fan blades, etc. The weight reduction efficiency of the structure increases with the increase of the size, which is usually in the range of 15–45%. In the manufacturing process, such parts require large area of diffusion bonding and have high requirement on the diffusion bonding technology and interface defect detection technology. Table 6.2 shows the comparison of the first 5-order natural frequencies of the solid and hollow sandwich structures.

With flexible manufacturing process, the four-layer plate structure can be manufactured into hollow sandwich structure with complex internal structure characteristics. It is divided into shell structure and shell-solid mixed structure, or H-shaped structure and X-shaped structure according to the internal geometrical characteristics. Currently, H-shaped structure has been widely used, while the application of X-shaped structure is rare in China.

The geometrical characteristics of the H-shaped four-layer plate structure are shown in Fig. 6.8a: the hollow area is formed by the skin and the reinforcing ribs perpendicular to the skin; the thickness of the skin is between 1.6 and 2.0 mm, and the reinforcing ribs has the similar thickness as the skin. This type of structure has high static bearing capacity, variable rigidity (by design), sound overall rigidity, and the advantage in weight reduction, which can easily achieve a weight reduction

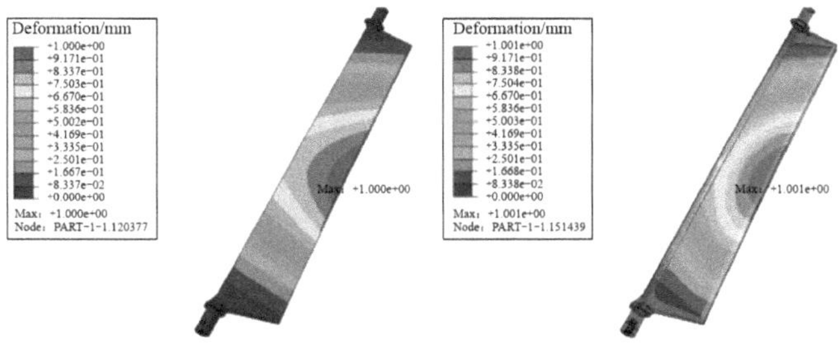

1th order vibration mode of the solid structure 1th order vibration mode of the two-layer plate structure

3th order vibration mode of the solid structure 3th order vibration mode of the two-layer plate structure

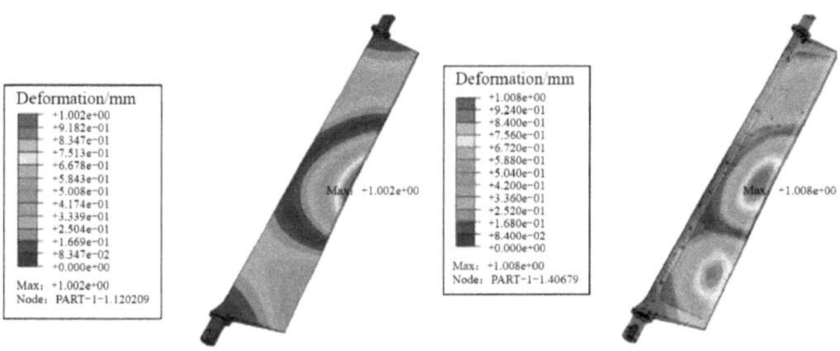

5th order vibration mode of the solid structure 5th order vibration mode of the two-layer plate structure

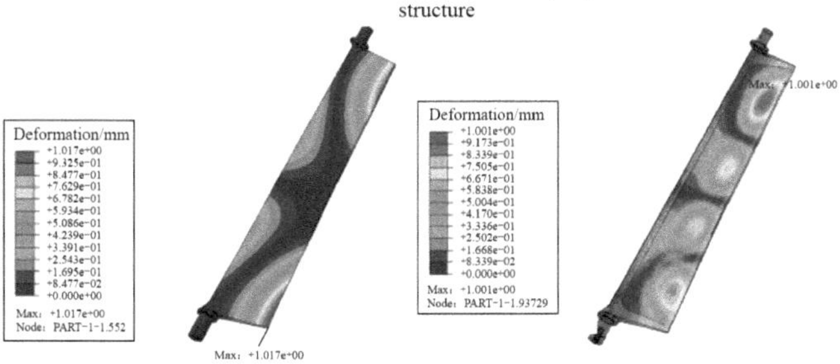

6th order vibration mode of the solid structure 6th order vibration mode of the two-layer plate structure

Fig. 6.5 The first 6-order model of vibration of the two-layer plate hollow structure

Table 6.1 Comparison of the first 6-order natural frequencies of the solid structure with two-layer plate hollow structure

Structure type	First 6-order natural frequencies/Hz					
	f_1	f_2	f_3	f_4	f_5	f_6
Solid structure	372.55	997.42	1255.6	1461.7	1945.8	2351.8
Two-layer plate hollow structure	447.87	1038	1209	1636.3	1734.3	2060.7

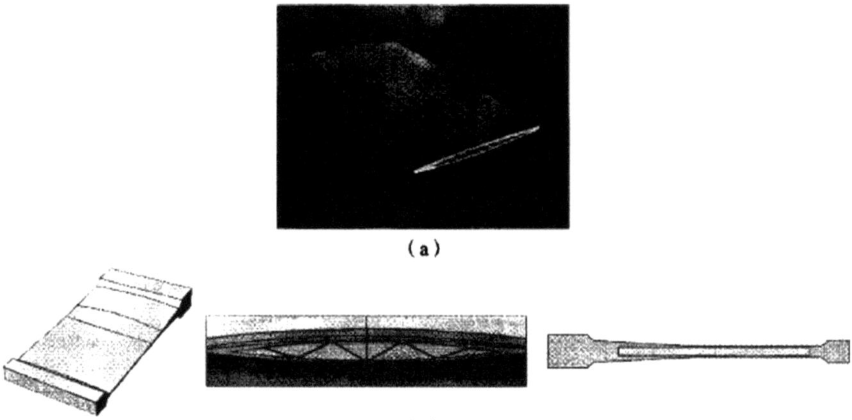

(a)

(b)

Fig. 6.6 Geometrical characteristics of the three-layer plate structure. **a** Skin-W-shaped truss reinforcement structure; **b** Characteristics of solid-shell mixed structure

efficiency of 20–50%. The major defect is the existence of inherent geometrical characteristics-triangle area, which is located in the joint of the rib with the skin and at 1/2 height of the stiffening rib, and is named because the geometry of the section is approximately a triangular (Fig. 6.8b). The side length of the triangle is generally between 0.1 and 2.0 mm and this is not good for the structure to bear the vibration load. Furthermore, due to the constraints of the manufacturing process, it is required to keep a large spacing between the stiffening ribs, resulting in poor local rigidity, dense high-order natural frequencies, and various types of local vibration modes (Fig. 6.9). Therefore, this type of structure is primarily used to bear static loads, and can also used to bear the structures with alternating loads but low life requirements, including rudder and wing surface. Table 6.3 shows the comparison of the first 16-order natural frequencies of the four-layer plate structure and the solid structure. Figure 6.10 shows the four-layer plate structure.

The geometrical characteristics of the X-shaped four-layer plate section structure are as shown in Fig. 6.11, consisting of skin and X-shaped reinforcement structure. This type of structure bearing has excellent static and dynamic bearing capacity, variable rigid (by design), and the overall rigidity is high, so it is especially suitable for high-thickness structure. The structure rigidity can be guaranteed by controlling

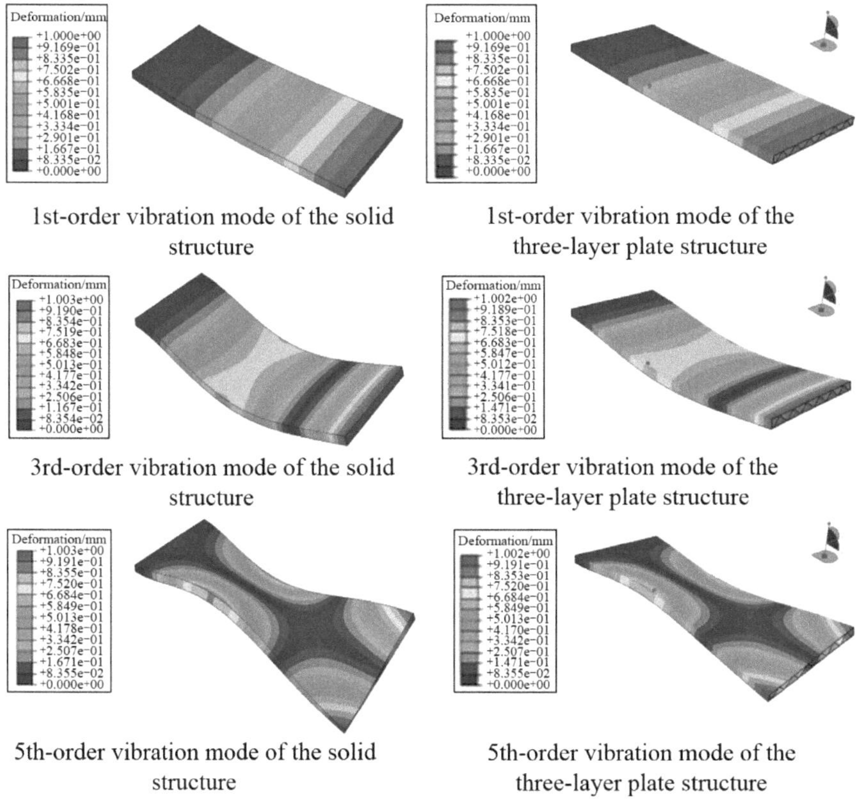

Fig. 6.7 Comparison of the first 5-order vibration modes of the solid and hollow sandwich structures

Table 6.2 Comparison of the first 5-order natural frequencies of the solid and hollow sandwich structures

Structure type	Natural frequency/Hz				
	f_1	f_2	f_3	f_4	f_5
Solid structure	61.74	302.90	383.22	735.78	966.66
Hollow sandwich structure	86.60	36 & 21	532.50	759.93	1166.2

the quantity of stiffening ribs. Without defects theoretically, this structure can be used to manufacture parts with high bearing capacity and high thickness. Similar to three-layer plate structure, this type of structure shows constraints in the ratio of the geometrical parameters of the skin and the truss. Though its skin thickness is generally higher than that of H-shaped structure, it can be reduced to a certain extent by optimizing the included angle of the truss and the skin angle or using other methods. However, the weight reduction efficiency of X-shaped structure is lower

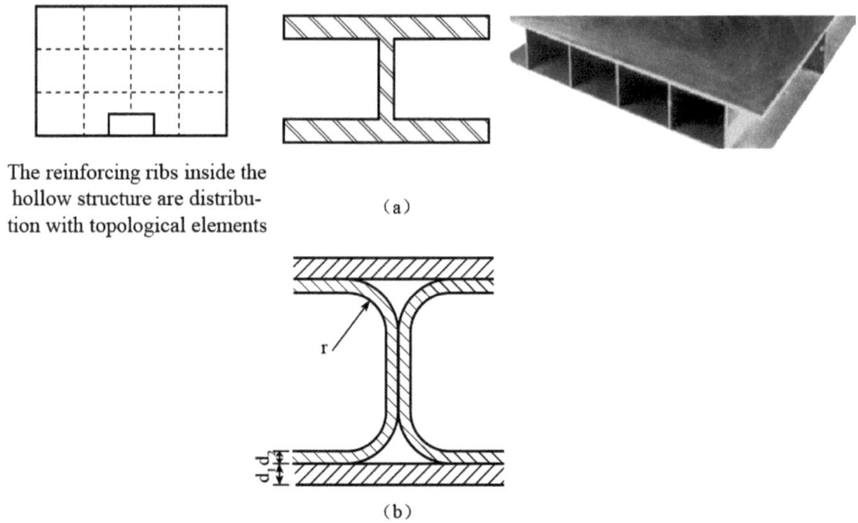

Fig. 6.8 Geometrical characteristics of H-shaped four-layer plate structure. **a** Geometric characteristics of the four-layer plate structure; **b** the inherent characteristic of the four-layer plate structure—triangle area

than the H-shaped structure, which is usually between 20 and 40%, but it involves relatively complex manufacturing process.

Based on the above analysis, the characteristics of geometry, load carrying capacity, and design method with the common SPF/DB structure can be summarized for reference in preparation of the structural design scheme, as shown in Table 6.4.

In recent years, new SPF/DB structures have emerged to further improve the bearing capacity and reduce the structure weight such as the lattice structure, as shown in Fig. 6.12. This type of structure is specially characterized by its multifunction. It can not only meet the bearing capacity and weight reduction functions of the traditional structure, but also have new functions such as heat insulation and vibration reduction, providing an effective attempt for the application field of the expanded titanium alloy lightweight structure.

6.2 Design Method of Superplastic Forming/Diffusion Bonding Structures

The design of SPF/DB structure is difficult mainly due to three reasons:

(1) It is very difficult to establish CAD models of parts. In designing this type of structure, there are various internal topological characteristics and geometrical shapes available for selection. geometric points are the characteristics controlling the lowest level of geometrical shapes, the structure design process generally

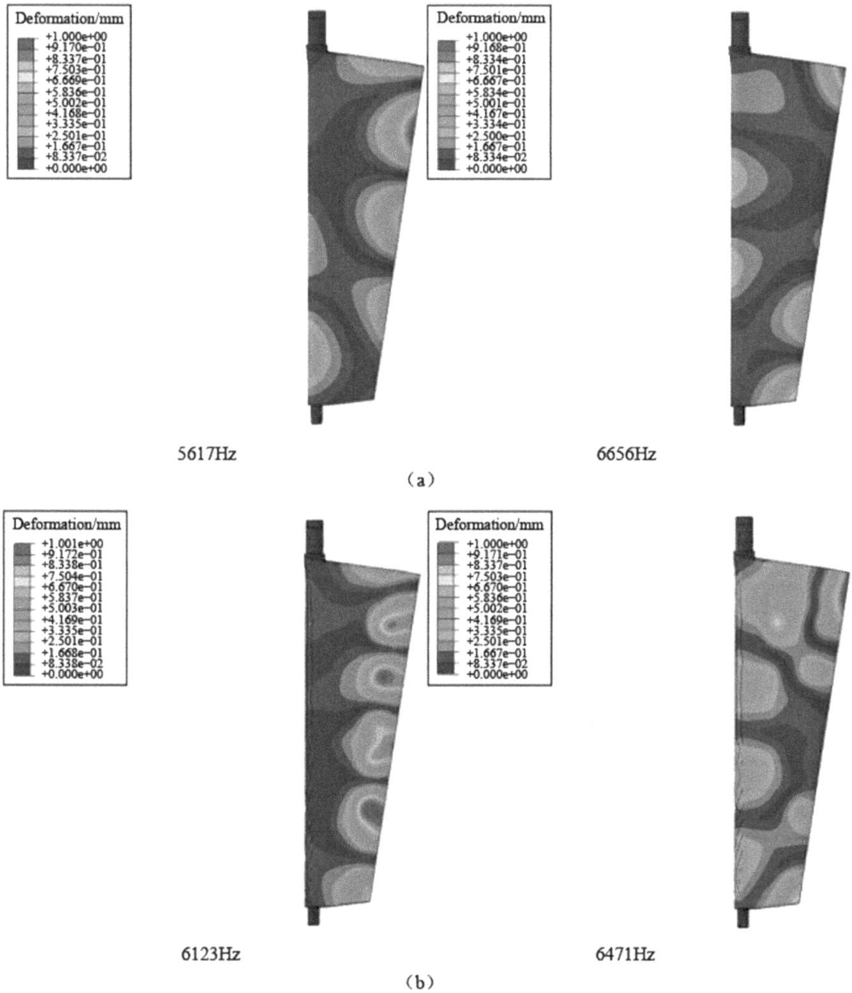

Fig. 6.9 Comparison of inherent frequencies between the four-layer plate structure and solid structure. **a** Solid structure; **b** Four-layer plate structure

follows the steps of forming a line from points, forming a plane from lines and enclosing a body from surfaces. A large number of geometric point coordinates are defined in the 3D virtual space of commercial CAD software, the position accuracy of the coordinates is difficult to control and the cycle is long. It is difficult to avoid the interference between the lines, planes and other characteristics when other geometrical characteristics are formed from the points gradually upward, so it is necessary to adjust the coordinates of geometric points one by one to prolong the CAD modeling cycle.

Table 6.3 Comparison of the first 16-order natural frequencies between four-layer plate structure and solid structure

Order	Natural frequency/Hz	
	Solid structure	Four-layer plate
f_1	431.91	548.03
f_2	978.71	1110.4
f_3	1354	1559.1
f_4	1837	1624
f_5	2273.6	2251.1
f_6	2797.8	3032.9
f_7	3852.2	3378.6
f_8	3901.3	3612.4
f_9	4597.1	4596.4
f_{10}	5332.9	4758.2
f_{11}	5681.3	4943.8
f_{12}	6657.4	6123.5
f_{13}	7328.1	6471.4
f_{14}	7377.9	7239.4
f_{15}	8451.5	7494.2
f_{16}	9165.9	7515.2

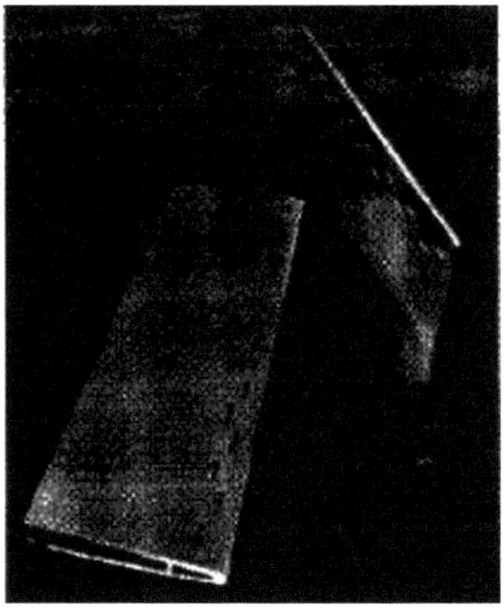

Fig. 6.10 Four-layer plate structure

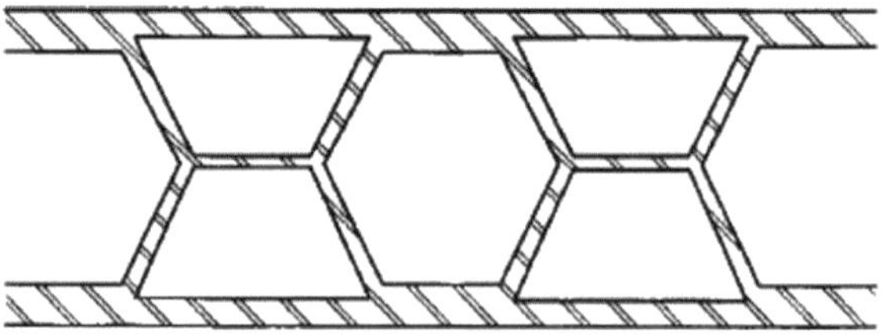

Fig. 6.11 Geometrical characteristics of X-shaped four-layer plate structure section

Table 6.4 Advantages and disadvantages of the SPF/DB structure

Structure type		Surface flatness	Structure rigidity	Weight reduction efficiency	Static load carrying property	Vibration load bearing property	Flexible layout	Easy-to-design
Single-layer plate		✓	×	○	×	×	✓	✓
Two-layer plate	Shell	○	×	○	×	×	✓	✓
	Mixed	✓	○	✓	○	×	✓	✓
Three-layer plate		✓	✓	○	✓	✓	V	×
Four-layer plate	H-shaped	✓	✓	✓	✓	×	✓	✓
	X-shaped	✓	✓	○	✓	✓	✓	×

Note: × indicates that the property is relatively poor among the four types of structures; ○ indicates that the property is moderate among the four types of structure. ✓ indicates that the property is relatively the best among the four types of structures.

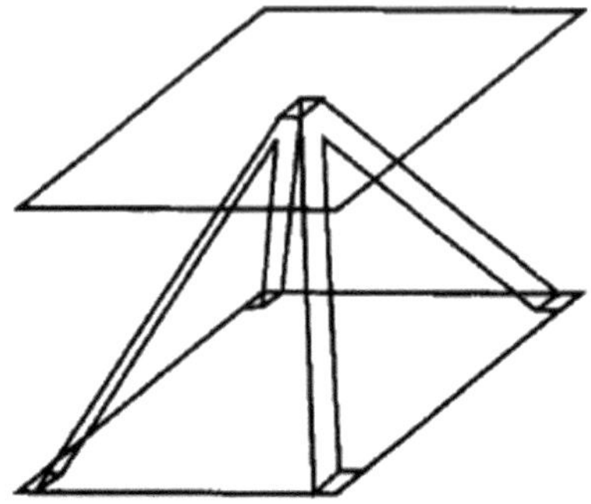

Fig. 6.12 Features of SPF/DB lattice structure

(2) It is very difficult to analyze the structural stress. The hollow sandwich structure has complex internal topological characteristics, which is not in compliance with the sweeping geometrical characteristics, and there are large difference in the dimensions of parts in different areas, for example, the structure is provided with the topological characteristics of solid and shell. In addition, high difficulty is observed in the quality control of CAD model. Correction or reconstruction of the local region of the CAD model always accompanied when establishing the finite element model, so it is difficult to control the element quality of the finite element model and the modeling cycle is long. Due to the above two reasons, high difficult is posed for accurately obtaining the structural stress distribution law.
(3) It is very difficult to assess the structural strength. The strength and failure mode of the parts are determined by the weakest area in the structure, it is necessary to take the surface and interior of the structure as the object and combine the relevant strength criteria to make a judgment in the process of determining the weak areas in the hollow sandwich structure; while the strength criteria should be developed based on material properties, geometrical characteristics, technological process as well as the effect of residual stress, so there is high difficulty and long cycle involved in the development of strength criteria through experimental verification and iterative optimization of part structure.

6.2.1 Structural Design Strategy

The commonly used structural design and analysis method are mainly divided into three steps: ① carry out shape design and internal structure design, and establish the structural geometrical model; ② reasonably simplify and disconcrete the geometrical model to build the finite element analysis model; ③ make response analysis and strength check, and feedback the structure designer for optimization. When designing the load-bearing and weight-sensitive hollow sandwich structure with this method, it founds difficulty in building the CAD model and finite element model with long cycle due to the complex geometrical characteristics of the structure. For the above, shortening the cycle of these two modeling processes will greatly promote the engineering application of the hollow sandwich structure.

In practice, the SPF/DB parts used for specific purposes are found with similar internal topological characteristics. With this basis, a new method can be adopted in the structural design of SPF/DB parts, that is, the structural design and optimization method based on finite element (FE) model. This design method divides the traditional design process into two stages, i.e. the outline design stage and the detailed design stage. Firstly, in the outline design stage, the similarity of the topological characteristics of the internal structure of the parts is used. For example, the hollow sandwich structure with hollow-solid hybrid and W-shaped internal truss has the macro sweeping characteristic in the extension direction of truss, while the section topological characteristics perpendicular to the sweeping direction of the truss are similar. This similarity is used to establish the finite element model containing the main geometrical characteristics of the structure, provide the load carrying performance analysis and carry out the iterative optimization of internal structural parameters until a reasonable selection interval of internal topological parameters is obtained; secondly, in the detailed design stage, a geometry model is established based on the control points determined by FE model, and additional feature are added to refine local characteristics to form a CAD model of part with complete geometrical characteristics. Then, virtual analysis is conducted based on this CAD model to verify the performance of the structural design parameters.

The idea based on the FE model design method allows to directly use the FE model having the main characteristics of the parts for optimizing the structural parameters, saving the time-consuming in CAD and FE modeling process. Beside, after the structural design scheme is determined, the node data in FE model will also be used as the basis for information input, inspection acceptance of the manufacturing process, thus to further shorten the manufacture process design cycle. The characteristics of the two design strategy are shown in Fig. 6.13.

The hollow sandwich structure is designed by using the core parametric design software. With long development period, the software needs to go through three stages to realize all the functions:

(1) In stage 1, it is required to have the ability of designing geometrical characteristics of hollow sandwich structure. According to the characteristics of hollow sandwich structure, the software can automatically complete the internal topology design based on the structural geometrical control parameters (including the surface control points data of several equal-height sections, wall thickness, the angle of the truss with skin, the boundary location of hollow area, etc.,) input by the designer under interactive conditions, and generate the internal geometrical characteristic control point cloud information; based on the requirements for interaction mode and load capacity analysis, a finite element model suitable for the analysis objective is generated. Three-layer plate hollow structure design software interface as shown in Fig. 5.40, the characteristics of control points in different regions generated during the modeling process as shown in Fig. 6.14. The software automatically generates the finite element models of

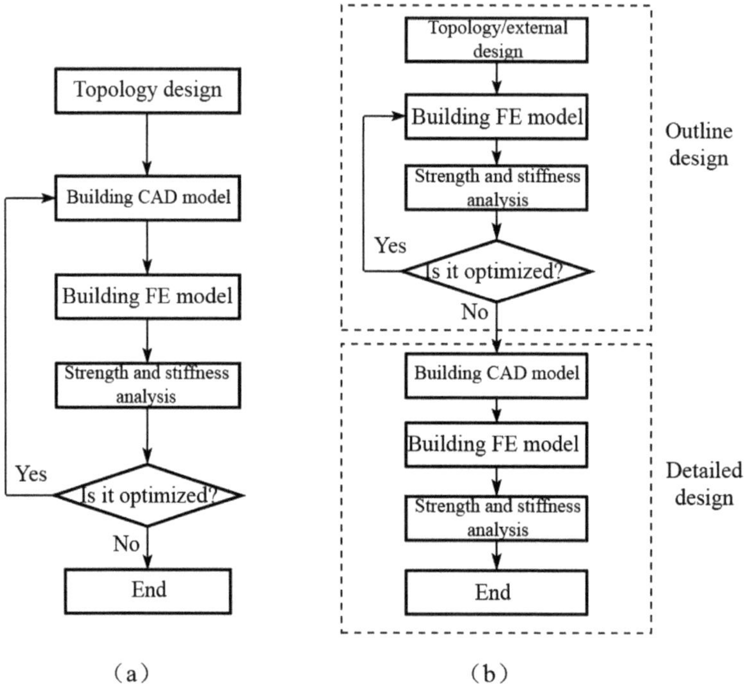

Fig. 6.13 Design method of the hollow sandwich structure. **a** Traditional design methods; **b** Design method of hollow sandwich structure

the hollow area and solid area on both sides based on the control parameters of external characteristics and internal geometrical characteristics; gradually completes the internal modeling of hollow sandwich structure according to the modeling process at the bottom of the interface. The built finite element models of hollow characteristic specimen of three-layer plate structure are shown in Fig. 6.15 respectively. The software can use 6-hedral element to disconcrete the parts, and meet the requirement of the desired element size, there are four layers of elements arranged along the direction of skin thickness of the hollow sandwich structure, and the first layer element is 0.2 mm thick.

(2) In stage 2, the constraints of the forming process and the bearing properties of the hollow sandwich structure on the geometrical design are integrated into the software to form the structure design software with the consideration of the forming process constraints, for example, ratio of skin thickness to truss thickness, the minimum/maximum angle between the truss and the skin, the minimum distance between parallel diffusion bonding interface, and the latter includes such as optimal angle range between the truss and skin, the minimum thickness of truss blanks, etc.

(3) In stage 3, the material short-term mechanical data and fatigue performance data obtained from the standard specimen and characteristic specimen are integrated

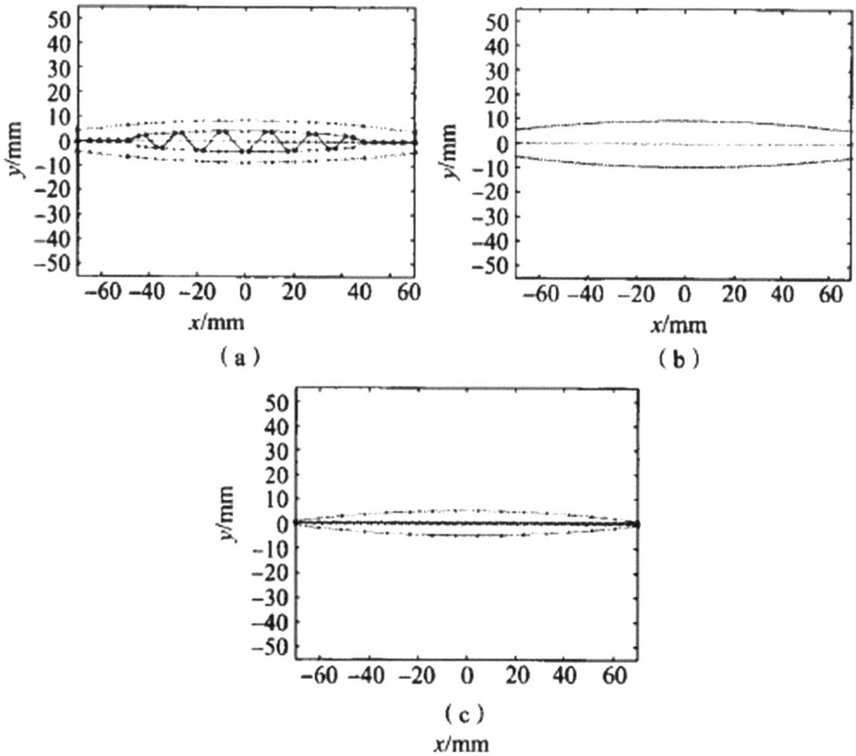

Fig. 6.14 Design of control points in hollow sandwich structure. **a** Control points in hollow area; **b** Control points in root solid area; **c** Control point in tip solid area

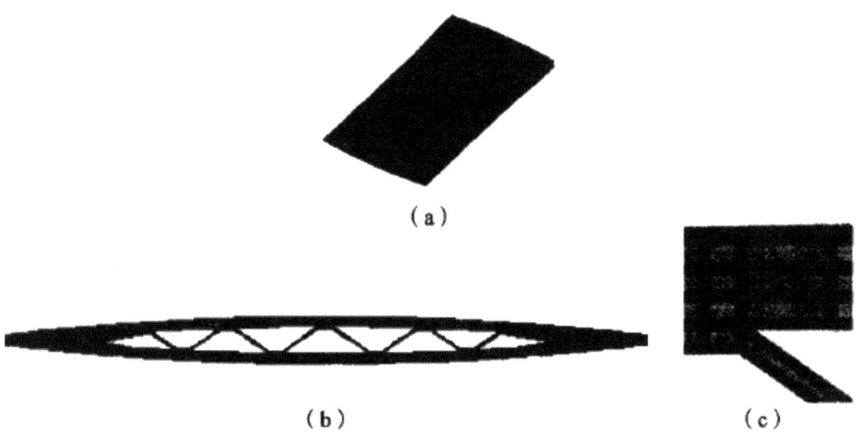

Fig. 6.15 Finite element model of hollow sandwich specimen. **a** Equal side view; **b** cross section of hollow sections; **c** detail of joint of skin and truss

into the software to enable prediction of the strength, damage and life prediction function of hollow sandwich structure with the consideration of the external load and residual stress.

6.2.2 Design Method of the Hollow Sandwich Alternative Load Carrying Structure

Compared with the traditional solid structure, the most significant topological characteristics of the hollow sandwich structure is the addition of a number of free surfaces, i.e. the interior free surface in the hollow sandwich structure, which enables the structure under the load to present the failure modes with diversified characteristics, the crack initiation location may initiate from the surface, subsurface or inner surface of the structure. Crack initiation location of the hollow sandwich structure is related to the stress distribution in the structure and the fatigue resistance performance of the structure, but not to the surface state, the distribution law of wall thickness, the angle between the truss and skin, surface roughness, diffusion bonding defects and other factors. It needs to take into consideration the influence of surface condition, internal stress level and the properties of materials, inner structural characteristics of the hollow sandwich structure, defects and residual stress. In the static bearing structure design, it is only required to consider the stress and material bearing capacity, while in the dynamic bearing structure design, all of the above factors must be considered, bringing in high difficulties.

Whether the theoretical crack initiation location can be controlled in the structure subsurface is the key to determine the design rationality of the dynamic load bearing structure. This is because the theory analysis and experiment results show that the hollow sandwich structure has the highest fatigue properties and longest service life when the crack initiation location is on the subsurface of the structure's out surface. The cause for this phenomenon is described as follows:

1. In case of the fatigue crack initiation on the outside surface of hollow sandwich structure, the performance of the structure is mainly decided by the material properties. When the fatigue resistance of structure is moderate, the reduction design of structure is conservative: this failure mode is commonly observed in hollow sandwich structures without surface strengthened; the service life of structure is mainly decided by the material properties, the vibration stress amplitude and roughness of the structure surface. Generally, the high-cycle fatigue properties of parts are close to those of standard specimens with the same surface state and similar stress gradient.
2. When crack initiation on the subsurface of hollow sandwich structure's outside surface, the structural performance is jointly dominated by material performance and surface strengthening level, in this condition, the structure has the best fatigue resistance: This failure mode is common found out in those structures with reasonable surface strengthening, good consistency between design and manufacturing of internal structure, and no significant stress concentration. Under

this condition, the high-cycle fatigue performance of the structure is close to the fatigue performance of the standard specimens strengthened with the same parameters.

3. Crack initiation occurs on the inner surface of the hollow sandwich structure, and the structure properties are dominated jointly by the material properties and internal stress. The weakest fatigue resistance is normally seen in the structures with radical structure reduction, less consistency in design and manufacture: the failure mode mainly occurs in the surface-strengthened structures, which involves two cases: first, the structure design is not reasonable, such as too small skin thickness, improper selection of surface strengthening parameters—high residual stress is resulted in the internal structure due to adoption of the upper limit of the effective strengthening interval, and crack initiation occurs from the inner surface of the hollow sandwich structure under the joint action of residual stress and vibration stress; second, the design and manufacture are not well matched, the internal geometrical dimension tolerance is out of the permissible range or there are presence of defects, thereby causing excessive stress concentration.

According to the fatigue failure mode and life control factors of the hollow sandwich structure, the strength prediction model of the hollow sandwich structure can be built based on Coffin life prediction formula and Goodman mean stress correction method.

Basic experiments are conducted to investigate the fatigue crack initiation from surface, subsurface and inner surface of the hollow sandwich structure. In case of crack initiation from the surface, it is possible to obtain the data of material properties under different stress ratios by using the surface polishing specimen; in case of crack initiation from the subsurface, it is possible to obtain the data of material properties under the conditions of symmetric cycle and different mean tensile stress by using the surface strengthening specimen when different shot peening parameters are used; in case of crack initiation from the inner surface, it is possible to obtain the data of material properties under different. In addition, the inner surface residual stress should be calculated by combining the surface strengthening parameters and structural geometrical parameters, and the law of influence on the fatigue properties of inner surface materials should be obtained by experimental test.

To develop the corresponding Coffin formula and Goodman correction method for the above three cases, it is necessary to perform the rotational bending fatigue test, axial fatigue test and the fatigue test with the mean stress not being zero. The strength correction method can be used to determine the influence of defects on the fatigue properties.

The axial-loading fatigue test data of standard specimens are shown in Fig. 6.16. The material used for preparation of the specimen is taken from the characteristic specimen of the hollow structure. According to the experimental results, the fatigue strength of the part material under axial loading and stress ratio R is 350 MPa, and the fatigue strength of the part material under axial loading and stress ratio is 300 MPa and 250 MPa respectively when the average tensile stress is 100 MPa and 200 MPa.

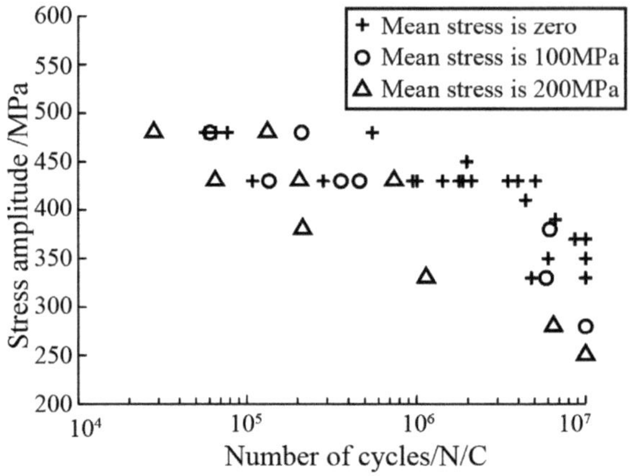

Fig. 6.16 Axial-loading fatigue test results of the smooth specimen made of TC4 alloy

Subject to surface strengthening, the inner surface of the hollow sandwich structure is under tensile residual stress and the residual stress is biaxial, at this time, the effect of inner surface residual stress on fatigue properties can be obtained by the fatigue test of characteristic specimen, such as the axial round bar specimen of the annular V groove, axial specimen of annular section. The specimens are shown in Fig. 6.17, and the annular section specimen after the fatigue test is shown in Fig. 6.18. The annual section specimens are used to simulate the biaxial residual stress state by means of applying axial tensile static load and the internal loading. The specimen can easily control the relative size of the orthogonal residual stresses, and enable to obtain the effect of residual stress on the fatigue properties of TC4 alloys when it is used to simulate the residual stress of the hollow sandwich structure with W-shaped truss, as shown in Fig. 6.19.

According to the test results of the standard specimen and biaxial fatigue specimen, the mean tensile stress causes reduction of the fatigue resistance of the material; the S–N curves of TC4 alloys obtained under different stress ratios are close to parallel lines, and the fatigue strength of the material under different mean stresses is shown in Table 6.5. Based on the test data, the corrected Goodman formula is used to create the effect of mean stress on the properties of TC4 alloys; for the fact that the internal residual stress is relatively small after the hollow blade surface is strengthened, the test data obtained under the mean stress of 100 and 200 MPa are selected to determine the corrected parameters of the mean stress, thus to obtain the Goodman mean stress correction parameter k = 0.79.

Goodman formula is

$$\frac{\Delta\sigma}{\sigma_{-1}} + \left(\frac{\sigma_{\mathrm{m}}}{\sigma_{\mathrm{b}}}\right)^{k} = 1$$

where: σ_{-1} is the fatigue strength of the material when the mean stress is zero; σ_{m} is the mean stress; σ_{b} is the strength limit of the material, which is constant and material dependent.

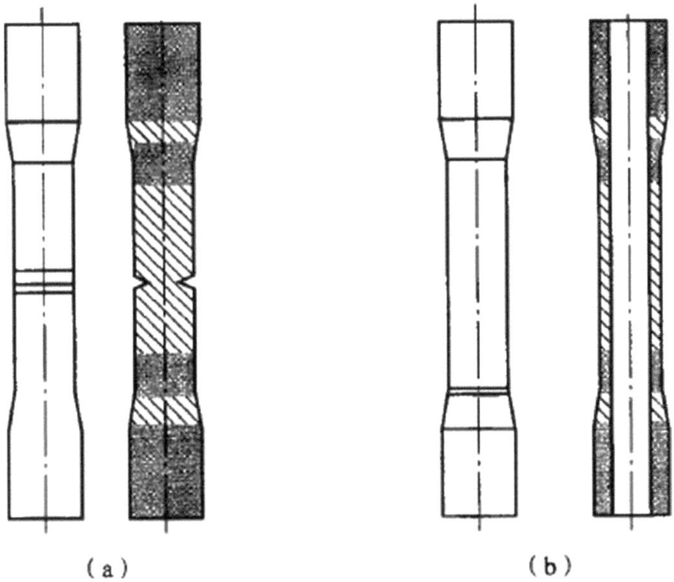

Fig. 6.17 Specimen for biaxial fatigue test. **a** Specimen with V-shaped groove; **b** Annular section specimen

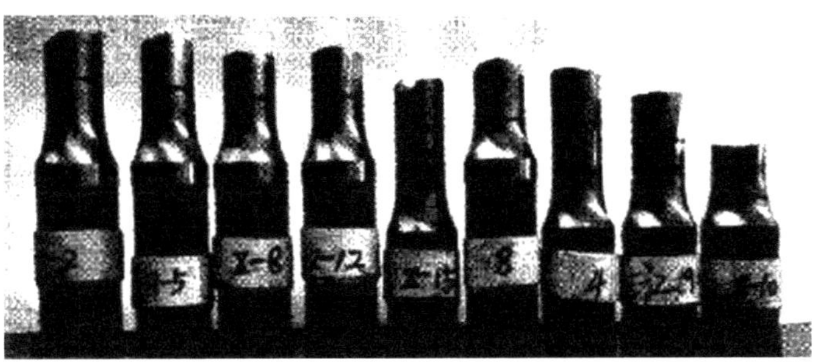

Fig. 6.18 Annular section specimen after fatigue test

Taking the W-shaped truss hollow sandwich structure as an example, brief analysis is made on the effect of wall thickness and surface strengthening parameters on the fatigue strength of the surface and inner surface of the rotating parts structure, and also on the fatigue strength of the structure under bending deformation, and the life of the structure under a certain stress level and the distribution law of damage. The external characteristics of W-shaped truss hollow sandwich parts are as shown in Fig. 6.20. There are four distribution rules of wall thickness, respectively, basic wall thickness distribution and the wall thickness distributions derived as per the proportion changes on that basis, and the wall thickness model includes 0.5 times, 0.8 times and 1.5 times. The said structures have the same outline, so increased wall

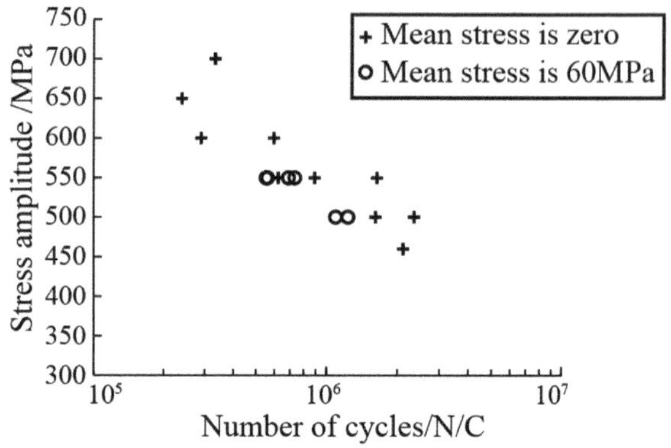

Fig. 6.19 Effect law of biaxial mean tensile stress on the material fatigue properties

Table 6.5 Fatigue strength of materials under different mean stresses

Data of smooth specimen				Data of surface strengthening specimen			
Mean stress (MPa)	0	100	200	Shotpeen intensity (mmN)	0.2	0.4	0.6
Fatigue strength (MPa)	360	300	256	Fatigue strength (MPa)	400	460	430

thickness means increased weight in parts. Figure 6.21 shows the change law of wall thickness with the W-shaped truss hollow sandwich parts.

Two strengthening processes and five sets of surface strengthening parameters are selected for analysis, as shown in Table 6.6. There are three sets of parameters selected for the wet shot peening process, with the shot strength of 0.3 mmN, 0.4 mmN and 0.5 mmN respectively, and the coverage rate is 200%. Two sets of parameters are selected for the dry shot peening process, with the shot strength of 0.15 mmA and 0.2 mmA respectively, and the coverage rate is 200%. The above surface strengthening parameters are all within the selected range of surface strengthening parameters determined by the standard bar fatigue test. In the selected range, increasing the

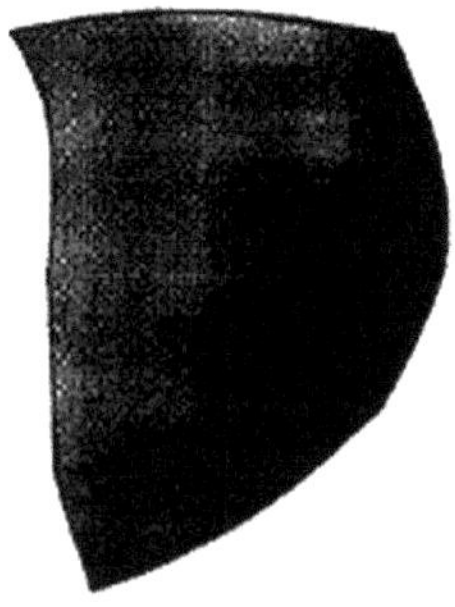

Fig. 6.20 External characteristics of the W-shaped truss hollow sandwich parts

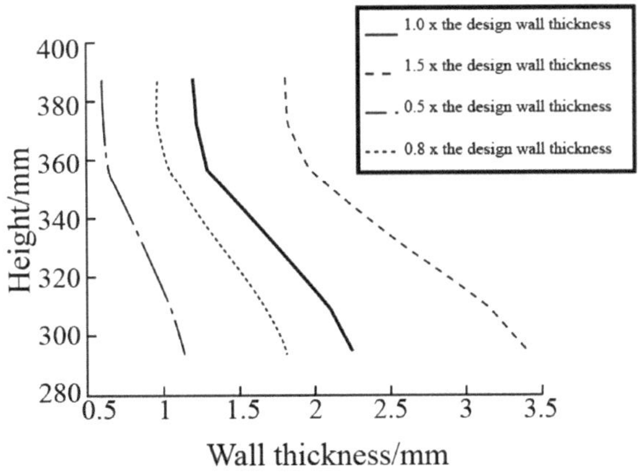

Fig. 6.21 Change law of the wall thickness of W-shaped truss hollow sandwich parts

shot strength can improve the surface strengthening level, and the fatigue strength of standard test bar increases accordingly. Therefore, the highest surface strengthening parameters are always adopted in the traditional structure, for example, the shot strength is taken as about 0.5 mmN for the wet shot peening; about 0.15 mmA for the dry shot peening.

The finite element model of W-shaped truss hollow sandwich structure is build to analyze its stress distribution under rotating condition at a speed of 10000 r/min. The displacement boundary condition is that the bottom face of the part is fixed, and while the rest area is kept free. In this case, the Mises stress contour of centrifugal loading is as shown in Fig. 6.22, the high stress area of centrifugal stress is located at the bottom, and the central area in the height direction of the parts is the high stress area of centrifugal stress. The stress distribution of the part under the 1st-order vibration mode is analyzed as shown in Fig. 6.23, which is in consistency with the stress distribution during assessment of its high cycle fatigue, representing one of the common deformation of parts during service period. In this state, the highest value point of vibration stress is located at the bottom of the part, high stress level is observed in the area below 2/3 area along the height direction, and there is small stress gradient in the height direction.

The effect of shot peening parameters on the deformation law of the W-shaped truss hollow sandwich structure with 0.8 times-wall thickness, is as shown in Fig. 6.24. As seen from the Figure, increasing the shotpeen intensity can lead to increase in the maximum deformation of the structure, but the deformation law remains. The maximum deformation of the W-shaped truss hollow sandwich structure occurs at the top front, and the trailing edge is observed with the maximum deformation. When 0.3 mmN process parameters are applied for the strengthening

Table 6.6 Surface strengthening parameters

Shotpeening intensity using wet blast/mmN			Shotpeening intensity in dry blast/mmA	
0.3	0.4	0.5	0.15	0.2

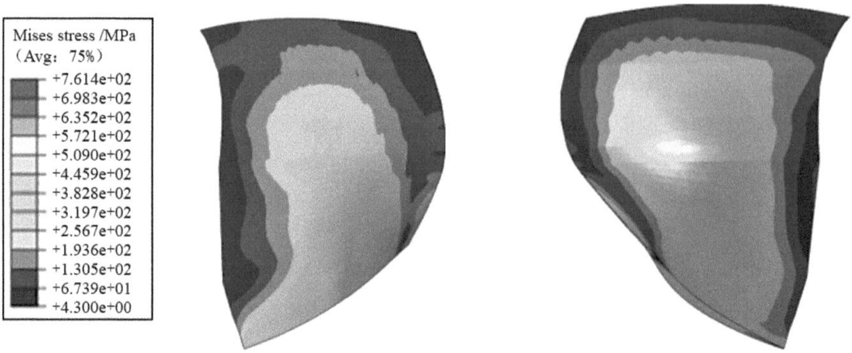

Fig. 6.22 Stress distribution law of W-shaped truss hollow sandwich structure under rotating condition

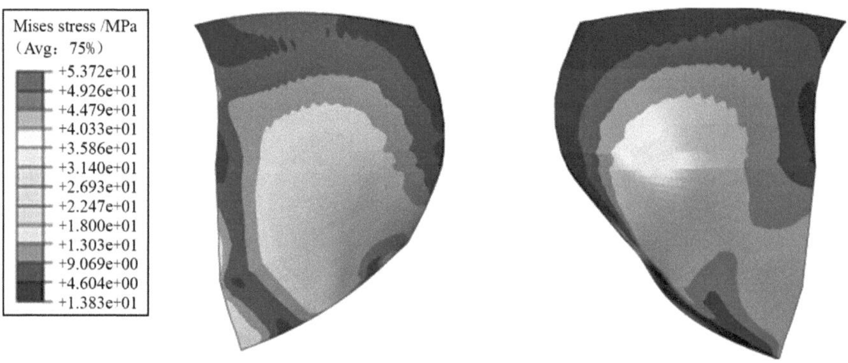

Fig. 6.23 Stress distribution of W-shaped truss hollow sandwich structure under the 1st-order vibration mode

process, the maximum deformation of the part is up to 0.302 mm; when 0.4 mmN process parameters are applied, the maximum deformation is 0.511 mm; and when 0.5 mmN process parameters are applied, the value is 0.597 mm.

According to the analysis, the deformation and residual stress of parts caused by shot peening parameters are different, but they follow the similar law. In case of 0.3 mmN shotpeen intensity and 0.8 times wall thickness parameters, the distribution law of residual stress is:

(1) The Mises stress distribution of the W-shaped truss hollow sandwich structure surface after surface strengthening is as shown in Fig. 6.25. For this structure, the rigid in hollow area is relatively weak, which is prone to deformation, so the residual stress in the area may be reduced. For this reason, the surface residual stress distribution characteristics of the parts is characterized by high stress in the solid area and low stress in the hollow area;

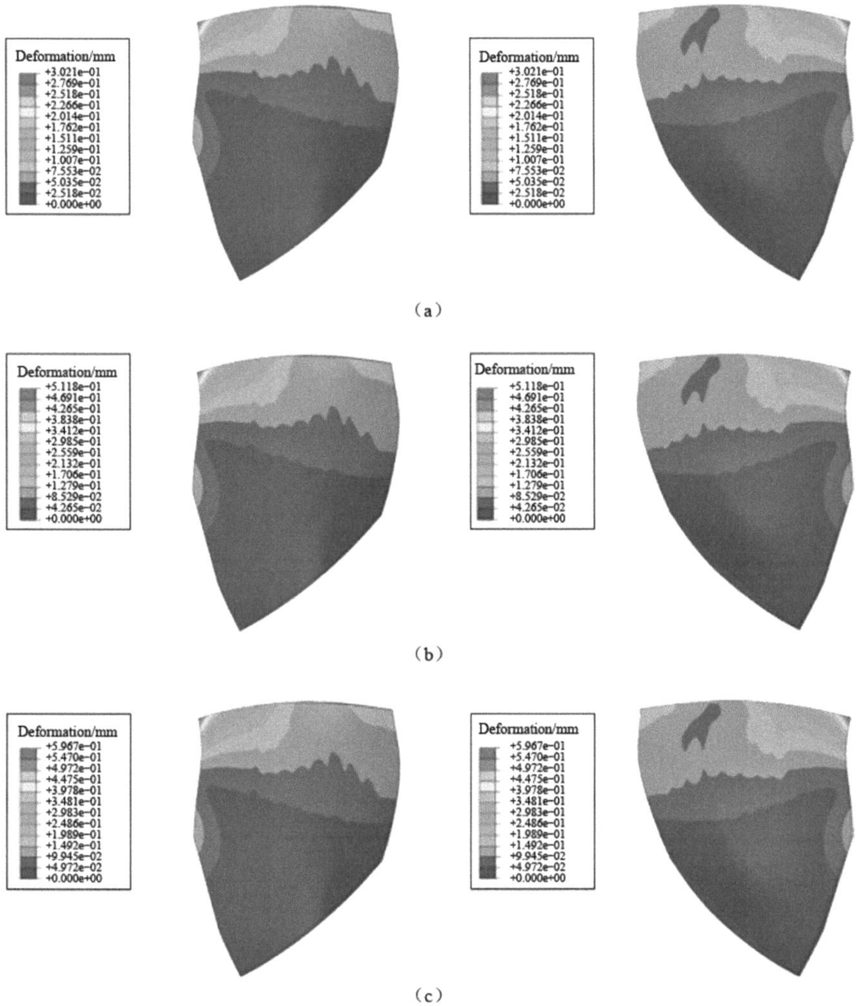

Fig. 6.24 Effect law of surface strengthening on deformation of the hollow sandwich structure. **a** Shot strength: 0.3 mmN, 0.8 times wall thickness; **b** Shot strength: 0.4 mmN, 0.8 times wall thickness; **c** Shot strength: 0.5 mmN, 0.8 times wall thickness

(2) The Mises stress distribution law of the W-shaped truss hollow sandwich structure inner surface after surface strengthening is as shown in Fig. 6.26. The residual stress on the inner surface of the hollow area presents a periodic distribution, and the residual stress in the interface area of the diffusion bonding is greater than that in other areas.

The effect law of surface strengthening parameters and wall thickness on the maximum residual stress of the inner surface of the hollow sandwich structure is shown in Fig. 6.27. As seen from the Figure, the residual stress on the inner surface

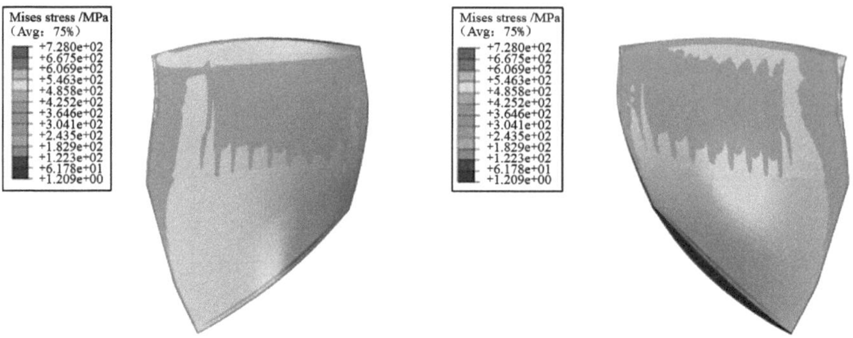

Fig. 6.25 Residual stress distribution on the external surface after surface strengthening. (Shot strength: 0.3 mmN, 0.8 times wall thickness)

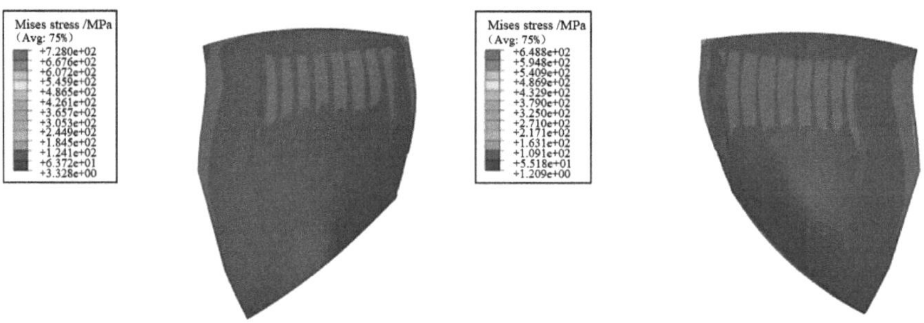

Fig. 6.26 Effect law of surface strengthening on the inner surface residual stress. (Shot strength: 0.3 mmN, 0.8 times wall thickness)

of the hollow sandwich structure increases with the decrease of wall thickness and the increase of shot strength. Taking the W-shaped truss hollow sandwich structure with 0.8 times-wall thickness as an example, its inner surface residual stress is tensile stress and the maximum residual stress is 45 MPa when the shot strength is 0.3 mmN; and the maximum residual stress on the inner surface reaches about 160 MPa when the shot strength becomes 0.2 mmA. Where the wall thickness is 0.5 times and the surface strengthening parameter is 0.2 mmA, the maximum residual stress on the inner surface will be close to 250 MPa.

Based on the life prediction model, the effect law of shotpeen intensity on structural damage of the W-shaped truss hollow sandwich structure is obtained under 1 time wall thickness condition, 1st-order bending vibration mode and the same amplitude, as well as the damage distribution of the concave surface, convex surface and inner surface of the convex surface, as shown in Figs. 6.28, 6.29 and 6.30. It can be seen from the Figure, when the shotpeen intensity is 0.3 mmN, the maximum damage is located in the convex surface and the middle span of W-shaped truss hollow sandwich structure, which is consistent with the maximum stress location of vibration. There is small damage with linear distribution on the inner surface on the

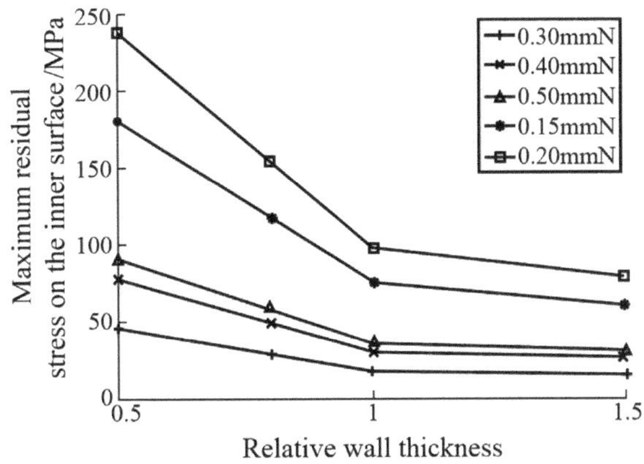

Fig. 6.27 Effects of relative wall thickness and shot strength on the maximum residual stress on the inner surface of the hollow sandwich structure

convex side of the part, which is the boundary position of the triangle formed by the diffusion bonding interface and skin and is related to the stress concentration at the intersection point of the truss and the skin. In addition, increasing shot strength will lead to decreased damage on the surface but increased damage on the inner surface. When the shot strength increases to 0.5 mmN, the damage at the intersection of the truss and the skin is significantly larger than that on the outer surface, this means that crack initiation will occur from the intersection of the truss and the skin.

Failure mode of the hollow sandwich structure is related to the geometrical parameters and surface strengthening parameters of the structure, and the former is in the upstream in the traditional part design process. Therefore, the surface strengthening process parameters are generally designed based on the determined geometrical parameters. The surface strengthening parameters of the hollow sandwich structure should be selected based on the strengthening parameter range obtained from the fatigue test data of standard specimens to allow the placement of theoretical weak position on the subsurface of the structure. Only when the above conditions are satisfied, the parts can be guaranteed with the highest fatigue resistance.

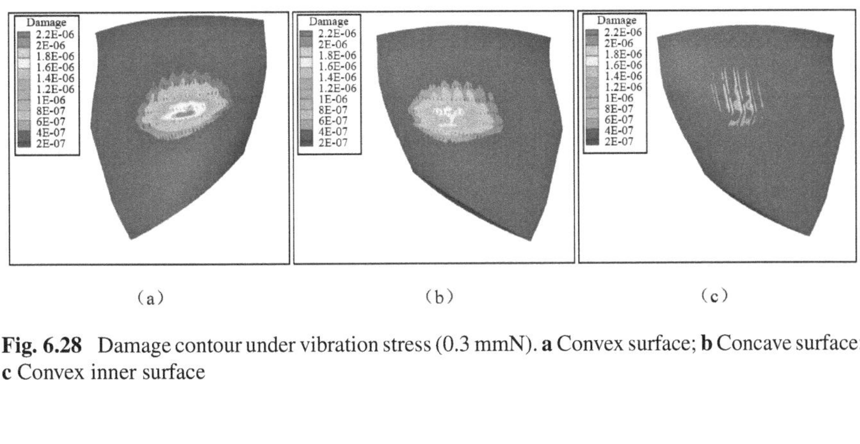

Fig. 6.28 Damage contour under vibration stress (0.3 mmN). **a** Convex surface; **b** Concave surface; **c** Convex inner surface

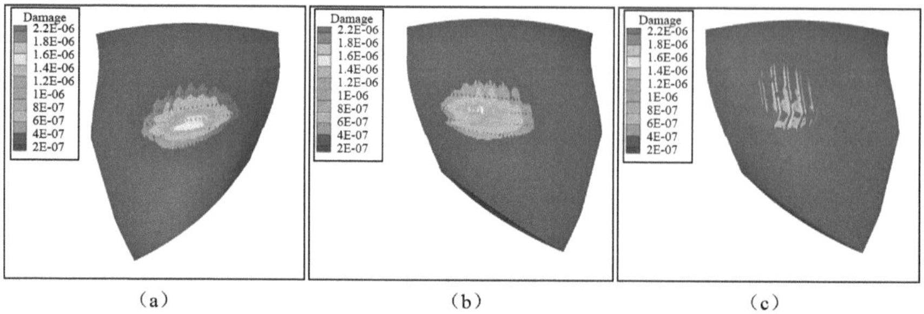

Fig. 6.29 Damage contour under vibration stress (0.4 mmN). **a** Convex surface; **b** Concave surface; **c** Convex inner surface

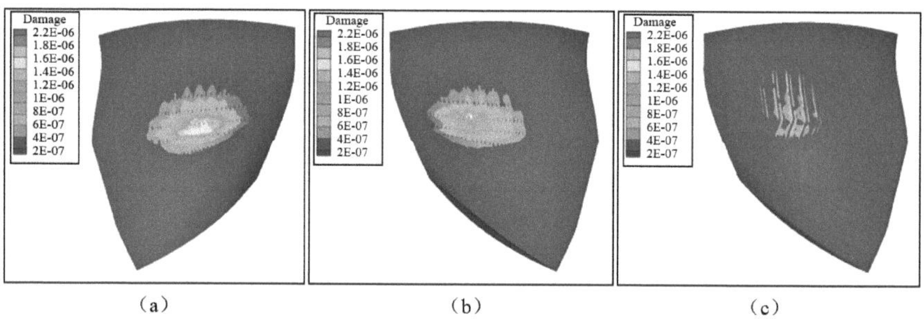

Fig. 6.30 Damage contour under vibration stress (0.5 mmN). **a** Convex surface; **b** Concave surface; **c** Convex inner surface

Figure 6.31 shows the reasons for placing the theoretical weak position of the hollow sandwich structure on the subsurface of the part's outside surface. The curves in the Figure shows the effect of surface strengthening parameters on the surface life and inner surface life of W-shaped truss hollow sandwich structure under 1.0 times wall thickness. From the Figure, it can be seen as follows:

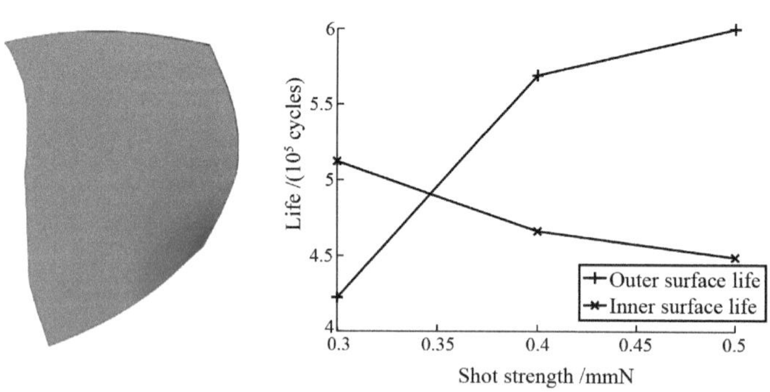

Fig. 6.31 Effect of the surface strengthening parameters on the inner and outer surface life of the three-layer plate structure

(1) Where the outside surface of the part is not strengthened or the level of strengthening is low, the inner surface and outer surface are in the similar state, i.e. the machining or polishing state, and the same applies to the material fatigue resistance of the two surfaces. Moreover, the theoretical design of the structure ensures that the surface has a higher normal alternating stress than the inside during service, and there is a small residual stress, which will not affect the fatigue resistance of the structure. As with the above characteristics, the surface and inside fatigue life is largely defendant on the amplitude of vibration stress and material properties. Therefore, the crack initiation time of the outer surface of the part is less than that of the inner surface, and the life of the part is controlled by the surface crack initiation time.
(2) By increasing the strengthening intensity of the part, it is possible to change the state of its outer surface and thereby improve the fatigue resistance of the material in this area, while the inner surface still remains the machining state. Therefore, it demonstrates that the part material presents a fatigue resistance higher than the inside, and the surface strengthening process cannot lead to change of the shape and load characteristics of the structure, and there is no change in the alternating stress of the outer surface, which still remains above the inside. However, after strengthening the compressive residual stress between the structure's outside surface and the subsurface can induce the tensile residual stress in the structure, while the amplitude of the alternating stress on the inner surface remains unchanged. As concluded from the above, the crack initiation time of the inner surface is shortened, and the variation increases with the increasing level of surface strengthening. Hence, Prompting the surface strengthening level contributes to prolonging the crack initiation time on the outer surface of the parts but shortening the crack initiation time on the inner surface. At a certain strengthening level, the overall life of the hollow sandwich structure is the longest when the life of the inner and outer surface is identical.
(3) With the further increasing of the surface shotpeen intensity, the fatigue resistance of the outside surface material continues to be strengthened and the same applies to the tensile state residual stress induced by this process within the structure, while the inner surface crack initiation time is shortened. Therefore, the life of the hollow sandwich structure decreases with the prompoting of surface strengthening level in this process.

The characteristics of the structure are demonstrated by A W-shaped truss hollow sandwich structure. The skin thickness of the part is 1.0 times the wall thickness, and the surface is subject to wet shotpeening. According to the fatigue test result of standard specimen, it concludes that the process allows an optional strengthening parameter range of 0.2–0.55 mmN for TC4 alloy. The effect of shotpeening on the life of parts is analyzed with the vibration stress amplitude of 500 MPa (Fig. 6.31), it can be seen that the best properties are be obtained when the shot strength is 0.35 mmN and the life of the outer surface life is the same as that of the inner surface. In the engineering practice, the shot strength selected is lower than the optimal solution, that is, keeping the fatigue resistance of the outer surface slightly lower than that of the inner surface. In this way, it is more convenient for nondestructive testing of the surface crack during the field service.

Due to complex internal geometrical characteristics of the hollow sandwich structure, there are a lot of simplification and abstraction work in the design of the structure. The structural strength evaluation criterion is coupled with the internal structure and process characteristics, which brings in great difficulty in establishing accurate criterion, so in the design and development process of a new structure, it is necessary to closely combine the theoretical design, process design and experimental evaluation.

6.3 Evaluation and Optimization Methods of the Quasi-static Bearing Properties of the Structure

In an early age, the hollow sandwich structure is primarily used for bearing static or quasi-static load, function as a secondary load-bearing structure, such as the stabilizer and rudder of the aircraft, engine strut, exhaustor convergence/divergence adjustment blade, etc. Concerning the completed quasi static load bearing experiments, the experiments are carried out on the stabilizer and rudder, with the aim of testing the aerodynamic load under the action of structure strength and stiffness. In case of bearing static or quasi-static loads, there are few factors affecting the failure of the hollow sandwich structures. Thank to the advancement of the virtual analysis technology, the prediction accuracy meets the design requirement of the static bearing structure, replacing the quasi-static load bearing verification experiment to a largest extent.

6.3.1 Quasi-static Bearing Performance Experiment

Aerodynamic force is known as the key load carried by the rudder and wing parts, which is perpendicular to the structure surface and distributed on the structure surface according to a certain law. There is small surface pressure generated by the aerodynamic load and it can not cause structural damage. Aerodynamic force can produce bending moment in the connection parts of the structure to form a high bending stress, and this area is generally the lap or transition area between a shell and a solid in a structure, geometrical changes can result in stress concentration, and mutual coupling of the two factors will have adverse effects on the bearing properties of the structure. Therefore, it is required to carry out experiment on the bearing properties of this type of structure. According to the common practice, the product of the safety factor and maximum design load is taken as the target value, and the quasi-static loading mode is adopted to apply the load on the parts to obtain the response law of the parts. For a long period of time, the experimenters have been devoted to seeking a loading method that is closer to the characteristics of pneumatic load.

Fig. 6.32 Static load carrying test of the hollow sandwich structure

In the previous experiments, tapes are bonded on the surface of the structure to transfer the tensile load, the loads are enabled to act uniformly on the surface of the specimen by setting multiple loading points, and special fixtures are used to ensure that the load uniformity is under control in case of high deformation in the structure; in recent years, an assessment method of airbag loading has been developed to achieve even distribution of the load on the surface of the tested structure.

The static load test of the hollow sandwich structure is shown in Fig. 6.32. As shown, one end of the structure is rigidly clamped, and four groups of tapes are installed on its surface to generate tensile load. The variation curve of displacement and strain of the hollow sandwich structure with load under quasi-static load (Fig. 6.33) is shown in Fig. 6.34. In the beginning of applying load, the deformation and load of the structure follows the linear change relationship, and when the load reaches a certain value, the deformation of part increases sharply with the increase of load until failure. The inspection results showed that the damage of the part is close to the clamping end, which is the transition area of the hollow area and solid zone in the structure and also the area with the maximum stress of the structure based on the theoretical analysis. According to the experiments, subject to the stress, the position where failure occurs is usually located in the area of maximum stress when the hollow sandwich structure is under static load.

6.3.2 *Quasi-static Load Failure Mode and Structural Optimization*

Failure characteristics of the rudder and wing components under quasi-static loading are shown in Fig. 6.35. The failure area is near the clamping end, located at the lap position between the hollow area and the solid area. Considering the clamping and loading characteristics of the structure in the process of quasi-static loading, the stress distribution is obtained by numerical method, as shown in Fig. 6.36. In the analysis, the structure is simplified into two parts: solid structure and shell structure.

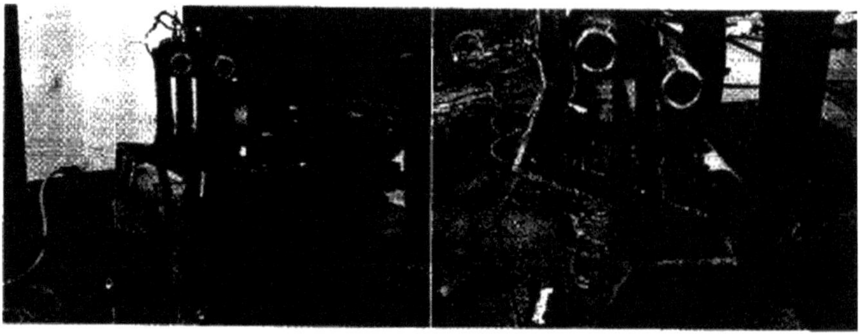

Fig. 6.33 Quasi-static tensile test of parts

The solid structure covers the pin hole and the nearby area, and the rest is the hollow area that deploys shell element for modeling. The analysis results show that the high stress area is near the pin hole, which is consistent with the failure position of the structure in the experiment. Hence, it is concluded that the failure mode of the hollow sandwich structure under static load is mainly determined by the macroscopic stress distribution.

According to the damage characteristics of the hollow sandwich structure, the main idea is proposed to improve the quasi static bearing properties through reduction of the stress concentration factor and the maximum stress level in the structure, for example, for the transition area between the entity and shell, the structure can be designed to the profile as shown in Fig. 6.37a when only the forming is considered but not the bearing characteristics. Following this design scheme, there may be large geometrical changes and high stress concentration. Through reduction of the thickness change gradient by way of step transition or fillet transition shown in Fig. 6.37b, it is possible to achieve reduce the local rigidity difference and the change rate of the bearing area of the structure, thereby effective enhancing the bearing properties of the structure.

6.4 Evaluation and Optimization Method of Alternating Load Bearing Properties of the Structure

In recent years, the applications of the hollow sandwich structure have gradually extended to the primary load carrying component, the dynamic load bearing structure, such as fan blade of high bypass-ratio turbofan engine, inlet guide van of engine, involving the risk of fatigue failure during working. The position with weak fatigue property is jointly decided by the structure design, process characteristics, residual stress, the manufacture and design conformance, as well as other factors. With increasingly higher indicators on the part structure and property requirements proposed in the design of aviation and aerospace vehicles, the load of the hollow

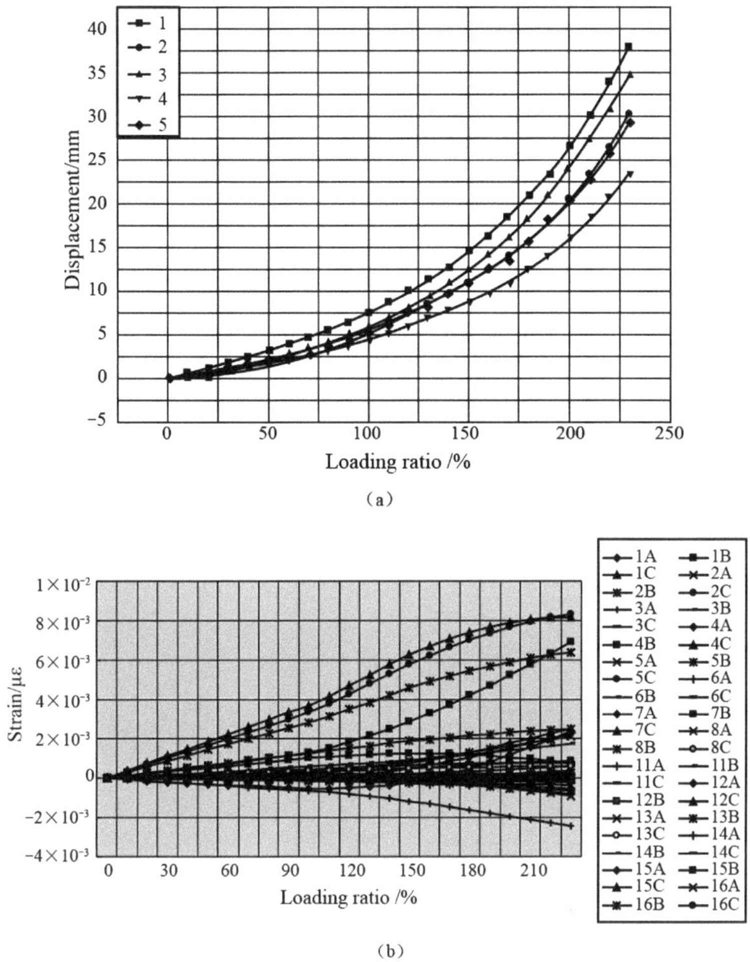

Fig. 6.34 Load-displacement law and load-strain law in the quasi-static tensile test. **a** Load-displacement curve; **b** Load-strain curve

Fig. 6.35 Failure characteristics of the hollow sandwich structure under quasi-static loading

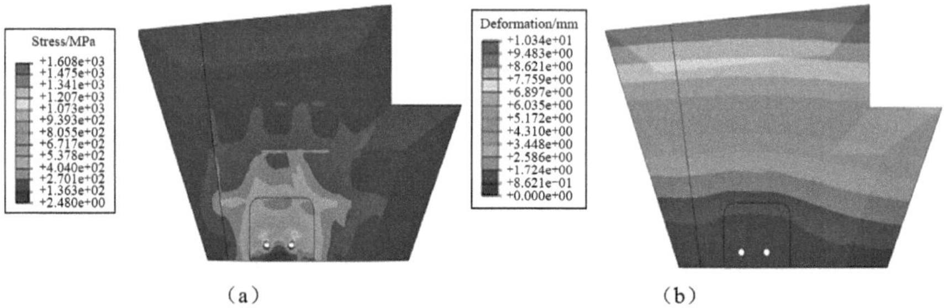

Fig. 6.36 Response of the hollow sandwich structure under static load. **a** Stress contour; **b** Deformation contour

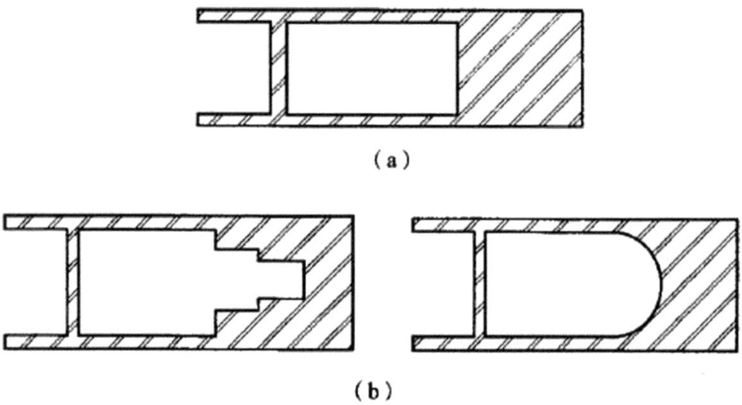

Fig. 6.37 Method of reducing the stress concentration of the hollow sandwich structure. **a** Structural design without considering the load-bearing requirement (sharp geometrical transition, high stress concentration); **b** Design with stress concentration reduction

sandwich structure is increased, and the effect of geometric discontinuity, unevenness bearing properties, vibration mode localization of vibration modes on fatigue properties becomes increasingly significant. Hence, for the advanced parts, the demand for adaptive theoretical and experimental methods of fatigue test is becoming more and more urgent.

6.4.1 Fatigue Properties Evaluation

There are three types of conventional methods for testing structural fatigue properties:

(1) Apply random vibration load to make the load present in the form of power spectral density. The experimental basis is GJB150-16-86 *Environmental test*

methods for military equipments—Vibration test, commonly used in the fatigue properties test of aircraft parts;

(2) Low-cycle fatigue test, applying load to the structure through the hydraulic drive mechanism;

(3) Apply sine-wave periodic load with a dwell frequency to make the load present in the form of stress amplitude. The experimental basis is HB5277-84 *Vibration fatigue test method for engine blades and materials* and this test method is applicable to test the S–N curve of structural fatigue properties and structural fatigue strength, commonly used in the testing the parts of aero-engines, also in the high-cycle fatigue test of the hollow sandwich parts.

For fatigue test of the hollow sandwich structure, the main equipment used includes excitation equipment, displacement sensor, acceleration sensor and strain gauge. The excitation equipment applies the load to the specimen, and among others, the most commonly used one is the electromagnetic vibration shaker (Fig. 6.38). With an excitation frequency of 10–3000 Hz, this type of equipment allows easy regulation. They can not only apply sine-wave periodic excitation of dwell frequency to the specimen, but also apply load according to the power spectrum density spectrum. The electromagnetic vibration shaker also has two main disadvantages:

(1) As the excitation frequency increases, the energy consumed by the moving parts and fixtures of the equipment itself increases quickly, so it is difficult to obtain large deformation of the structure and the stress amplitude is small;

(2) The upper limit of equipment frequency is relatively low, with the excitation frequency not exceeding 10 kHz.

In recent years, by virtue of the rapid response characteristics of magnetostrictive materials, a loading device supporting an excitation frequency of 25 kHz has been

Fig. 6.38 Electromagnetic vibration shaker

developed (Fig. 6.39). This type of excitation device provides relatively small thrust, not suitable for the fatigue test of large parts. At present, it is commonly used to test the vibration fatigue properties of aeroengine high-pressure compressor blades.

When performing fatigue test on the fan blades of the aeroengine medium and high bypass-ratio aeroengine, it is required to fix the specimen with a firm fixture weighting about hundreds of kilograms. When performing the vibration figure test in high-order vibration mode on the structure, there is large energy consumption on the working fixture, and the vibration stress on the specimen is relatively small. To address this problem, small fixed specimens, a loading device based on pneumatic principle is developed to directly transfer the load to the blade through the high-speed airflow so that higher stress level can be obtained by the blades when they vibrate under high-order vibration mode, thereby eliminating the long test cycle of such blades in the traditional testing methods. The loading device is shown in Fig. 6.40.

Fig. 6.39 Magnetostrictive loading device

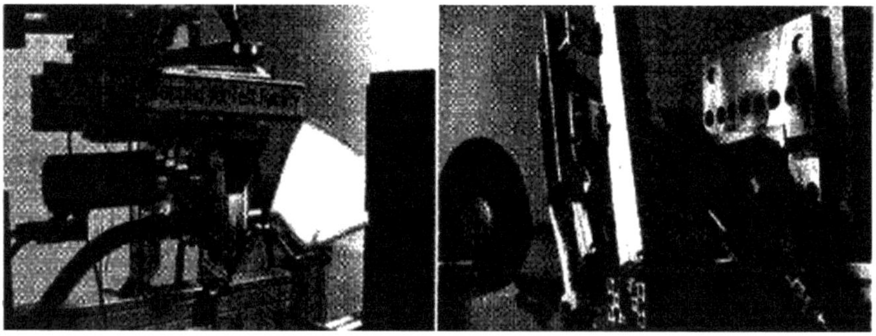

Fig. 6.40 Air excitation fatigue test device

During the fatigue test, the displacement sensor and acceleration sensor are used for real-time monitoring of the specimen condition and feeding back the information to the excitation equipment to form a closed-loop negative feedback, so as to complete control of the specimen condition. As relatively mature equipment, the acceleration sensor is commonly used in random vibration test; the displacement sensor is normally seen in the fatigue test at fixed frequency. Traditionally, eddy current displacement sensor is used to measure the displacement data, it has relatively small range, the theoretical working distance of the probe is shorter, and the phenomenon easily occurs, such as sensor specimen wear or waveform distortion. Therefore, this type of sensors are being gradually replaced by the laser displacement sensor. The digital laser displacement sensor has high measurement accuracy and strong anti-jamming, and the distance from the probe and the measured part can reach over100 mm. Among others, the overall performance of the laser displacement sensor based on the triangle ranging is more balanced. The electric laser displacement sensor is shown in Fig. 6.41.

In the fatigue test, the fixture is used to connect the specimen with the excitation equipment and transfer the load. Therefore, it is required to have light weight and good rigidity, and free of resonance frequency in the frequency range of the test.

For the random vibration test, the fixture used has a simple structure, and can be simply fixed by pressure plate or by a special-designed fixture, there is no strict limit for the installation position of parts to the extent that the load can be applied to the structure in three directions, so the fixture design is relatively simple. Figure 6.42 shows the random vibration test fixture for a certain type of rudder, allowing installation of four parts at a time.

In the sine-wave load fatigue test at dwell frequency, the resonance principle is followed to generate stable alternating stress in the specimen, for example, in the fatigue test of various aeroengine blades. Generally, the steady-state alternating stress on the structure is required to be no less than 300 MPa. When the titanium alloy

Fig. 6.41 Laser displacement sensor

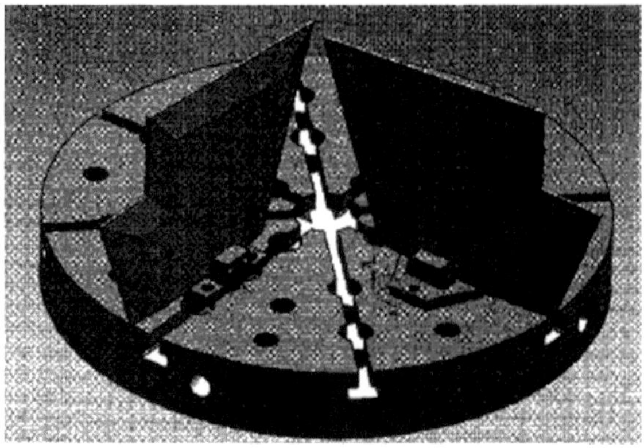

Fig. 6.42 Random vibration test fixture

hollow sandwich parts are used in the manufacture of the low-pressure compressor rotors or stators, the structural size is relatively large, and the amplitude is high under the required test stress level. For example, for the blades with a span 250 mm, the maximum amplitude is about 25 mm if the stress amplitude of the tested position is 340 MPa; for high bypass-ratio engine fan blades with a span of 780 mm, the maximum amplitude is about 60 mm if the stress amplitude of the tested position is 340 MPa.

The highly deformed specimen will result in high-load fixture. In case of poor rigidity and deformation of the fixture, it is less effective in transferring the load. It can be started from two aspects to seek improvement of the rigidity and bearing capacity of fixtures:

(1) Optimize the shape of the fixture, for example, by increasing the contact area between the fixture and the excitation device, as shown in Fig. 6.43a.
(2) Select proper fixed position. In case of small-size and light-weight specimen, it can be fixed on the vertical bench of the vibration shaker; If in case of a large-size specimen with the center of gravity of the structure far away from the fixture, the specimen can be fixed vertically on the horizontal sliding bench of the vibration shaker, as shown in Fig. 6.43b.

6.4.2 Fatigue Failure Mode and Optimization of the Structure

When the SPF/DB structure is used to carry alternating loads, the fatigue property of the structure is top issue to be solved. H-shaped hollow sandwich structure has medium alternating load bearing capacity, suitable for the service environment with low dynamic load level, such as the inlet guide van of engine, etc. Such parts have moderate requirements on the surface integrity of the structure, surface polishing is generally adopted and the surface roughness is around 0.4. Under laboratory load conditions, the fatigue strength of the structure can generally reach 300 MPa. Figure 6.44a shows the vibration fatigue test of an engine's inlet guide van, and

(a) (b)

Fig. 6.43 Clamping the specimens on the vertical bench and horizontal sliding bench (It is helpful to significantly improve the fixture rigidity and reduce the stress by increasing the contact size between the fixture and vibration shaker in the deformation plane of the specimen.). **a** The specimen is clamped on a vertical bench; **b** The specimen is clamped on the horizontal sliding bench

the failure modes are shown in Fig. 6.44b, c. Comparing with the stress contour, it can be seen that the failure position shown in Fig. 6.44b is located in the maximum stress area in the structure, which is a normal failure; the failure position shown in Fig. 6.44c is located in the hollow-solid transition area, where the vibration stress level is low and the failure is abnormal.

In case of normal failure of the H-shaped hollow sandwich structure normal fails, Photos of the crack source area and the location photos of crack initiation are shown in Fig. 6.45, when the crack initiation occurs on the surface of structure, there is a close relationship between the fatigue life of the specimen and the surface quality of the area; and Fig. 6.45b shows the crack initiation from the cut signs of the surface machining. Therefore, it is possible to effectively improve the fatigue life by improving the surface integrity of the structure.

As shown in Fig. 6.46 photo of fatigue crack initiation location of H-shaped hollow sandwich structure during abnormal failure, the crack initiation occurs in the "triangle area" of the structure. There are complex mechanisms for crack initiation in the triangle area, including the stress concentration caused by material discontinuity, stress concentration caused by geometrical changes, and non-uniform residual stress caused by fluctuation of the manufacturing process. “Triangle area” is an natural characteristic of the H-shaped hollow sandwich structure, which is related to the forming process. The triangle size can be effectively controlled by optimizing the manufacturing parameters, thereby improving the fatigue properties of the structure.

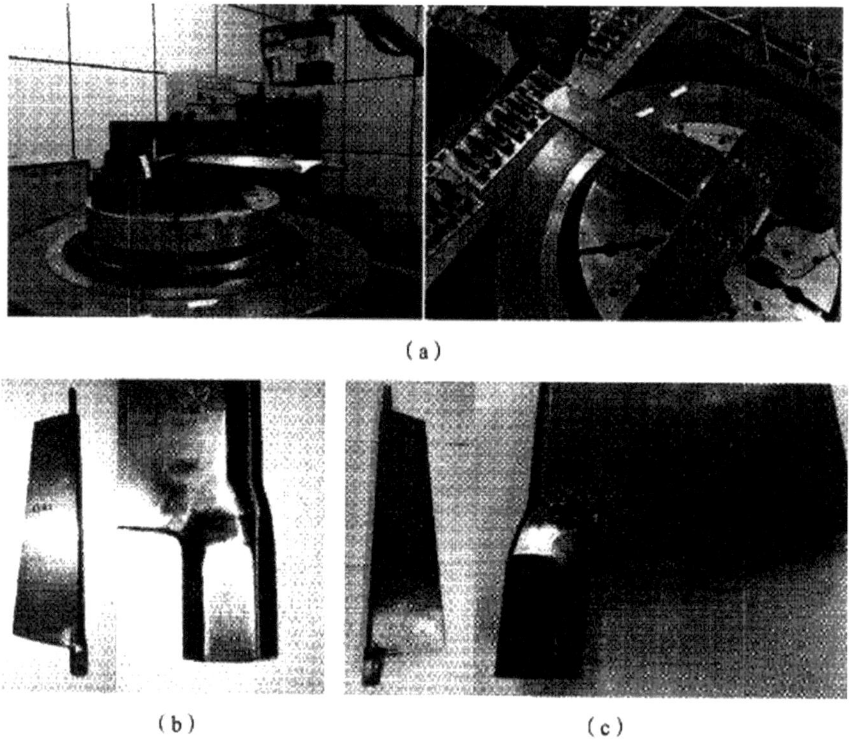

Fig. 6.44 Vibration fatigue test characteristics and failure characteristics of H-shaped four-layer plate structure. **a** Characteristics of vibration fatigue test; **b** crack initiation at the clamping portion; **c** crack initiation at the blade body

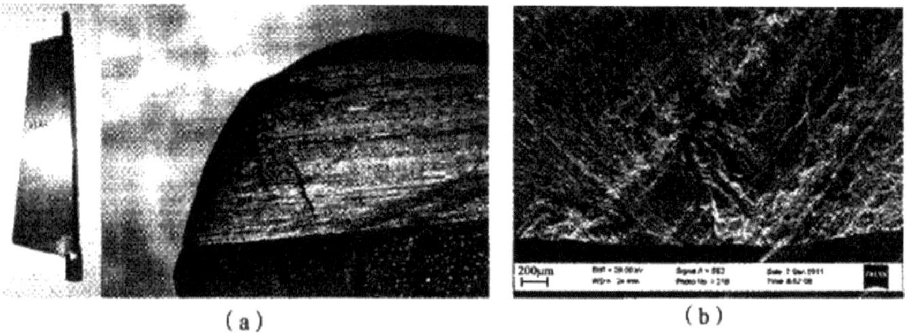

Fig. 6.45 Photo of fatigue crack initiation location of H-shaped hollow sandwich structure during abnormal failure. **a** Macro photos of the fracture; **b** SEM photo of the fracture

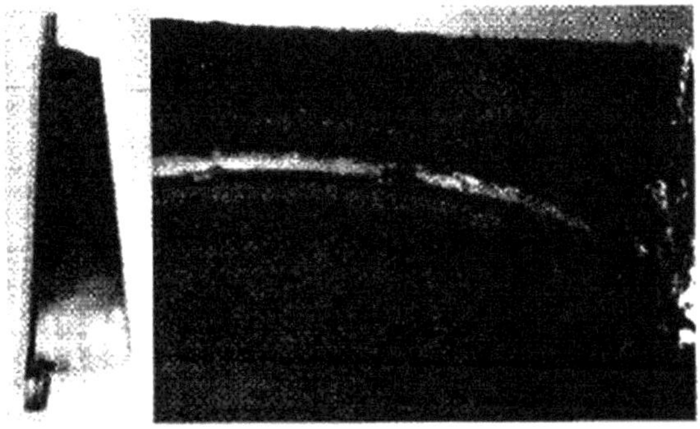

(a)

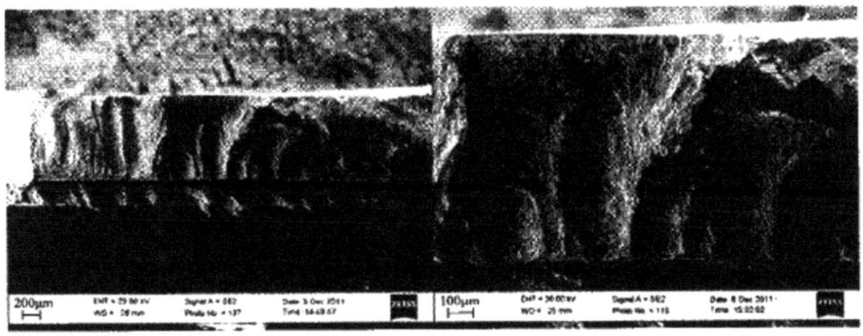

(b)

Fig. 6.46 Photo of fatigue crack initiation location of the H-shaped hollow sandwich structure during abnormal failure. **a** Optical microscrope photo of the fracture; **b** SEM photo of the fracture

From the view of the manufacturing process, there are two ways to improve the fatigue properties of the H-shaped hollow sandwich structure:

(1) Optimize the process parameters to reduce the triangle area size, for example, by increasing the forming temperature and prolonging the gas pressure holding time in the forming process, which is suitable for controlling the triangle area size in the H-shaped hollow sandwich structure. Reduction of the triangle area size can lead to reduction of the stress that causes the triangle area to continue to deform. From formula (6.1) and formula (6.2), it can be seen the stress to reduce the size of the triangle area present a linear change with the deformation of the triangle area. Taking the H-shaped four-layer plate structure for an example, the initial thickness of the internal blank is usually 0.8 mm, and the gas pressure is 3.0 MPa; for TC4 alloys, when the triangle radius is reduced to 0.1 mm at about 900 °C, the strain rate of the material will be reduced below 10^{-5}/s, that is, the triangle size cannot be further reduced. It is realizable to reduce the creep resistance and the triangle area size by increasing the forming temperature, but

there may be deterioration of material properties at the same time. Figure 6.47 shows the local geometrical model and stress model of the triangle area in H-shaped hollow sandwich structure.

$$\sigma = \frac{P \times r}{d_2} \tag{6.1}$$

$$\sigma = K\dot{\varepsilon}^m \varepsilon^n \tag{6.2}$$

(2) Optimizing the structure shape of the blank, for example, profile changing the transition form between the hollow area and solid area of the structure from vertical transition to fillet transition or elliptical transition (Fig. 6.48), can effectively reduce the triangle size of the hollow-solid transition position in the H-shaped hollow sandwich structure. Theoretically, the elliptic fillet transition can achieve complete elimination of the "triangle" on the hollow-solid boundary, as shown in Fig. 6.48. Therefore, the above improvement measures are most effective ones for the triangle size control of the hollow-solid transition area in H-shaped hollow sandwich structure.

It is possible to eliminate the natural structural characteristic of "triangle area" in H-shaped hollow sandwich structure by using other internal topology structures, for example, the "X-shaped" structure, as shown in Fig. 6.49. The manufacturing process of X-shaped internal structure is similar to the H-shaped structure, while this allows to completely eliminate the "triangle" characteristics, and it is suitable for use in manufacture of thick parts. In addition, this process has limitation on the cooperative relationship between the thickness of the inner and outer blanks and the angle between the truss and blanks, for example, when the angle between the truss and the skin is designed to be 45°, the ratio of inner blank thickness to outer blank thickness is required to be less than or equal to 0.33, otherwise, surface forming quality control of parts will be affected. Therefore, the weight reduction efficiency of the X-shaped structure is lower than that of H-shaped structure; if the skin material

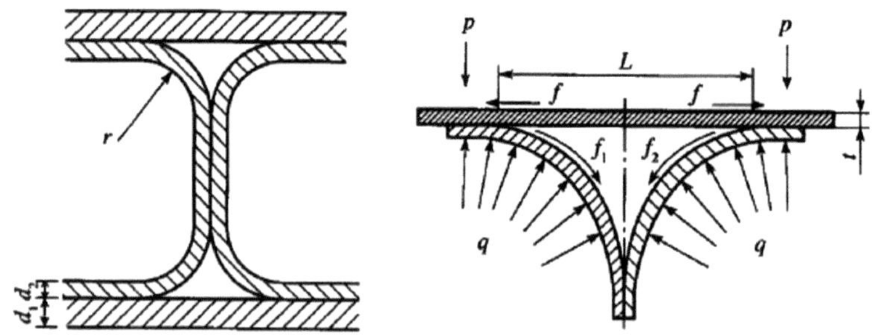

(a) (b)

Fig. 6.47 Local geometrical model and stress model of the triangle area in H-shaped hollow sandwich structure. **a** Geometrical model; **b** Stress model

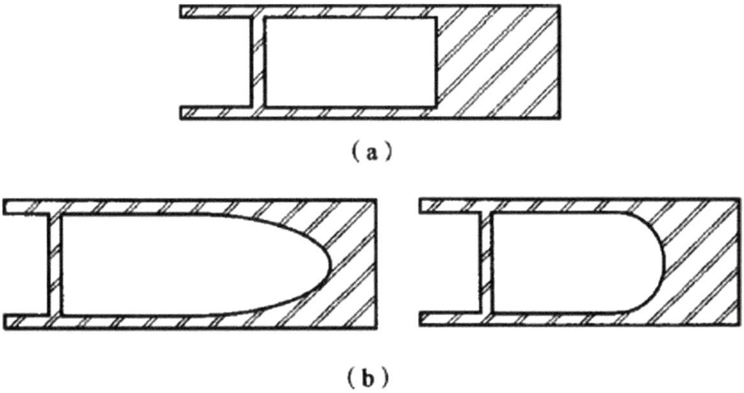

Fig. 6.48 Method of reducing stress level of the H-shaped hollow sandwich structure. **a** Typical section of H-shaped four-layer plate structure; **b** Reduction of the triangle size of the hollow-solid transition position

of X-shaped structure is removed by the subsequent auxiliary process after the inner structure forming, the weight reduction efficiency of the X-shaped structure can remain at the same level as that of the H-shaped structure.

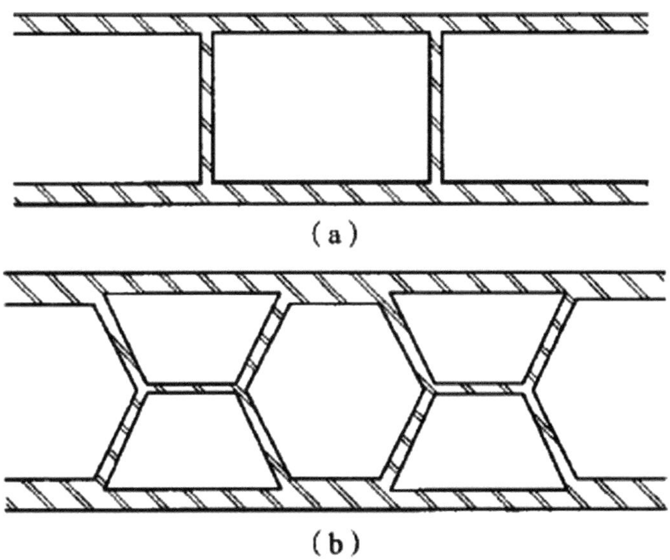

Fig. 6.49 Method of reducing the stress level of the four-layer plate structure. **a** Typical section of H-shaped four-layer plate structure; **b** X-shaped four-layer plate structure

For the titanium alloy W-shaped three-layer hollow plate structure, the fatigue crack initiation occurs on the surface, subsurface and inner surface. The crack initiation characteristics of a certain type of blade under laboratory environment are shown in Fig. 6.50.

Abrasive belt grinding is a traditional method used to improve the surface quality of the structure. For titanium alloy W-shaped structure, the abrasive belt grinding requires the surface roughness Ra to be more than 0.4, and the fatigue crack initiation surface characteristics of the titanium alloy W-shaped three-layer plate structure during surface grinding is as shown in Fig. 6.51. As seen from the Figure, crack initiation occurs on the surface of the structure, and in the vibration fatigue test, it is deemed as normal failure if the maximum vibration stress is detected on the surface of the structure; however, in case of poor integration of the structural surface, improper local grinding can lead to presence of grinding/machining signs on the structural surface, as shown in Fig. 6.51b, which will reduce the fatigue properties of the structure. For the above, the targeted measure is to improve the control requirements of surface roughness to avoid non-compliance of the local roughness with the

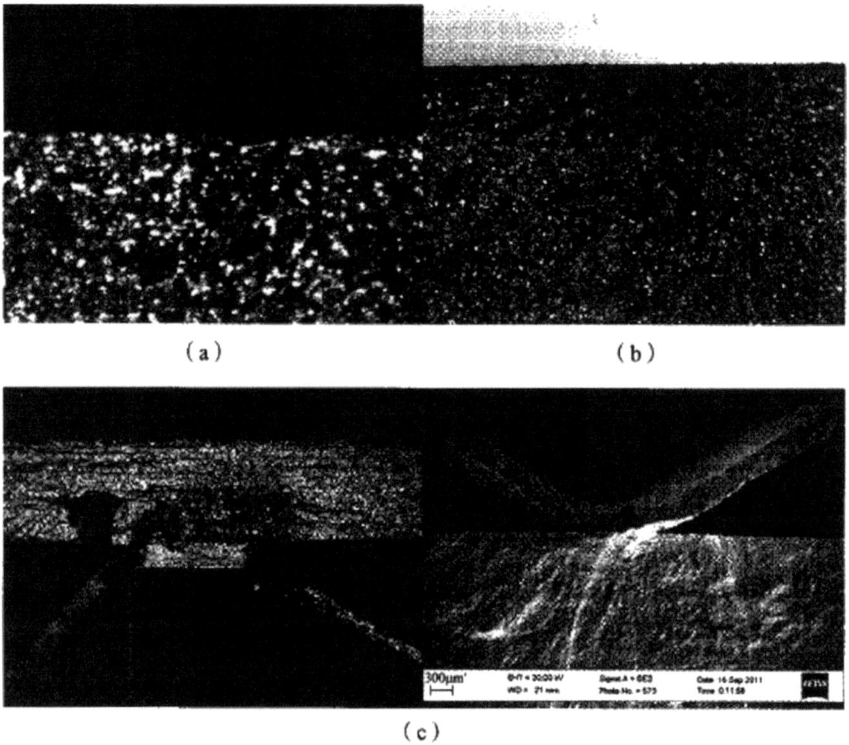

Fig. 6.50 Fatigue failure characteristics of blades under laboratory conditions. **a** Crack initiation from the surface of the structure; **b** Crack initiation from subsurface; **c** Crack initiation from the inner surface

(a) (b)

Fig. 6.51 Characteristics of fatigue crack surface initiation in TI-alloy W-shaped three-layer plate structure subjected to surface grinding. **a** Initiation on the surface of the structure; **b** Crack initiation from the machining signs

requirements; and it also requires the grinding direction of the abrasive belt to be parallel to the direction of the principal stress under the structure.

In the engineering application of traditional solid structure, surface strengthening has been used as an effective way to improve the fatigue properties, and the common strengthening process includes rolling strengthening, shot peening, laser shotpeening, etc. It is generally believed that the strengthening process introduces a compressive residual stress field between the outer surface and the subsurface of the structure, which acts to increase the crack initiation time and prolong the micro crack expanding time, ultimately delivering improved overall fatigue life of the structure. Application of these methods are also seen in the titanium alloy W-shaped three-layer plate structure. In order to obtain the effect of different strengthening degree/ process parameters on the fatigue properties of the structure, the hollow-characteristic specimens (Fig. 6.52) are adopted or real parts. In Fig. 6.52, specimens for surface grinding, strong surface strengthening and moderate surface strengthening are given from left to right, and there are different initiation locations of fatigue crack in three types of specimens, respectively the outer surface, subsurface, and inner surface of the structure, as shown in Fig. 6.53; the fatigue life of the threes is shown in Fig. 6.54, it can be seen that the structure has the longest life under the moderate shot strength.

The internal geometry characteristics of titanium alloy W-shaped three-layer plate structure is guaranteed by the manufacture process, and forming precision significantly affects the local stress, causing deviation of the stress level of the internal structure from the theoretical design. If the stress reaches a certain level, it may cause fatigue crack initiation from the inner side of the structure. The internal crack initiation caused by manufacturing deviation as is shown in Fig. 6.55. The said manufacturing deviations may include: ① Skin thickness deviates from the theoretical design; ② Width of the diffusion bonding interface is less than the theoretical design; ③ Angle between the truss and skin deviates from the theoretical design;

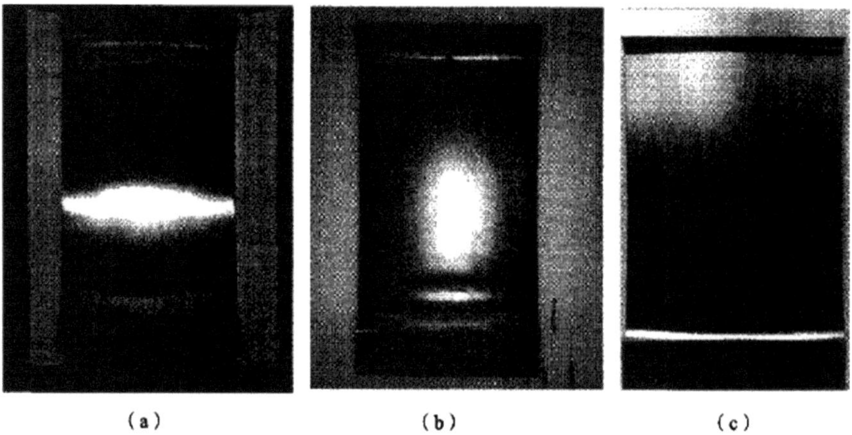

Fig. 6.52 Titanium alloy W-shaped hollow sandwich structure specimen. a Surface polishing; b Strong surface strengthening; c Moderate surface strengthening

④ Interface mis-bonding; ⑤ Local distortion of truss occurs. The above deviations requires proper control during part manufacturing.

The fatigue life of titanium alloy W-shaped three-layer plate structure is closely related to the structure design, manufacturing process. In terms of structure design, it is necessary to combine the characteristics of manufacturing process and accessibility to reasonably control the distribution of materials and thereby achieve reasonable stress distribution in the structure, making the thick part in the structure able to bear more load; while the relatively weak part, such as the diffusion bonding interface, is arranged in a relatively low stress level; with regard to the manufacturing process, it is important to improve the consistency of design and manufacture, control the geometrical size of the structure and actively simplify the stress concentration of the structure in order to create an idea structure featuring long life, good bearing properties and low weight. Although there are many crack initiation locations in the titanium alloy hollow structure, it is possible to obtain the rule between the crack initiation location and the structure life by the statistical method (Fig. 6.56). As shown in the Figure, the structure with crack initiation on its subsurface, has relatively good fatigue properties. This law may be used as a direction for optimization of hollow structure/process design of titanium alloys.

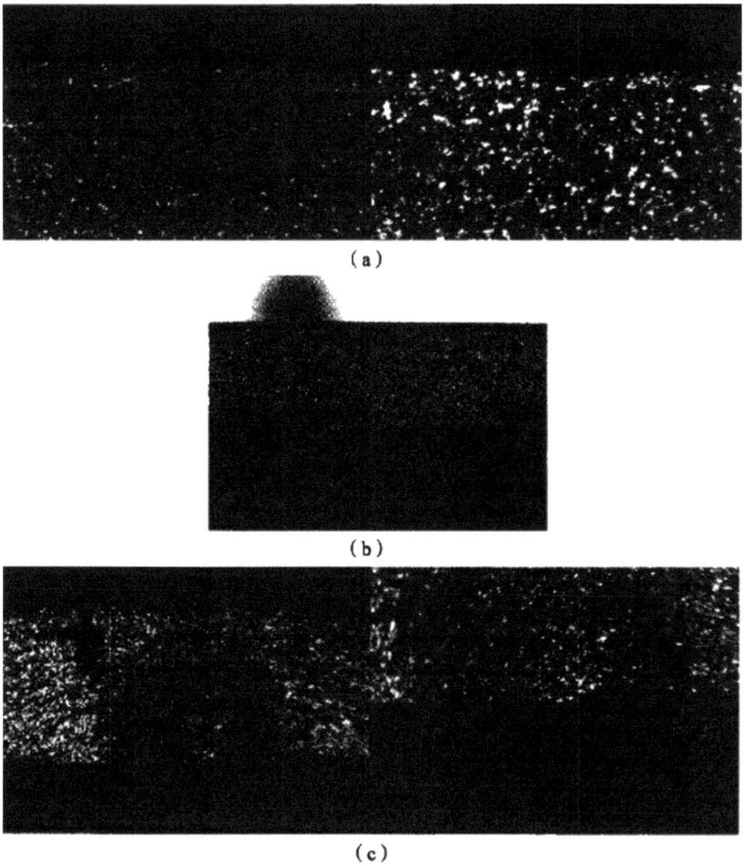

Fig. 6.53 Effect of surface strengthening on fatigue crack initiation of the titanium alloy W-shaped hollow sandwich structure. **a** Effective strengthening is not obtained (for crack initiation on the external surface); **b** Surface strengthening degree is within a certain range (for crack initiation on the subsurface); **c** Surface strengthening is excessive (for crack initiation on the inner surface)

Fig. 6.54 Comparison of the fatigue life of specimen under different surface states

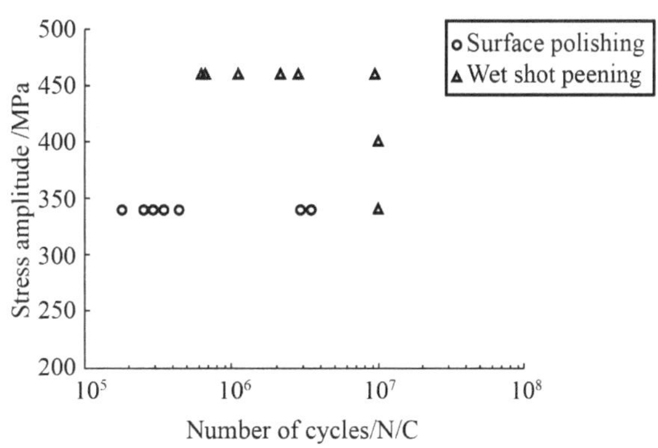

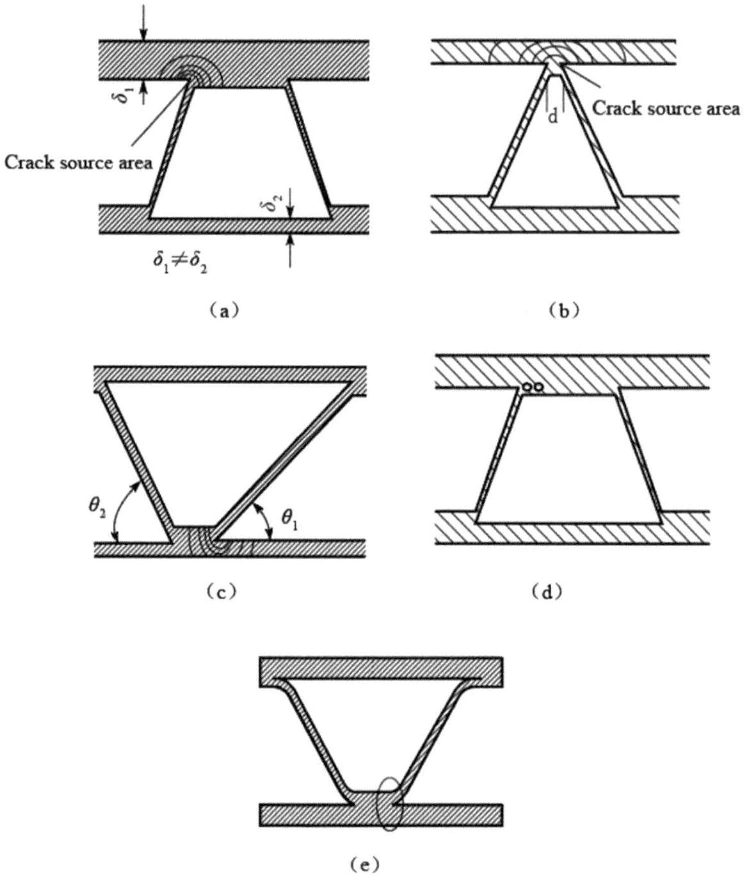

Fig. 6.55 Crack initiation caused by internal local stress concentration. **a** Skin thickness deviates from the theoretical design; **b** Width of the diffusion bonding interface is less than the theoretical design; **c** Angle between the truss and skin deviates from the theoretical design; **d** Diffusion bonding interface fails; **e** Local distortion of truss occurs

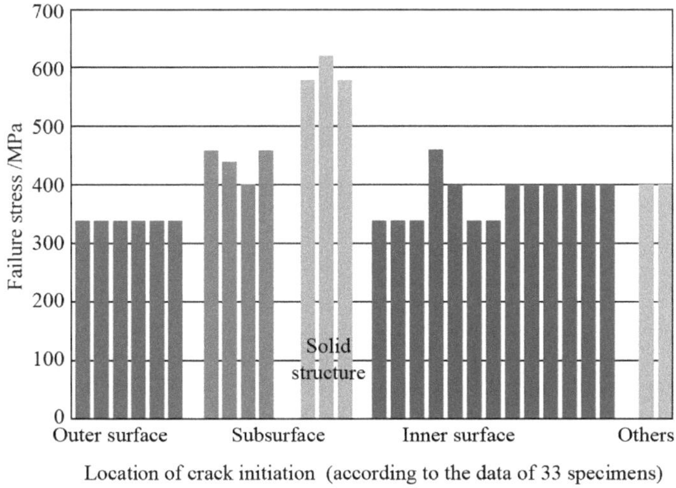

Fig. 6.56 Law of crack initiation with the fatigue life of the hollow sandwich structure

References

1. David Serra. 2008. Superplastic forming applications on aero engines: A review of ITP manufacturing processes. In *6th EUROSPF Conference.*
2. John Whurr. 2013. *Future Civil Aeroengine Architectures & Technologies.*
3. Ritzmann, Swen, and Scott Courtney. 2016. Blade runners. *Fatigue Testing*, 62–64.
4. Wendenburg, Erk, and Grégory Pugliesi. 2012. *High Cycle Fatigue Testing by Means of Chopped Air Excitation*